Quadratic Algebras

University
LECTURE
Series

Volume 37

Quadratic Algebras

Alexander Polishchuk
Leonid Positselski

American Mathematical Society
Providence, Rhode Island

2000 *Mathematics Subject Classification.* Primary 16S37, 16S15, 16E05, 16E30, 16E45, 16W50, 13P10, 60G10.

For additional information and updates on this book, visit
www.ams.org/bookpages/ulect-37

Library of Congress Cataloging-in-Publication Data

Polishchuk, Alexander, 1971–
Quadratic algebras / Alexander Polishchuk, Leonid Positselski.
p. cm. — (University lecture series, ISSN 1047-3998 ; v. 37)
Includes bibliographical references.
ISBN 0-8218-3834-2 (acid-free paper)
1. Quadratic fields. 2. Associative rings. 3. Commutative rings. 4. Stochastic processes. I. Positselski, Leonid, 1973- II. Title. III. University lecture series (Providence, R.I.) ; 37.

QA247.P596 2005
512′.4—dc22 2005048198

∞ The paper used in this book is acid-free and falls within the guidelines
established to ensure permanence and durability.
Visit the AMS home page at http://www.ams.org/

10 9 8 7 6 5 4 3 2 1 10 09 08 07 06 05

Contents

Introduction

The goal of this book is to introduce the reader to some recent developments in the study of associative algebras defined by quadratic relations. More precisely, we are interested in (not necessarily commutative) algebras over a field that can be presented using a finite number of generators and (possibly nonhomogeneous) quadratic relations. This book is devoted to some aspects of the theory of such algebras, mostly evolving around the notions of Koszul algebra and Koszul duality. Its content is a mixture of known results with a few original results that we circulated since 1994 as a preprint of the same title.

One of the original motivations for the study of quadratic algebras came from the theory of quantum groups (see [**43, 77**]). Namely, quadratic algebras provide a convenient framework for "noncommutative spaces" on which quantum groups act (see [**78**]). One of the basic problems that arose in this context is how to control the growth of a quadratic algebra (e.g., measured by Hilbert series). A related question is whether there are generalizations of the Poincaré-Birkhoff-Witt theorem (for universal enveloping algebras) to more general quadratic algebras. The core of this book is our attempt to present some partial solutions. It turns out that one can shed some light on questions of this kind using the remarkable notion of Koszul algebra introduced by S. Priddy [**104**]. In fact, the study of this notion brought some dramatic changes to the area. Loosely speaking, our experience shows that general quadratic algebras behave as badly as possible, while for Koszul algebras the situation is usually much nicer. As we hope to convince the reader, the study of Hilbert series provides a good illustration of this principle.

Perhaps one of the important features of the theory of Koszul algebras is duality: for each Koszul algebra there is a dual Koszul algebra (roughly speaking, it is obtained by passing to the dual space of generators and the orthogonal space of quadratic relations). This often leads to remarkable connections between seemingly unrelated problems. For example, Koszul duality of the symmetric algebra and the exterior algebra underlies the famous description of coherent sheaves on projective spaces in terms of modules over the exterior algebra due to J. Bernstein, I. Gelfand and S. Gelfand [**27**]. More generally, in a number of situations one can prove an equivalence of derived categories of modules over Koszul dual algebras (see [**23, 11, 24, 51**]). This topic is beyond the scope of our book although we will discuss some more elementary aspects of Koszul duality.

The notion of Koszulness also proved to be a really impressive prediction tool. In many examples a few observations may suggest that some quadratic algebra is Koszul. Then this conjecture turns out to be related to some important and nontrivial features of the setting. It is also quite amazing that many important quadratic algebras naturally arising in various fields of mathematics are Koszul. Examples known to us arise in the following areas:

(i) algebraic geometry—certain homogeneous coordinate algebras are Koszul (see [**29, 37, 39, 67, 50, 72, 73, 89, 96**]);
(ii) representation theory—certain subcategories of the category $\mathcal{O}$ for a semisimple complex Lie algebra are governed by Koszul algebras (see [**19, 24**]);
(iii) noncommutative geometry—the Koszulness condition arises naturally in the theory of exceptional collections; the algebras describing certain noncommutative deformations of projective spaces are Koszul (see [**30, 31, 117**]);
(iv) topology—Steenrod algebra, cohomology algebras of formal rational $K[\pi, 1]$-spaces, holonomy algebras of supersolvable hyperplane arrangements, as well as some algebras related to configuration spaces of surfaces are Koszul; the category of perverse sheaves on a triangulated space is equivalent to modules over a Koszul algebra (see [**104, 88, 113, 28, 97, 127**]);
(v) number theory—the Milnor K-theory ring of any field (tensored with $\mathbb{Z}/l\mathbb{Z}$ for a prime l) is conjectured to be Koszul—this is a strengthening of the Bloch-Kato conjecture relating Milnor K-theory with Galois cohomology (see [**103, 102**]);
(vi) noncommutative algebra—the universal algebra generated by pseudoroots of a noncommutative polynomial is Koszul (see [**111, 93**]).

Checking the Koszul property usually requires some effort and the methods of proof vary from one case to another. Although we do not try to give a systematic exposition of these methods here, the reader will find a few sample techniques for checking Koszulness (mostly in chapter 2).

As we have already mentioned, one of the central questions studied in our book is how to generalize the Poincaré-Birkhoff-Witt-theorem (PBW-theorem) to quadratic algebras. Recall that the classical PBW-theorem for the universal enveloping algebra $U\mathfrak{g}$ of a Lie algebra $\mathfrak{g}$ can be formulated in two different ways. In the first formulation one starts with a basis of $\mathfrak{g}$ and then the theorem states that certain standard monomials in basis elements form a basis of $U\mathfrak{g}$. Another formulation simply asserts that the associated graded algebra of $U\mathfrak{g}$ with respect to the standard filtration coincides with the symmetric algebra $\mathbb{S}\mathfrak{g}$. Thus, the first way to generalize the PBW-theorem to other algebras is to modify the notion of standard monomials. Assume that we have a graded quadratic algebra (i.e., quadratic relations are homogeneous). Then using lexicographical order on the set of all monomials in generators one can define a certain set of standard monomials (depending on quadratic relations). The analogue of the PBW-theorem in this case states that if the standard monomials form a basis in the grading component of degree 3 then the same is also true for all grading components (so that we get a *PBW-basis* in our algebra). This theorem is a particular case of the so-called diamond lemma in the theory of Gröbner bases developed in works on combinatorial algebra in the late 70s (see [**26, 35, 36**]). Note that the universal enveloping algebra $U\mathfrak{g}$ can be homogenized by adding an extra central generator, so that the classical PBW-theorem would fit into this context.

Before stating the second generalization of the PBW-theorem let us say a few words about the terminology adopted in the book. We use the term "quadratic algebra" only in reference to algebras defined by homogeneous quadratic relations (because with the exception of chapter 5 we consider only such algebras). Assigning degree 1 to each generator one can view a quadratic algebra as a graded algebra $A = \bigoplus_{n\geqslant 0} A_n$ such that A_0 is the ground field and A is the quotient of the tensor algebra of A_1 by an ideal generated in degree 2. Note that sometimes (e.g., in applications to representation theory) it is necessary to consider more general quadratic algebras such that A_0 is not necessarily equal to the ground field but rather is a semisimple algebra. We will briefly discuss algebras of this kind in section 9 of chapter 2.

Our second generalization of the PBW-theorem deals with a "nonhomogeneous quadratic algebra", i.e., an algebra with a finite number of generators and nonhomogeneous quadratic defining relations. If A is such an algebra then one can consider the natural filtration on A determined by the set of generators. Let us denote by $\mathrm{gr}A$ the associated graded algebra. On the other hand, one can truncate the relations in A leaving only their homogeneous quadratic parts. Let $A^{(0)}$ be the obtained quadratic algebra. The nonhomogeneous PBW-theorem states that the natural map $A^{(0)} \to \mathrm{gr}A$ is an isomorphism provided $A^{(0)}$ is Koszul and a certain self-consistence condition is satisfied (this result was proved independently by A. Braverman and D. Gaitsgory [**33**]). This self-consistency condition is obtained by looking at expressions of degree 3 in generators. In the case $A = U\mathfrak{g}$ it coincides with the Jacobi identity for the Lie bracket on $\mathfrak{g}$.

It is interesting that the notion of Koszulness appears also in the first generalization of the PBW-theorem: quadratic algebras having a basis of standard monomials, called *PBW-algebras*, are always Koszul (this observation goes back to S. Priddy [**104**]). However, the converse is not true: Koszul algebras are not necessarily PBW (see section 3 of chapter 4). In fact, the class of PBW-algebras is substantially smaller than that of Koszul algebras and is much easier to study. For example, the set of PBW-algebras with a given number of generators is constructible in Zariski topology while the set of Koszul algebras is often not constructible (see section 3 of chapter 4 and section 6 of chapter 6). On the other hand, there are many parallel results for both classes of algebras. Firstly, both properties can be formulated in terms of distributivity of certain lattices of vector spaces. Secondly, various natural operations with quadratic algebras, such as quadratic duality, free product, tensor product, Segre product and Veronese powers preserve both classes. The comparison between the classes of Koszul and PBW-algebras is also an important part of the present work. In our experience PBW-algebras often provide a good testing ground for guessing the general pattern that might hold for all Koszul algebras. Usually there is no problem with proving that a pattern holds for PBW-algebras; however, the case of Koszul algebras is often much harder (if at all accessible).

One of the most striking properties of Koszul algebras is the following.

Koszul Deformation Principle (V. Drinfeld [**43**]). *If a formal family of graded quadratic algebras $A(t)$ is flat in the grading components of degree $\leqslant 3$ and the algebra $A(0)$ is Koszul then the family is flat in all degrees.*

More precisely, a similar statement holds for local deformations (in Zariski topology) if we consider only a finite number of grading components (see Theorem 2.1 of chapter 6). The second version of the PBW-theorem considered above can be easily deduced from this principle. Another unexpected consequence that we derive from it is the finiteness of the number of Hilbert series of Koszul algebras with a fixed number of generators (the analogous statement for quadratic algebras is wrong). We conjecture that Hilbert series of Koszul algebras enjoy several interesting properties that can be easily checked for PBW-algebras (although we prove that these two sets of Hilbert series are different). For example, we conjecture that the Hilbert series of a Koszul algebra is always rational.

The study of Hilbert series of Koszul algebras led to the discovery in [**100**] of an interesting connection with the theory of discrete stochastic processes. Namely, to every Koszul algebra A one can associate a *one-dependent* stationary stochastic sequence of 0's and 1's. It is convenient to encode probabilities of various events in such a process by a linear functional $\phi : \mathbb{R}\{x_0, x_1\} \to \mathbb{R}$ on the free algebra in two variables, taking nonnegative values on all monomials and satisfying $\phi(1) = 1$. Then the condition of one-dependence is equivalent to the equation

$$\phi(f \cdot (x_0 + x_1) \cdot g) = \phi(f)\phi(g),$$

where $f, g \in \mathbb{R}\{x_0, x_1\}$. Abusing the terminology we call such a functional ϕ a *one-dependent process*. It is easy to see that ϕ is uniquely determined by the values $(\phi(x_1^n))$. Now the one-dependent process associated with a Koszul algebra A is defined by

$$\phi_A(x_1^{n-1}) = a_n / a_1^n,$$

where $a_n = \dim A_n$. Nonnegativity of values of ϕ on all monomials is equivalent to a certain system of polynomial inequalities for the numbers a_n. The fact that these inequalities are indeed satisfied for a Koszul algebra seems to be a remarkable coincidence. However, the analogy between the two theories does not end here. It turns out that under this correspondence the subclass of PBW-algebras maps to the set of so-called two-block-factor processes. The relation between all one-dependent processes and the subclass of two-block-factors was intensively studied in the 90s after it was proved in [**2**] that a one-dependent process does not have to be a two-block factor (see [**1, 118, 122**]). This topic seems to be surprisingly similar to the relation between Koszul and PBW-algebras. Motivated by this analogy we conjecture that the Hilbert series associated with every one-dependent process admits a meromorphic continuation to the entire complex plane. Rationality of Hilbert series of Koszul algebras would follow from this (by a theorem of E. Borel [**32**]). We also observe that the polynomial inequalities satisfied by the numbers $(\phi(x_1^n))$ form a subset in the well-known system of inequalities defining the notion of a *totally positive sequence* (also known as *Polya frequency sequence*). It is known that the generating series of a totally positive sequence admits a meromorphic continuation (see [**71**]). This can be considered as another hint in favor of our conjecture.

Here is the more detailed outline of the content of the book.

Chapter 1 contains some basic definitions and results concerning cohomology of graded algebras, quadratic algebras and distributivity of lattices. In particular, in section 2 we define quadratic duality for quadratic algebras and quadratic modules (we use the term "Koszul duality" when referring to this duality in the case of Koszul algebras and Koszul modules).

In chapter 2 we describe various equivalent definitions of Koszulness, including Backelin's criterion in terms of distributivity of lattices (see [**15**]). We give similar equivalent definitions for a related notion of n-Koszulness that has an advantage of being defined by a finite number of conditions. We also show that many results about quadratic and Koszul algebras have natural analogues for quadratic and Koszul *modules*. In section 5 we consider the problem of preservation of Koszulness under homomorphisms of various types between graded algebras, generalizing some results of Backelin and Fröberg [**20**]. In section 7 we give examples of projective varieties with Koszul homogeneous coordinate algebras. In section 8 we explain how to associate to a Koszul algebra A a (graded) *infinitesimal bialgebra* (or ϵ*-bialgebra*)

V_A. This construction can be viewed as a categorification of the one-dependent process ϕ_A associated with A, because the values of ϕ_A on monomials are given by dimensions of certain multigrading components of V_A. In section 9 we consider some generalizations of the notion of Koszulness including an important case of graded algebras $A = \bigoplus_{n\geqslant 0} A_n$ such that A_0 is a semisimple algebra (in the rest of the book we assume that A_0 is the ground field). We also give an interpretation of Koszul algebras in terms of monoidal functors from a certain universal (nonunital) monoidal category.

In chapter 3 we consider several natural operations on quadratic algebras and modules that preserve Koszulness and discuss the behavior of Hilbert series under these operations. Following [**20**] we consider free sums, free products, along with several types of tensor products, the Segre product $A \circ B$, the dual operation "black circle product" $A \bullet B$ and Veronese powers $A^{(n)}$. The operation $A \bullet B$ is also closely related to the internal cohomomorphism operation introduced by Manin (see [**77, 79**]). We prove that if one of the algebras is Koszul then the Hilbert series of $A \bullet B$ can be computed in terms of those of A and B and show that this is impossible if both algebras are not Koszul. An interesting application of these operations is given in section 5, where we show, following D. Piontkovskii [**92**], that Koszulness of a quadratic algebra A cannot be determined from the knowledge of the Hilbert series of A and $A^!$.

Chapter 4 is devoted to PBW-algebras. We start by giving a proof of the PBW-theorem for quadratic algebras that gives a criterion for the existence of a PBW-basis (as we have mentioned before, this is really a particular case of the diamond lemma). Then we prove that PBW-algebras are Koszul and give a criterion of the PBW-property in terms of distributivity of lattices in the spirit of Backelin's criterion of Koszulness. We also check that the class of PBW-algebras is stable under quadratic duality and under all operations considered in chapter 3. Then we discuss Hilbert series of PBW-algebras. We show that the Hilbert series of a PBW-algebra is a generating function for the number of paths in a finite oriented graph and hence is rational. In section 7 we prove a generalization of the PBW-theorem involving filtrations with values in an ordered semigroup. In section 8 we consider commutative PBW-algebras. We prove that they are Koszul and compute their Hilbert series. We also present some examples showing that the sets of Hilbert series of PBW-algebras and Koszul algebras are different. In section 9 we discuss a generalization of the classes of Koszul and PBW-algebras from graded algebras to $\mathbb{Z}$-algebras. In section 11 we consider 3-dimensional elliptic Sklyanin algebras. We prove that they are Koszul but do not admit a PBW-basis even viewed as $\mathbb{Z}$-algebras.

In chapter 5 we consider nonhomogeneous quadratic algebras. For these algebras we prove in section 2 the PBW-theorem involving an analogue of the Jacobi identity and Koszulness of the corresponding homogeneous quadratic algebra. We also prove in section 3 a version of this theorem for nonhomogeneous quadratic modules. In section 4 we consider an analogue of quadratic duality for the nonhomogeneous case. It turns out that the dual object to a nonhomogeneous quadratic algebra is a so-called CDG-algebra (curved DG-algebra). In section 5 we give some examples of nonhomogeneous quadratic algebras and modules. In particular, we list all solutions of the analogue of the Jacobi identity in the case of the quadratic relations corresponding to a free commutatative superalgebra, and consider an example related to the PBW-theorem for quantum universal enveloping algebras (Example 6). The remainder of this chapter is devoted to various cohomological calculations related to nonhomogeneous quadratic duality.

Chapter 6 is devoted to the Koszul Deformation Principle for quadratic algebras and some of its consequences, such as finiteness of the number of Hilbert series of Koszul algebras with a fixed number of generators. Furthermore, in section 3 we give an explicit bound on this number and in section 7 we prove that the number of such Hilbert series is finite even if the ground field is allowed to vary. In section 4 we discuss some results on generic algebras among quadratic algebras with a given number of generators and relations. In section 5 we consider examples of possible Hilbert series for algebras with a small number of generators and relations. Section 6 contains counterexamples from [**56**] showing that the set of Koszul algebras is not constructible and that the set of Hilbert series of quadratic algebras with a given number of generators is infinite.

In chapter 7 we explain the connection between Koszul algebras and one-dependent processes. We start by formulating several conjectures on Hilbert series of Koszul algebras, such as the rationality conjecture. Then we derive a system of polynomial inequalities satisfied by the numbers $a_n = \dim A_n$ for a Koszul algebra A. The polynomials of a_n appearing in these inequalities express the dimensions of multigrading components of the ϵ-bialgebra V_A. Then we show that these inequalities allow one to associate a one-dependent process to the sequence (a_n). We show that Koszul duality corresponds to the natural duality on one-dependent processes and also introduce analogues of some other operations on Koszul algebras for one-dependent processes. In section 5 we show that the one-dependent process associated with a PBW-algebra is a two-block-factor and that every two-block-factor can be approximated by those obtained from PBW-algebras. In section 7 we review the notion of a Hilbert space representation of a one-dependent process due to V. de Valk [**121**]. In section 8 we discuss the conjecture that the Hilbert series of a one-dependent process can be extended meromorphically to the entire complex plane. We show that this series always admits a meromorphic continuation to the disk $|z| < 2$ (it converges for $|z| < 1$) and prove the conjecture for two-block-factor processes. In section 9 we give a construction due to B. Tsirelson of a one-dependent process associated with an arbitrary quadratic algebra and a Hermitian form on the space of generators. In section 10 we consider an analogue for Koszul *modules* of the construction of a one-dependent process from a Koszul algebra.

In the Appendix we recall some definitions concerning DG-algebras, DG-modules and Massey products.

Acknowledgments. First, we would like to thank A. Vaintrob whose question about possible generalizations of the PBW-theorem to quadratic algebras started this work in 1991. Also, we are grateful to J. Backelin, A. Braverman, J. Bernstein, P. Etingof, V. Ginzburg, V. Ostrik, D. Piontkovskii, J.-E. Roos, A. Schwarz, B. Shelton, B. Tsirelson, and S. Yuzvinsky for many interesting discussions and suggestions. Special thanks are due to J. Backelin for pointing out several mistakes in the manuscript. Finally, we are grateful to the referee for many useful suggestions.

CHAPTER 1

Preliminaries

In this chapter after setting up notation and reviewing some homological algebra (including the bar construction) we give basic definitions concerning quadratic algebras and quadratic duality. We review some general results on cohomology of graded algebras and modules in sections 3–5. Sections 6 and 7 contain some basic results on distributive lattices.

0. Conventions and notation

Throughout this work, by an *algebra* A we mean an associative algebra with unit $1_A \in A$ over a fixed ground field $\Bbbk$. The unit acts identically on all our A-*modules.* A *graded algebra* is a graded vector space $A = \bigoplus_{i\in\mathbb{Z}} A_i$ with an algebra structure such that $A_i \cdot A_j \subset A_{i+j}$ and $1_A \in A_0$. A *graded (left) A-module* is a graded vector space $M = \bigoplus_{i\in\mathbb{Z}} M_i$ with an A-module structure such that $A_i \cdot M_j \subset M_{i+j}$. For a graded A-module M we denote by $M(n)$ the same module with shifted grading: $M(n)_i = M_{i+n}$.

An *augmented algebra* A is an algebra equipped with a direct sum decomposition $A = \Bbbk \oplus A_+$ such that $\Bbbk = \Bbbk \cdot 1_A$ is the line spanned by 1_A and A_+ is a two-sided ideal in A. The ground field $\Bbbk$ is equipped with the left and right A-module structures via the augmentation.

Vector spaces are usually assumed to be finite-dimensional. Graded vector spaces $V = \bigoplus_{n\in\mathbb{Z}} V_n$ are assumed to be *locally finite-dimensional*, i.e., to have finite-dimensional grading components. For such a graded vector space V we denote by V^* the graded dual vector space with components $(V^*)_n = V^*_{-n}$.

We say that a graded vector space $V = \bigoplus_n V_n$ has *polynomial growth* if there exists a constant $C > 0$ and a positive integer d such that $\dim V_n \leqslant C \cdot n^d$ for $n \gg 0$. On the other hand, if there exists a constant $c > 1$ such that $\dim V_n \geqslant c^n$ for $n \gg 0$, then we say that V has *exponential growth.*

Starting from section 2 of chapter 1 we assume that our graded algebras $A = \bigoplus_i A_i$ are locally finite-dimensional with $A_i = 0$ for $i < 0$ and $A_0 = \Bbbk \cdot 1_A$. We always equip an algebra A of this kind with the natural augmentation such that $A_+ = \bigoplus_{i=1}^{\infty} A_i$. We assume our graded A-modules $M = \bigoplus_i M_i$ to be locally finite-dimensional and bounded below (i.e., with $M_i = 0$ for $i \ll 0$). By a *nonnegatively graded* A-module we mean a graded A-module M such that $M_i = 0$ for $i < 0$.

Many of our definitions (and results) for algebras have analogues for modules over algebras. Sometimes, we will use the letter "(M)" to label such analogues.

For a vector space V, we denote by $\mathbb{T}(V) = \bigoplus_{i=0}^{\infty} \mathbb{T}^i(V)$, where $\mathbb{T}^i(V) = V^{\otimes i}$, the free associative algebra (tensor algebra) generated by V. The free commutative (symmetric) algebra generated by V is denoted $\mathbb{S}(V) = \bigoplus_{i=0}^{\infty} \mathbb{S}^i(V)$, and the free skew-commutative (exterior) algebra is denoted $\bigwedge(V) = \bigoplus_{i=0}^{\infty} \bigwedge^i(V)$. The free associative (noncommutative polynomial) algebra generated by a set of variables $x_1, \dots, x_m$ over a field $\Bbbk$ is denoted $\Bbbk\{x_1, \dots, x_m\}$ and the free commutative (polynomial) algebra generated by $x_1, \dots, x_m$ is denoted $\Bbbk[x_1, \dots, x_m]$.

For vector spaces V and W over $\Bbbk$ (or algebras, or modules over them) the tensor product over $\Bbbk$ is denoted simply by $V \otimes W$.

For a set of vectors v_i in a vector space V, we denote by $\langle v_i : i \in I \rangle \subset V$ their linear span. For a vector space W, we denote by $\mathbb{G}(W) = \coprod_{u=0}^{\dim W} \mathbb{G}_u(W)$ the Grassmannian variety of all vector subspaces $U \subset W$, where $u = \dim U$. By $\mathbb{P}^1_{\Bbbk} = \Bbbk \cup \{\infty\}$ we denote the projective line over $\Bbbk$.

With the exception of section 7 of chapter 6 we consider algebraic varieties over a fixed field $\Bbbk$. We say that x is a *point* of a variety X and we write $x \in X$ if x is a $\overline{\Bbbk}$-valued point of X, where $\overline{\Bbbk}$ is an algebraic closure of $\Bbbk$. When using topological notions for algebraic varieties we always mean these notions with respect to Zariski topology.

For a real number $x \in \mathbb{R}$, the symbol $\lceil x \rceil$ means the minimal integer which is not smaller than x, while $\lfloor x \rfloor$ is the maximal integer not greater than x. With the exception of sections 5 and 7 of chapter 7 we denote by $[m, n]$ the segment of integer numbers between m and n.

We use both chain and cochain complexes. In the former case we use lower indexing (as in $C_\bullet$) and the differential is lowering the degree by one: $C_i \to C_{i-1}$. In the latter case we use upper indexing (as in $K^\bullet$) and the differential is raising the degree by one: $K^i \to K^{i+1}$. For a chain complex of graded vector spaces $V_\bullet$ (where the differential preserves the grading) we denote by $H_{ij}V_\bullet$ the component of grading j in the i-th homology.

1. Bar constructions

In this section we consider bar and cobar constructions that provide explicit realizations of the functors Tor and Ext. Although we are mostly interested in the graded case we also consider these functors for nongraded algebras and modules (to be used in chapter 5). The reader can skip this chapter and refer to it when necessary.

For an algebra A and a pair of A-modules M, N we denote by $\mathrm{HOM}_A(M, N)$ the space of A-module homomorphisms and by $\mathrm{EXT}^i_A(M, N)$ the corresponding extension spaces (defined as derived functors of $\mathrm{HOM}_A(M, N)$).

Assume that A is augmented: $A = \Bbbk \oplus A_+$. Then for every left A-module M we have the following resolution by free A-modules, called the (*normalized*) *bar-resolution* $\widetilde{\mathcal{B}ar}_\bullet(A, M)$:

$$\cdots \longrightarrow A \otimes A_+ \otimes A_+ \otimes M \longrightarrow A \otimes A_+ \otimes M \longrightarrow A \otimes M \longrightarrow 0,$$

where $\widetilde{\mathcal{B}ar}_i(A, M) = A \otimes A_+^{\otimes i} \otimes M$ and the differential is given by

$$\partial(a_0 \otimes \cdots \otimes a_i \otimes m) = \sum_{s=1}^{i} (-1)^s a_0 \otimes \cdots \otimes a_{s-1}a_s \otimes \cdots \otimes a_i \otimes m + (-1)^{i+1} a_0 \otimes \cdots \otimes a_i m.$$

We view $\widetilde{\mathcal{B}ar}_i(A, M)$ as an A-module by letting A act on the left. (There is a similar construction for a nonaugmented algebra A, with A_+ replaced by A everywhere, called the *nonnormalized bar-resolution*.)

Since $\widetilde{\mathcal{B}ar}_\bullet(A, M)$ is a free resolution of M, we have

$$\mathrm{EXT}^i_A(M, N) = H^i(\mathrm{HOM}_A(\widetilde{\mathcal{B}ar}_\bullet(A, M), N)).$$

In other words, $\mathrm{EXT}^i_A(M, N)$ can be computed as the i-th cohomology of the *cobar-complex*

$$\mathcal{COB}^\bullet(A, M, N) = \mathrm{HOM}_A(\widetilde{\mathcal{B}ar}_\bullet(A, M), N).$$

Note that $\mathcal{COB}^i(A, M, N) \simeq \mathrm{HOM}_{\Bbbk}(A_+^{\otimes i} \otimes M, N)$.

Now assume that A is a *graded* algebra A and M, N is a pair of *graded* A-modules M. We will denote by $\mathrm{Ext}^i_A(M, N) = \bigoplus_{j\in\mathbb{Z}} \mathrm{Ext}^{ij}_A(M, N)$ the derived functor of the graded homomorphisms functor $\mathrm{Ext}^0_A(M, N) = \mathrm{Hom}_A(M, N) = \bigoplus_{j\in\mathbb{Z}} \mathrm{Hom}^j_A(M, N)$, where $\mathrm{Hom}^j_A(M, N)$ is the space of all homomorphisms mapping M_k to N_{k-j}. The first grading i is called the *homological grading* and the second j is called the *internal* one. As above we can compute these spaces using the graded cobar-complex:

$$\mathrm{Ext}^i_A(M, N) = H^i(\mathcal{C}ob^\bullet(A, M, N)),$$

where

$$\mathcal{C}ob^\bullet(A, M, N) = \mathrm{Hom}_A(\widetilde{\mathcal{B}ar}_\bullet(A, M), N) \simeq \mathrm{Hom}_{\Bbbk}(A_+^{\otimes\bullet} \otimes M, N).$$

This identification is compatible with internal gradings.

For an algebra A, a right A-module R, and a left A-module L, we denote by $\mathrm{Tor}^A_i(R, L)$ the derived functors of the tensor product over A, so that $\mathrm{Tor}^A_0(R, L) = R \otimes_A L$. If A is a graded algebra and the modules R and L are graded then the spaces $\mathrm{Tor}^A_i(R, L)$ acquire the corresponding *internal grading* induced by the grading of $R \otimes_A L$:

$$\mathrm{Tor}^A_i(R, L) = \bigoplus_{j\in\mathbb{Z}} \mathrm{Tor}^A_{ij}(R, L).$$

For example, $\mathrm{Tor}^A_{0,j} = (R \otimes_A L)_j$ is spanned by elements $x \otimes y$, where $x \in R_k$ and $y \in L_{j-k}$. Using the bar resolution above we can compute these spaces as homology of the *bar-complex*:

$$\mathrm{Tor}^A_i(R, L) = H_i(\mathcal{B}ar_\bullet(R, A, L)),$$

where

$$\mathcal{B}ar_\bullet(R, A, L) = R \otimes_A \widetilde{\mathcal{B}ar}_\bullet(A, L).$$

Note the following duality between bar and cobar complexes:

$$\mathcal{COB}^\bullet(A, M, R^\vee) = \mathcal{B}ar_\bullet(R, A, M)^\vee,$$

where for a vector space V we set $V^\vee = \mathrm{HOM}_{\Bbbk}(V, \Bbbk)$ (in the left-hand side we use the natural structure of a left A-module on $R^\vee$). This leads to the corresponding duality between homology:

$$\mathrm{EXT}^i_A(M, R^\vee) = \mathrm{Tor}^A_i(R, M)^\vee.$$

In the graded case we have a similar duality using graded duals:

$$\mathcal{C}ob^\bullet(A, M, R^*) = \mathcal{B}ar_\bullet(R, A, M)^*$$
$$\mathrm{Ext}^{ij}_A(M, R^*) = \mathrm{Tor}^A_{ij}(R, M)^*,$$

where R^* is graded by $(R^*)_k = (R_{-k})^*$.

Let A be a graded algebra, M be a graded left A-module. The spaces $\mathrm{Ext}_A^{ij}(M,\Bbbk)$ and $\mathrm{Tor}_{ij}^A(\Bbbk, M)$ are called respectively *cohomology* and *homology* spaces of the module M. A particular case of the above duality is

$$\mathrm{Ext}_A^{ij}(M,\Bbbk) = \mathrm{Tor}_{ij}^A(\Bbbk, M)^*.$$

We denote the relevant bar and cobar complexes by

$$\mathcal{B}ar_\bullet(A, M) = \mathcal{B}ar_\bullet(\Bbbk, A, M),$$
$$\mathcal{C}ob^\bullet(A, M) = \mathcal{C}ob^\bullet(A, M, \Bbbk) = \mathcal{B}ar_\bullet(A, M)^*.$$

In the case $M = \Bbbk$ we set

$$\mathcal{B}ar_\bullet(A) = \mathcal{B}ar_\bullet(A, \Bbbk),$$
$$\mathcal{C}ob^\bullet(A) = \mathcal{C}ob^\bullet(A, \Bbbk) = \mathcal{B}ar_\bullet(A)^*,$$

where $\mathcal{B}ar_i(A) = A_+^{\otimes i}$.

The Yoneda multiplication (composition) on the Ext-spaces can be described as follows. Let M, N, P be left A-modules and

$$c' : A_+^{\otimes i'} \otimes N \longrightarrow P, \quad c'' : A_+^{\otimes i''} \otimes M \longrightarrow N$$

be cocycles representing some classes $\xi' \in \mathrm{EXT}_A^{i'}(N, P)$ and $\xi'' \in \mathrm{EXT}_A^{i''}(M, N)$ respectively. Starting from c'', define the corresponding morphism of the resolutions

$$\tilde{c}'' : \widetilde{\mathcal{B}ar}_{\bullet+i''}(A, M) \longrightarrow \widetilde{\mathcal{B}ar}_\bullet(A, N)$$

as

$$\tilde{c}''_i : A \otimes A_+^{\otimes i} \otimes A_+^{\otimes i''} \otimes M \longrightarrow A \otimes A_+^{\otimes i} \otimes N$$
$$a_0 \otimes x \otimes y \otimes m \longmapsto a_0 \otimes x \otimes c''(y \otimes m).$$

It is easy to verify that $\tilde{c}''$ is a morphism of complexes of A-modules and the composition $\widetilde{\mathcal{B}ar}_{i''}(A, M) \longrightarrow \widetilde{\mathcal{B}ar}_0(A, N) \longrightarrow P$ coincides with the original cocycle c''. Therefore, the composition class $\xi' \circ \xi'' \in \mathrm{EXT}_A^{i'+i''}(M, N)$ is represented by the cocycle $c' \circ \tilde{c}'' : \widetilde{\mathcal{B}ar}_{i'+i''}(A, M) \longrightarrow \widetilde{\mathcal{B}ar}_{i'}(A, N) \longrightarrow P$, which is given explicitly by

$$c' \circ c'' : A_+^{\otimes i'} \otimes A_+^{\otimes i''} \otimes M \longrightarrow P$$
$$x \otimes y \otimes m \longmapsto c'(x \otimes c''(y \otimes m)).$$

The same computation works for graded Ext-spaces.

The dual *comultiplication* structure on the Tor-spaces can be explicitly described as follows. Let P be a finite-dimensional left A-module. Then we have a natural identity element $\mathrm{id}_P \in P \otimes_\Bbbk P^*$. Therefore, for every pair (R, L), where R is a right A-module and L is a left A-module, we have a natural map of vector spaces, functorial in R and L

$$R \otimes_A L \longrightarrow (R \otimes_A P) \otimes_\Bbbk (P^* \otimes_A L)$$

and hence the induced map

$$\Delta : \mathrm{Tor}_i^A(R, L) \longrightarrow \mathrm{Tor}_{i'}^A(R, P) \otimes_\Bbbk \mathrm{Tor}_{i''}^A(P^*, L), \qquad i' + i'' = i.$$

This map is dual to the Yoneda product

$$\mathrm{EXT}_A^{i'}(P, R^\vee) \otimes \mathrm{EXT}_A^{i''}(L, P) \longrightarrow \mathrm{EXT}_A^i(L, R^\vee).$$

On the level of bar-complexes it is given by the formula

$$\begin{aligned}\Delta\colon R\otimes A_+^{\otimes i'}\otimes A_+^{\otimes i''}\otimes L &\longrightarrow (R\otimes A_+^{\otimes i'}\otimes P)\otimes(P^*\otimes A_+^{\otimes i''}\otimes L)\\ r\otimes x\otimes y\otimes l &\longmapsto r\otimes x\otimes \mathrm{id}_P\otimes y\otimes l.\end{aligned}$$

The Yoneda multiplication defines a graded algebra structure on $\mathrm{EXT}_A^*(\Bbbk,\Bbbk) = \bigoplus_{i=0}^\infty \mathrm{EXT}_A^i(\Bbbk,\Bbbk)$ and a structure of a graded left $\mathrm{EXT}_A^*(\Bbbk,\Bbbk)$-module on $\mathrm{EXT}_A^*(M,\Bbbk)=\bigoplus_{i=0}^\infty \mathrm{EXT}_A^i(M,\Bbbk)$. Analogously, the above construction (for $P = R = \Bbbk$ and L equal to $\Bbbk$ or M) makes $\mathrm{Tor}_*^A(\Bbbk,\Bbbk)=\bigoplus_{i=0}^\infty \mathrm{Tor}_i^A(\Bbbk,\Bbbk)$ into a *graded coalgebra* and $\mathrm{Tor}_*^A(\Bbbk,M)=\bigoplus_{i=0}^\infty \mathrm{Tor}_i^A(\Bbbk,M)$ a *graded left comodule* over it. The same structures appear for graded Ext and Tor. Moreover, in this case there is an additional internal grading, so that we get a bigraded algebra and bigraded module (resp., bigraded coalgebra and bigraded comodule).

Now assume that A and M are locally finite-dimensional with $A_i=0$ for $i<0$ and $M_i=0$ for $i\ll 0$. Then we have natural isomorphisms

$$\mathcal{C}ob^i(A)=(A_+^{\otimes i})^*=A_+^{*\otimes i},\qquad \mathcal{C}ob^i(A,M)=(A_+^{\otimes i}\otimes M)^*=A_+^{*\otimes i}\otimes M^*,$$

where the first tensor component in $A_+^{\otimes i}$ is coupling with the first tensor component in $A_+^{*\otimes i}$, etc. The cobar-differential on $\mathcal{C}ob^\bullet(A,M)$ takes the form

$$d(f_1\otimes\cdots\otimes f_i\otimes g)=$$
$$\sum_{s=1}^{i}(-1)^{s-1}f_1\otimes\cdots\otimes d_{1,A}(f_s)\otimes\cdots\otimes f_i\otimes g+(-1)^i f_1\otimes\cdots\otimes f_i\otimes d_{1,M}(g),$$

where $f_s\in A_+^*$, $g\in M^*$, and the map $d_{1,A}\colon A_+^*\longrightarrow A_+^*\otimes A_+^*$ (resp., $d_{1,M}\colon M^*\longrightarrow A_+^*\otimes M^*$) is dual to the multiplication $A_+\otimes A_+\longrightarrow A_+$ (resp., the action $A_+\otimes M\longrightarrow M$). A similar formula holds for the differential on $\mathcal{C}ob^\bullet(A)$. The algebra structure on $\mathcal{C}ob^\bullet(A)$ and the left $\mathcal{C}ob^\bullet(A)$-module structure on $\mathcal{C}ob^\bullet(A,M)$ are given by the formulas

$$\begin{aligned}(f_1'\otimes\cdots\otimes f_{i'}')(f_1''\otimes\cdots\otimes f_{i''}'')&=f_1'\otimes\cdots\otimes f_{i'}'\otimes f_1''\otimes\cdots\otimes f_{i''}''\\ (f_1\otimes\cdots\otimes f_i)(f_1'\otimes\cdots\otimes f_{i'}'\otimes g)&=f_1\otimes\cdots\otimes f_i\otimes f_1'\otimes\cdots\otimes f_{i'}'\otimes g\end{aligned}$$

and are easily verified to be compatible with the differential d by the usual Leibniz rules, so that $\mathcal{C}ob^\bullet(A)$ is a DG-algebra and $\mathcal{C}ob^\bullet(A,M)$ is a DG-module over it (see Appendix). Disregarding the differential, we can say that $\mathcal{C}ob(A)=\mathbb{T}(A_+^*)$ and $\mathcal{C}ob(A,M)=\mathcal{C}ob(A)\otimes M^*$ is the free associative graded algebra generated by A_+^* and the free graded left $\mathcal{C}ob(A)$-module generated by M^*. The induced multiplicative structures on the cohomology spaces $\mathrm{Ext}_A^*(\Bbbk,\Bbbk)$ and $\mathrm{Ext}_A^*(M,\Bbbk)$ are immediately found to coincide with the Yoneda multiplication computed above.

Analogously, the coalgebra and comodule structures on the spaces $\mathrm{Tor}_*^A(\Bbbk,\Bbbk)$ and $\mathrm{Tor}_*^A(\Bbbk,M)$ are induced by the *DG-coalgebra* and *DG-comodule* structures on the complexes $\mathcal{B}ar_\bullet(A)$ and $\mathcal{B}ar_\bullet(A,M)$ defined above.

In particular, we see that for the opposite graded algebra A^{op} the Ext-spaces are the same as for A: $\mathrm{Ext}_{A^{\mathrm{op}}}^{ij}(\Bbbk,\Bbbk)=\mathrm{Ext}_A^{ij}(\Bbbk,\Bbbk)$, and the Yoneda multiplication just changes to the opposite one. (Another way to deduce the first property is to note that $\mathrm{Ext}_A^{ij}(\Bbbk,\Bbbk)=\mathrm{Tor}_{ij}^A(\Bbbk,\Bbbk)^*$ and the latter spaces for A and A^{op} coincide by the definition.)

2. Quadratic algebras and modules

Recall that we assume our graded algebras A (resp., graded A-modules M) to be locally finite-dimensional with $A_i = 0$ for $i < 0$ and $A_0 = \mathbb{k}$ (resp., locally finite-dimensional and bounded below).

Definition 1. A graded algebra A is called *one-generated* if the natural map $p\colon \mathbb{T}(A_1) \longrightarrow A$ from the tensor algebra generated by A_1 is surjective. A one-generated algebra is called *quadratic* if the kernel $J_A = \ker p$ is generated as a two-sided ideal in $\mathbb{T}(A_1)$ by its subspace $I_A = J_A \cap \mathbb{T}^2(A_1) \subset A_1 \otimes A_1$. Therefore, a quadratic algebra A is determined by a vector space of generators $V = A_1$ and an arbitrary subspace of quadratic relations $I \subset V \otimes V$. We denote this by $A = \{V, I\}$.

For every graded algebra A, there is a uniquely defined quadratic algebra $\mathrm{q}A$ together with an algebra homomorphism $\mathrm{q}A \longrightarrow A$ which is an isomorphism in degree 1 and a monomorphism in degree 2. Namely, in the above notation one has $\mathrm{q}A = \{A_1, I_A\}$. We call this algebra the *quadratic part* of A.

The *quadratic dual algebra* to a quadratic algebra $A = \{V, I\}$ is defined by $A^! = \{V^*, I^\perp\}$, where V^* is the dual vector space to V and $I^\perp \subset V^* \otimes V^*$ is the orthogonal complement to I with respect to the natural pairing $\langle v_1 \otimes v_2,\, v_1^* \otimes v_2^* \rangle = \langle v_1, v_1^* \rangle \langle v_2, v_2^* \rangle$ between $V \otimes V$ and $V^* \otimes V^*$.

Definition 2(M). Let $A = \{V, I\}$ be a quadratic algebra. A left A-module M is called *quadratic* if $M_i = 0$ for $i < 0$, the natural map $A \otimes M_0 \longrightarrow M$ is surjective, and its kernel J_M is generated as an A-submodule in $A \otimes M_0$ by the subspace $K_M = J_M \cap A_1 \otimes M_0$. In other words, a quadratic A-module is determined by a vector space of generators $H = M_0$ and an arbitrary subspace of relations $K \subset V \otimes H$. We denote this by $M = \langle H, K \rangle = \langle H, K \rangle_A$.

Note that we *do not* consider quadratic modules over nonquadratic algebras.

For every graded algebra A and a nonnegatively graded A-module M there is a natural quadratic module $\mathrm{q}_A M$ over the quadratic algebra $\mathrm{q}A$ together with a morphism $\mathrm{q}_A M \longrightarrow M$ of modules over $\mathrm{q}A$ which is an isomorphism in degree 0 and a monomorphism in degree 1. In other words, one has $\mathrm{q}_A M = \langle M_0, K_M \rangle_{\mathrm{q}A}$. We call this module the *quadratic part* of M.

For a quadratic module $M = \langle H, K \rangle_A$ over a quadratic algebra $A = \{V, I\}$ we define the *quadratic dual module* $M^! = M^!_A$ over the quadratic algebra $A^!$ dual to A by $M^!_A = \langle H^*, K^\perp \rangle_{A^!}$, where $K^\perp \subset V^* \otimes H^*$ is the orthogonal complement to $K \subset V \otimes H$.

Examples. 1. If $A = \mathbb{S}(V)$ is the symmetric algebra then $A^! = \bigwedge(V^*)$ is the exterior algebra with the dual space of generators.
2. If $A = \mathbb{T}(V)$ is the tensor algebra then $A^! = \mathbb{k} \oplus V^*$ is an algebra with $A^!_i = 0$ for $i \geqslant 2$.
3. For any quadratic algebra A the duality for modules sends free A-modules to trivial $A^!$-modules and vice versa: $\mathbb{k}^!_A = A^!$ and $A^!_A = \mathbb{k}$.
4. If A is a commutative quadratic algebra then quadratic relations in $A^!$ are linear combinations of odd supercommutators of generators. Hence, $A^!$ is isomorphic to the universal enveloping algebra $U(\mathfrak{g})$ of a graded Lie superalgebra $\mathfrak{g}$ (where $\mathfrak{g}$ is quadratic, i.e., generated by $\mathfrak{g}_1$ with defining relation of degree 2).
5. Similarly to the previous example, the dual of a skew-commutative quadratic algebra is the universal enveloping algebra of a usual Lie algebra (also quadratic). For example, let V be a finite-dimensional vector space with a skew-symmetric form

$\omega \neq 0$. Then we can consider the quadratic algebra $A = \bigwedge(V^*)/(\omega)$. The dual quadratic algebra $A^!$ is isomorphic to the universal enveloping algebra $U(\mathfrak{h})$, where $\mathfrak{h} = \mathfrak{h}_\omega$ is the graded Lie algebra generated by $\mathfrak{h}_1 = V$ and by one more element c of degree 2 with defining relations $[v, v'] = \omega(v, v')c$. It is easy to see that if the rank of ω is $\geqslant 4$ then these relations imply that c is central, so in this case $\mathfrak{h}$ is the Heisenberg central extension of V by $\Bbbk$ associated with ω. For example, if ω is nondegenerate and $\dim V = 2n \geqslant 4$, then $\mathfrak{h}$ is the standard Heisenberg Lie algebra and A can be presented as follows: it is generated by $2n$ elements $x_1, \dots, x_n, y_1, \dots, y_n$ with the relations

$$[x_i, x_j] = [y_i, y_j] = 0 \text{ for } 1 \leqslant i, j \leqslant n,$$
$$[x_i, y_j] = 0 \text{ for } i \neq j,$$
$$[x_1, y_1] = [x_2, y_2] = \dots = [x_n, y_n].$$

6. If M is a graded module over a graded algebra A then we can introduce an algebra structure on $A_M = A \oplus M(-1)$ by the rule $(a, m)(a', m') = (aa', am')$. It is easy to see that if A is a quadratic algebra and M is a quadratic A-module then A_M is again a quadratic algebra. This observation can often be used to deduce module analogues of results on quadratic algebras.

3. Diagonal cohomology

The following result which is essentially due to S. Priddy and C. Löfwall [**104, 75**], gives a cohomological interpretation of quadratic duality.

PROPOSITION 3.1. *For any graded algebra A and a nonnegatively graded A-module M all nonzero cohomology spaces $\mathrm{Ext}_A^{ij}(M, \Bbbk)$ are concentrated in the region $i \leqslant j$. The diagonal subalgebra $\bigoplus \mathrm{Ext}_A^{ii}(\Bbbk, \Bbbk)$ of the algebra $\bigoplus \mathrm{Ext}_A^{ij}(\Bbbk, \Bbbk)$ is always a quadratic algebra and the diagonal submodule $\bigoplus \mathrm{Ext}_A^{ii}(M, \Bbbk)$ is a quadratic module over it. More precisely,*

$$\bigoplus \mathrm{Ext}_A^{ii}(\Bbbk, \Bbbk) \simeq (\mathrm{q}A)^!, \qquad \bigoplus \mathrm{Ext}_A^{ii}(M, \Bbbk) \simeq (\mathrm{q}_A M)^!_{(\mathrm{q}A)}.$$

Proof: As we have seen in section 2, these Ext-spaces can be computed as the cohomology of the cobar-complexes

$$\mathcal{C}ob^{ij}(A) = \sum_{k_1 + \cdots + k_i = j,\ k_s \geqslant 1} A_{k_1}^* \otimes \cdots \otimes A_{k_i}^*$$
$$\mathcal{C}ob^{ij}(A, M) = \sum_{k_1 + \cdots + k_i + l = j,\ k_s \geqslant 1,\ l \geqslant 0} A_{k_1}^* \otimes \cdots \otimes A_{k_i}^* \otimes M_l^*.$$

All the assertions follow easily from this. □

Note that if A is one-generated then the diagonal part of $\mathrm{Ext}_A^*(\Bbbk, \Bbbk)$ coincides with the subalgebra generated by $\mathrm{Ext}_A^1(\Bbbk, \Bbbk)$. To generate non-diagonal pieces one needs higher Massey products (see [**82**] and Appendix).

4. Minimal resolutions

In this section we prove that every graded module (bounded below) over a (nonnegatively) graded algebra admits a minimal free resolution, unique up to an isomorphism.

We start with the following version of Nakayama's lemma for noncommutative graded algebras.

LEMMA 4.1. *Let A be a graded algebra and M a graded A-module. Then a graded vector subspace $X \subset M$ generates M as an A-module iff the composition $X \longrightarrow M \longrightarrow \Bbbk \otimes_A M$ is surjective.*

Proof: X generates M iff the natural morphism $f : A \otimes X \longrightarrow M$ is surjective. It is clear that this implies surjectivity of the map $\overline{f} : X \longrightarrow \Bbbk \otimes_A M$. Conversely, assume that $\overline{f}$ is surjective and let us show that f is surjective. We can argue by induction in n that the degree n component f_n is surjective. This is trivially true for $n \ll 0$. Assume that f_i is surjective for all $i < n$. Given an element $m \in M_n$ there exists $x \in X_n$ such that $m - x \in A_+ M$. Hence, by the assumption $m - x$ belongs to the image of f, so m is also in the image of f. □

It follows immediately from the above lemma that inside any generating subspace $X \subset M$ one can find a smaller generating subspace $X' \subset X$ such that the map $X' \longrightarrow \Bbbk \otimes_A M$ is an isomorphism.

Definition. A bounded above complex of free graded A-modules

$$\cdots \longrightarrow P_2 \longrightarrow P_1 \longrightarrow P_0 \longrightarrow 0$$

is called *minimal* if all the induced maps $\Bbbk \otimes_A P_{i+1} \longrightarrow \Bbbk \otimes_A P_i$ vanish.

For every graded A-module M, one can construct a minimal free graded A-module resolution of M in the following way. Using Lemma 4.1 choose a generating subspace $X_0 \subset M$ such that $X_0 \simeq \Bbbk \otimes_A M$. Consider the corresponding morphism $A \otimes X_0 \longrightarrow M$ and let $M_1 \subset A \otimes X_0$ be its kernel. Choose a generating subspace $X_1 \subset M_1$ such that $X_1 \simeq \Bbbk \otimes_A M_1$. It is easy to see that continuing in this manner we will find a minimal resolution of M consisting of the modules $P_i = A \otimes X_i$. Since any two free resolutions of the same module M are connected by a chain map inducing the identity on M, it follows that a minimal resolution is unique up to a nonunique isomorphism.

In fact, the following more general statement is true.

PROPOSITION 4.2. *Every bounded above complex $F_\bullet$ of free graded A-modules admits a decomposition $F_\bullet = P_\bullet \oplus T_\bullet$ into the direct sum of two subcomplexes of free graded A-modules, where the complex $P_\bullet$ is minimal and the complex $T_\bullet$ is acyclic.*

Proof: Assume that our complex

$$\cdots \longrightarrow F_{i+1} \xrightarrow{d_{i+1}} F_i \xrightarrow{d_i} F_{i-1} \longrightarrow \cdots$$

satisfies $\overline{d}_i = \Bbbk \otimes d_i = 0$ for all $i \leqslant n$ for some $n \in \mathbb{Z}$. This is certainly true for sufficiently small n since $F_\bullet$ is bounded above. We are going to show that in this case there exists a decomposition $F_\bullet \simeq F'_\bullet \oplus T_\bullet$ into a direct sum of subcomplexes of free graded modules, where $T_\bullet$ is an acyclic complex concentrated in degrees $n+1$ and n, and the differentials d'_i in $F'_\bullet$ satisfy $\overline{d'}_i = 0$ for $i \leqslant n+1$. Iteration of this process would give the required decomposition. To construct $F'_\bullet$ and $T_\bullet$ as above let us consider the A-module $M = F_n / d_{n+1}(F_{n+1})$. Let $F_i = A \otimes X_i$ for $i \in \mathbb{Z}$. Since X_n generates M there exists a graded subspace $Y \subset X_n$ such that the map $Y \longrightarrow \Bbbk \otimes_A M$ is an isomorphism. Set $F'_n = Y \otimes A$. We claim that the composed map $f : F_{n+1} \longrightarrow F_n \longrightarrow X_n / Y \otimes A$ is surjective. Indeed, the map $F'_n \longrightarrow M$ is surjective by Lemma 4.1, hence $F_n = F'_n + d_{n+1}(F_{n+1})$ and therefore the map $f : F_{n+1} \longrightarrow X_n / Y \otimes A$ is surjective. Let $F'_{n+1} \subset F_{n+1}$ be the kernel of f. Note that $F'_{n+1} = \overline{d}_{n+1}^{-1}(Y) \otimes A$. Let us set $F'_i = F_i$ for $i \neq n, n+1$. Since the map

$F_{n+2} \longrightarrow F_{n+1}$ factors through F'_{n+1} we obtain a subcomplex $F'_\bullet \subset F_\bullet$. Moreover, we have a split exact triple of complexes

$$\begin{array}{ccccccccccc}
\cdots & \longrightarrow & F'_{n+2} & \longrightarrow & F'_{n+1} & \xrightarrow{d'_{n+1}} & F'_n & \longrightarrow & F'_{n-1} & \longrightarrow & \cdots \\
 & & \downarrow & & \downarrow & & \downarrow & & \downarrow & & \\
\cdots & \longrightarrow & F_{n+2} & \longrightarrow & F_{n+1} & \xrightarrow{d_{n+1}} & F_n & \longrightarrow & F_{n-1} & \longrightarrow & \cdots \\
 & & \downarrow & & \downarrow & & \downarrow & & \downarrow & & \\
\cdots & \longrightarrow & 0 & \longrightarrow & X_n/Y \otimes A & \xrightarrow{\mathrm{id}} & X_n/Y \otimes A & \longrightarrow & 0 & \longrightarrow & \cdots
\end{array}$$

We claim that $\overline{d'}_{n+1} = 0$. Indeed, by the choice of Y we have a direct sum decomposition $X_n = \overline{d}_{n+1}(X_{n+1}) \oplus Y$. Therefore, $\overline{d}^{-1}_{n+1}(Y)$ coincides with the kernel of $\overline{d}_{i+1}$ and the map

$$\Bbbk \otimes_A F'_{n+1} = \overline{d}^{-1}_{n+1}(Y) \longrightarrow Y = \Bbbk \otimes_A F'_n$$

is zero. It is also clear that $\overline{d'}_n = 0$ and $\overline{d'}_i = 0$ for $i < n$. It remains to choose a free submodule $T_{i+1} \subset F_{i+1}$ that projects isomorphically to $X/Y \otimes A$ and to set $T_i = d_{i+1}(T_{i+1})$. □

Remark. The decomposition considered in the above proposition is unique up to an automorphism of $F_\bullet$. One can also check that a morphism between minimal complexes inducing an isomorphism on homology is an isomorphism itself.

The following class of minimal resolutions will play an important role for the notion of Koszulness.

Definition. A resolution

$$\cdots \longrightarrow P_2 \longrightarrow P_1 \longrightarrow P_0 \longrightarrow M \longrightarrow 0$$

of a graded A-module M by free graded A-modules is called a *linear free resolution* if each P_i is generated in degree i.

A linear free resolution can be written in the form

$$\cdots \longrightarrow V_2 \otimes A(-2) \longrightarrow V_1 \otimes A(-1) \longrightarrow V_0 \otimes A \longrightarrow M \longrightarrow 0,$$

where V_i are vector spaces (of degree zero).

It is clear that linear free resolutions are minimal. Note also that M admits a linear free resolution iff $\mathrm{Tor}^A_{ij}(\Bbbk, M) = 0$ for $i \neq j$ (equivalently, $\mathrm{Ext}^{ij}_A(M, \Bbbk) = 0$ for $i \neq j$). Finally, we observe that if a module M admits a linear free resolution then it is unique up to *unique* isomorphism. Indeed, since P_i is generated in degree i, it follows that an endomorphism of $P_\bullet$ is determined by its action on $\Bbbk \otimes_A P_\bullet$, i.e., on the spaces $\mathrm{Tor}^A_i(\Bbbk, M)$. But this action is trivial for any endomorphism of $P_\bullet$ inducing the identity on M.

In fact, as we will see in section 3 for every nonnegatively graded A-module M one can construct a certain complex of free A-modules that will coincide with the linear free resolution when M admits one.

5. Low-dimensional cohomology

In this section we prove the well-known result that the spaces $\mathrm{Tor}^A_1(\Bbbk, \Bbbk)$ and $\mathrm{Tor}^A_2(\Bbbk, \Bbbk)$ describe respectively minimal homogeneous generators and relations of a graded algebra A, while the spaces $\operatorname{Tor}_0^A(\Bbbk, M)$ and $\operatorname{Tor}_1^A(\Bbbk, M)$ correspond to minimal generators and relations of a graded A-module M.

PROPOSITION 5.1. *Let A be a graded algebra. Assume that a graded A-module M is represented as the cokernel of a morphism of free graded A-modules,*

$$A \otimes Y \longrightarrow A \otimes X \longrightarrow M \longrightarrow 0,$$

where $X = \bigoplus_{j\in\mathbb{Z}} X_j$ and $Y = \bigoplus_{j\in\mathbb{Z}} Y_j$ are graded vector spaces. Then the spaces $\operatorname{Tor}_{0,j}^A(\Bbbk, M) = (M/A_+M)_j$ are naturally identified with certain quotients of the spaces X_j, while the spaces $\operatorname{Tor}_{1,j}^A(\Bbbk, M)$ are subquotients of the spaces Y_j. Furthermore, for any module M there exists a presentation of the above type such that we have natural isomorphisms $X_j \simeq \operatorname{Tor}_{0,j}^A(\Bbbk, M)$ and $Y_j \simeq \operatorname{Tor}_{1,j}^A(\Bbbk, M)$.

Proof: Let us extend the resolution one step further to the left, that is, choose a morphism $A \otimes Z \longrightarrow A \otimes Y$ such that the resulting complex is exact at the term $A \otimes Y$. Tensoring with the right A-module $\Bbbk$, we get a complex of graded vector spaces $Z \xrightarrow{\psi} Y \xrightarrow{\varphi} X$. By the definition, we have $\operatorname{Tor}_0^A(\Bbbk, M) \simeq \operatorname{coker}\varphi$ and $\operatorname{Tor}_1^A(\Bbbk, M) \simeq \ker\varphi/\operatorname{im}\psi$. As was explained in section 4 one can choose a representing morphism $A \otimes Y \longrightarrow A \otimes X$ for M such that $\varphi = \psi = 0$. This proves the last statement. □

PROPOSITION 5.2. *Assume that a graded algebra A is represented as the quotient of the free algebra $\mathbb{T}(X)$ generated by a graded vector space X by the ideal generated by a graded subspace $Y \subset \mathbb{T}(X)$,*

$$\mathbb{T}(X) \otimes Y \otimes \mathbb{T}(X) \longrightarrow \mathbb{T}(X) \longrightarrow A \longrightarrow 0.$$

Then the spaces $\operatorname{Tor}_{1,j}^A(\Bbbk, \Bbbk) = (A_+/A_+^2)_j$ are naturally identified with certain quotients of the spaces X_j, while the spaces $\operatorname{Tor}_{2,j}^A(\Bbbk, \Bbbk)$ are subquotients of the spaces Y_j. For every graded algebra A there exists a presentation of this type such that $X_j \simeq \operatorname{Tor}_{1,j}^A(\Bbbk, \Bbbk)$ and $Y_j \simeq \operatorname{Tor}_{2,j}^A(\Bbbk, \Bbbk)$.

Proof: More generally, assume that an augmented algebra A is the quotient algebra of an augmented algebra B by a two-sided ideal J. Then we claim that there is a natural five-term exact sequence

$$\operatorname{Tor}_2^B(\Bbbk, \Bbbk) \longrightarrow \operatorname{Tor}_2^A(\Bbbk, \Bbbk) \longrightarrow J/(B_+J + JB_+) \\ \longrightarrow \operatorname{Tor}_1^B(\Bbbk, \Bbbk) \longrightarrow \operatorname{Tor}_1^A(\Bbbk, \Bbbk) \longrightarrow 0.$$

Indeed, it is obtained from the following exact triple of complexes, two of which are just initial fragments of the bar-complexes for A and B:

$$\begin{array}{ccccccccc}
0 & \longrightarrow & J & \longrightarrow & B_+ & \longrightarrow & A_+ & \longrightarrow & 0 \\
 & & \uparrow & & \uparrow & & \uparrow & & \\
0 & \longrightarrow & B_+ \otimes J \oplus J \otimes B_+ & \longrightarrow & B_+ \otimes B_+ & \longrightarrow & A_+ \otimes A_+ & \longrightarrow & 0 \\
 & & \uparrow & & \uparrow & & \uparrow & & \\
0 & \longrightarrow & \text{kernel} & \longrightarrow & B_+ \otimes B_+ \otimes B_+ & \longrightarrow & A_+ \otimes A_+ \otimes A_+ & \longrightarrow & 0
\end{array}$$

In the case $B = \mathbb{T}(X)$ and $J = (Y)$, we have $\operatorname{Tor}_1^B(\Bbbk, \Bbbk) = X$ and $\operatorname{Tor}_2^B(\Bbbk, \Bbbk) = 0$. On the other hand, the space Y maps surjectively onto $J/(B_+J + JB_+)$, so the first assertion follows.

It remains to construct a presentation of A with the required minimality property. First, arguing as in Lemma 4.1, it is easy to check that a homogeneous vector subspace $X \subset A_+$ generates the algebra A iff it maps surjectively on the quotient space A_+/A_+^2. Therefore, one can choose a generating subspace X such that it maps isomorphically to this quotient. Then it follows from the above five-term exact sequence that $\mathrm{Tor}_2^A(\Bbbk, \Bbbk) \simeq J/(B_+J + JB_+)$. It remains to use Lemma 4.1 for the B-B-bimodule J to find an appropriate minimal relations space Y. □

Using duality between $\mathrm{Tor}_{ij}^A(\Bbbk, M)$ and $\mathrm{Ext}_A^{ij}(M, \Bbbk)$ we deduce the following

COROLLARY 5.3. *Let A be a graded algebra, M be a nonnegatively graded A-module.*
(i) A is one-generated iff $\mathrm{Ext}_A^{1,j}(\Bbbk, \Bbbk) = 0$ for $j > 1$;
(ii) A is quadratic iff $\mathrm{Ext}_A^{i,j}(\Bbbk, \Bbbk) = 0$ for $j > i$ and $i = 1, 2$.
(iiM) Assume that A is quadratic. Then M is quadratic iff $\mathrm{Ext}_A^{i,j}(M, \Bbbk) = 0$ for $j > i$ and $i = 0, 1$. □

Alternatively, the relation between minimal generators and relations and Tor-spaces can be described as follows. For a graded algebra A and integer $k \geqslant 1$ we denote by $A^{\langle k \rangle}$ the graded algebra with the same generators and relations of degree $< k$ as in A, and with no generators or defining relations of degree $\geqslant k$. Thus, we have $\mathrm{Tor}_{1,j}^{A^{\langle k \rangle}}(\Bbbk, \Bbbk) = \mathrm{Tor}_{2,j}^{A^{\langle k \rangle}}(\Bbbk, \Bbbk) = 0$ for $j \geqslant k$ and there is a morphism of graded algebras $A^{\langle k \rangle} \longrightarrow A$ which is an isomorphism in degree $< k$. Analogously, for a graded A-module M and integer k denote by $M^{\langle k \rangle}$ the graded A-module with $\mathrm{Tor}_{0,j}^A(M^{\langle k \rangle}, \Bbbk) = \mathrm{Tor}_{1,j}^A(M^{\langle k \rangle}, \Bbbk) = 0$ equipped with a morphism of graded modules $M^{\langle k \rangle} \longrightarrow M$ which is an isomorphism in degree $< k$. Then there are natural exact sequences

$$0 \longrightarrow \mathrm{Tor}_{2,k}^A(\Bbbk, \Bbbk) \longrightarrow A_k^{\langle k \rangle} \longrightarrow A_k \longrightarrow \mathrm{Tor}_{1,k}^A(\Bbbk, \Bbbk) \longrightarrow 0$$

$$0 \longrightarrow \mathrm{Tor}_{1,k}^A(M, \Bbbk) \longrightarrow M_k^{\langle k \rangle} \longrightarrow M_k \longrightarrow \mathrm{Tor}_{0,k}^A(M, \Bbbk) \longrightarrow 0.$$

6. Lattices and distributivity

Here we present some results from the general lattice theory. Most of them are due to B. Jónsson [**69**] and Musti–Buttafuoco [**86**].

A *lattice* is a discrete set Ω endowed with two idempotent (i.e., $a \star a = a$), commutative, and associative binary operations $\wedge, \vee \colon \Omega \times \Omega \longrightarrow \Omega$ satisfying the following *absorption identities*:

$$a \wedge (a \vee b) = a, \quad (a \wedge b) \vee b = b.$$

We write $a \leqslant b$, if the equivalent conditions $a \wedge b = a$ or $a \vee b = b$ hold. The dual lattice Ω° is obtained by switching the operations $\wedge$ and $\vee$.

A lattice is called *distributive* if it satisfies the following *distributivity identity*:

$$a \wedge (b \vee c) = (a \wedge b) \vee (a \wedge c).$$

We will see that dual lattices Ω and Ω° are distributive simultaneously. However, the distributivity condition above for *fixed* elements a, b, c is not self-dual in general, except in the case of a *modular lattice*.

A lattice is called *modular*, if the distributivity identity holds for any triple of its elements a, b, c such that $a \geqslant c$. Equivalently, one should have

$$a \geqslant c \implies a \wedge (b \vee c) = (a \wedge b) \vee c.$$

This condition is clearly self-dual.

LEMMA 6.1. [**69**] *For any triple of elements x, y, z of a modular lattice Ω, the following two dual distributivity conditions are equivalent and do not change after permuting the elements x, y, z:*

$$x \wedge (y \vee z) = (x \wedge y) \vee (x \wedge z), \quad x \vee (y \wedge z) = (x \vee y) \wedge (x \vee z).$$

Proof [**69**]: If the first equation holds, then by modularity and absorption we have

$$(z \vee x) \wedge (z \vee y) = z \vee (x \wedge (z \vee y)) = z \vee (x \wedge z) \vee (x \wedge y) = z \vee (x \wedge y).$$

The other implications can be obtained by duality and permutation of x, y, z. □

From now on we assume our lattices to be modular. A triple of elements x, y, z of a lattice Ω is called *distributive* if it satisfies the equivalent conditions of the above lemma.

A *sublattice* of a lattice Ω is a subset closed under both operations $\vee$ and $\wedge$. *The sublattice generated by a subset* $\mathcal{X} \subset \Omega$ consists of all elements of Ω that can be obtained from the elements of $\mathcal{X}$ using these operations. Note that a finitely generated distributive lattice is finite. The analogous statement for modular lattices is not true [**64**].

Our main goal is to find a system of equations on the elements of $\mathcal{X}$ that would guarantee distributivity of the sublattice generated by $\mathcal{X}$. Temporarily, let us make the following definition.

Definition. A subset $\mathcal{X} \subset \Omega$ is called *distributive* if for any pair of finite subsets $\mathcal{Y}$, $\mathcal{Z} \subset \mathcal{X}$ and any element $x \in \mathcal{X}$ the triple

$$\bigvee_{y \in \mathcal{Y}} y, \quad x, \quad \bigwedge_{z \in \mathcal{Z}} z$$

is distributive. It follows easily from modularity that it is enough to consider non-intersecting x, $\mathcal{Y}$, and $\mathcal{Z} \subset \mathcal{X}$.

Note that if $\mathcal{X}$ is distributive then for any pair of finite subsets $\mathcal{Y}$, $\mathcal{Z} \subset \mathcal{X}$ one has

$$\Big(\bigvee_{y \in \mathcal{Y}} y\Big) \wedge \Big(\bigwedge_{z \in \mathcal{Z}} z\Big) = \bigvee_{y \in \mathcal{Y}} \Big(y \wedge \bigwedge_{z \in \mathcal{Z}} z\Big).$$

We are going to prove that this condition is actually sufficient for distributivity of the sublattice generated by $\mathcal{X}$ (see Theorem 6.3 below). The proof will be based on the following technical result.

PROPOSITION 6.2. *Let $\mathcal{X} \in \Omega$ be an N-element subset in a modular lattice Ω such that any of its proper subsets is distributive. Then for every $2 \leqslant k \leqslant N-1$ the following two conditions are equivalent:*

(a) *For any decomposition $\mathcal{X} = \mathcal{Y}' \sqcup \mathcal{Y}'' \sqcup \mathcal{Z}$ of the set $\mathcal{X}$ into a disjoint union of nonempty subsets $\mathcal{Y}'$, $\mathcal{Y}''$, $\mathcal{Z}$ such that $\#\mathcal{Y}' + \#\mathcal{Y}'' = k$, the triple*

$$\bigvee_{y' \in \mathcal{Y}'} y', \quad \bigvee_{y'' \in \mathcal{Y}''} y'', \quad \bigwedge_{z \in \mathcal{Z}} z$$

is distributive.

(b) *For any decomposition $\mathcal{X} = \mathcal{Y} \sqcup \mathcal{Z}' \sqcup \mathcal{Z}''$ of the set $\mathcal{X}$ into a disjoint union of nonempty subsets $\mathcal{Y}$, $\mathcal{Z}'$, $\mathcal{Z}''$ such that $\#\mathcal{Y} = k-1$, the triple*

$$\bigvee_{y \in \mathcal{Y}} y, \quad \bigwedge_{z' \in \mathcal{Z}'} z', \quad \bigwedge_{z'' \in \mathcal{Z}''} z''$$

is distributive.

Moreover, if one of these distributivity conditions holds for some choice of a decomposition then it holds for all decompositions, i.e., (a) and (b) are satisfied (for fixed k). Finally, $\mathcal{X}$ is distributive iff these conditions are satisfied for all $2 \leqslant k \leqslant N-1$.

Proof: First, we note that for $\mathcal{Y}' = \mathcal{Y}$, $\mathcal{Z} = \mathcal{Z}''$ and $\mathcal{Y}'' = \mathcal{Z}' = \{x\}$ corresponding conditions (a) and (b) coincide. Using the assumption of distributivity for proper subsets of $\mathcal{X}$ let us show that condition (a) does not change after rearranging elements between the subsets $\mathcal{Y}'$ and $\mathcal{Y}''$, having the subset $\mathcal{Z}$ fixed. Indeed, the left-hand side of the distributivity equation

$$\Big(\bigvee\nolimits_{y' \in \mathcal{Y}'} y' \vee \bigvee\nolimits_{y'' \in \mathcal{Y}''} y''\Big) \wedge \bigwedge\nolimits_{z \in \mathcal{Z}} z = \bigvee\nolimits_{y \in \mathcal{Y}' \sqcup \mathcal{Y}''} y \wedge \bigwedge\nolimits_{z \in \mathcal{Z}} z$$

does not change under such a rearranging. For the right-hand side we have

$$\Big(\bigvee\nolimits_{y' \in \mathcal{Y}'} y' \wedge \bigwedge\nolimits_{z \in \mathcal{Z}} z\Big) \vee \Big(\bigvee\nolimits_{y'' \in \mathcal{Y}''} y'' \wedge \bigwedge\nolimits_{z \in \mathcal{Z}} z\Big)$$
$$= \bigvee\nolimits_{y' \in \mathcal{Y}'} \Big(y' \wedge \bigwedge\nolimits_{z \in \mathcal{Z}} z\Big) \vee \bigvee\nolimits_{y'' \in \mathcal{Y}''} \Big(y'' \wedge \bigwedge\nolimits_{z \in \mathcal{Z}} z\Big) = \bigvee\nolimits_{y \in \mathcal{Y}' \sqcup \mathcal{Y}''} \Big(y \wedge \bigwedge\nolimits_{z \in \mathcal{Z}} z\Big)$$

by the distributivity of $\mathcal{Y}' \sqcup \mathcal{Z}$ and $\mathcal{Y}'' \sqcup \mathcal{Z}$. Analogously, one can show that condition (b) depends only on the subset $\mathcal{Y}$. Using these equivalences one can rearrange any decomposition (a) or (b) of the set $\mathcal{X}$ into any other. For the proof of the last assertion note that if $\mathcal{X}$ is distributive then condition (a) holds for any decomposition with $\mathcal{Y}''$ consisting of one element. □

THEOREM 6.3. [**86**] *A subset $x_1, \dots, x_N \in \Omega$ of a modular lattice Ω generates a distributive sublattice if and only if for any sequence of indices $1 \leqslant i_1 < \cdots < i_l \leqslant N$ and any number $2 \leqslant k \leqslant l-1$ the triple*

$$x_{i_1} \vee \cdots \vee x_{i_{k-1}},\ x_{i_k},\ x_{i_{k+1}} \wedge \cdots \wedge x_{i_l}$$

is distributive. In particular, if any proper subset of $x_1, \dots, x_N$ generates a distributive sublattice, then the whole set $x_1, \dots, x_N$ has the same property iff for any $2 \leqslant k \leqslant N-1$ the triple

$$x_1 \vee \cdots \vee x_{k-1},\ x_k,\ x_{k+1} \wedge \cdots \wedge x_N$$

is distributive.

Proof: The "only if" part is trivial, so we only have to prove the "if" part. Arguing by induction in N we can assume that any proper subset of $x_1, \dots, x_N$ generates a distributive sublattice. Applying Proposition 6.2 we immediately see that the set $x_1, \dots, x_N$ is distributive (in the sense of the definition given above). It is enough to prove that any subset $f_1, \dots, f_s$ of the sublattice generated by $x_1, \dots, x_N$ is distributive (in fact, it is enough to prove this for all three-element subsets).

Let us define the *complexity* of an element $f \in \Omega$ of the sublattice generated by $x_1, \dots, x_N$, as the minimal number of the operation signs $\wedge$ and $\vee$ used in an expression of f in terms of x_i. The complexity of a subset $f_1, \dots, f_s$ is by definition the sum of complexities of all f_t. We use induction in the complexity of a subset $f_1, \dots, f_s$ and for subsets of the same complexity the induction in s.

So suppose we are given a subset $f_1, \dots, f_s$ and $f_1 = g \vee h$ is the first step of a minimal expression for f_1. By the induction assumption, the set $g, h, f_2, \dots, f_s$ is distributive and any proper subset of $f_1, \dots, f_s$ is also distributive. It remains to observe that condition (a) of Proposition 6.2 holds for any decomposition $\{f_1, \dots, f_s\} = \mathcal{Y}' \sqcup \mathcal{Y}'' \sqcup \mathcal{Z}$ with $f_1 \in \mathcal{Y}'$ because it reduces to a similar condition for the distributive set $g, h, f_2, \dots, f_s$. □

The above theorem shows that a subset of a modular lattice is distributive iff it generates a distributive sublattice. Here are some other corollaries.

COROLLARY 6.4. **[69]** *Let Ω be a modular lattice and $\mathcal{X} \subset \Omega$ be a subset. Then the following two conditions are equivalent:*

(a) *the set $\mathcal{X}$ is distributive;*

(b) *any subset $\mathcal{X}' \subset \mathcal{X}$ containing no pairs of elements x_1, x_2 such that $x_1 \leqslant x_2$, is distributive.*

Proof: Assume that (b) holds. Clearly, we can also assume that $\mathcal{X}$ is finite and (by induction) that all its proper subsets are distributive. If $\mathcal{X}$ contains no elements $x_1 \leqslant x_2$, there is nothing to prove. Otherwise, let $x_1 \leqslant x_2$ be such a pair. Then distributivity of $\mathcal{X}$ follows from the fact that condition (a) of Proposition 6.2 holds for all decompositions with $x_1, x_2 \in \mathcal{Y}'$ (where $k > 2$) and for all decompositions with $\mathcal{Y}' = \{x_1\}$ and $\mathcal{Y}'' = \{x_2\}$ ($k = 2$). □

COROLLARY 6.5. *Let $u, x_1, \dots, x_N$ be elements of a modular lattice Ω. Then the following conditions are equivalent:*

(a) *the set $u, x_1, \dots, x_N$ is distributive;*

(b) *both sets $x_1, \dots, x_N$ and $u \wedge x_1, \dots, u \wedge x_N$ are distributive and the equation*

$$u \wedge \left(\bigvee_{i\in I} x_i\right) = \bigvee_{i\in I} (u \wedge x_i)$$

holds for any subset $I \subset [1, N]$;

(b*) *both sets $x_1, \dots, x_N$ and $u \vee x_1, \dots, u \vee x_N$ are distributive and the equation*

$$u \vee \left(\bigwedge_{i\in I} x_i\right) = \bigwedge_{i\in I} (u \vee x_i)$$

holds for any subset $I \subset [1, N]$;

(c) *both sets $u \wedge x_1, \dots, u \wedge x_N$ and $u \vee x_1, \dots, u \vee x_N$ are distributive and all triples u, x_i, x_j are distributive for $1 \leqslant i < j \leqslant N$.*

Proof: (b) $\Longrightarrow$ (a): By induction, we can assume that any proper subset of the set $u, x_1, \dots, x_N$ is distributive. Now the condition of Theorem 6.3 for $x_{N+1} = u$ and $k = N$ essentially coincides with the equation in (b), while the conditions for $k < N$ can be expressed in terms of $u \wedge x_1, \dots, u \wedge x_N$.

(c) $\Longrightarrow$ (a): By induction, we can assume that the union of any proper subset of $x_1, \dots, x_N$ with u is distributive. Then the last condition of (c) is only needed in the trivial case $N = 2$. As we have noticed above the conditions of Theorem 6.3 applied to $x_1, \dots, x_N, u$, for $k < N$ can be expressed in terms of $u \wedge x_1, \dots, u \wedge x_N$. Similarly, the conditions of this theorem applied to $u, x_1, \dots, x_N$, for $k > 2$ can be expressed in terms of $u \vee x_1, \dots, u \vee x_N$. Therefore, condition (a) of Proposition 6.2 holds for all k. □

COROLLARY 6.6. *Let $x_1, \dots, x_N$ and $u_0 \leqslant \dots \leqslant u_l$ be elements of a modular lattice Ω such that $u_0 \leqslant x_i \leqslant u_l$ for all $1 \leqslant i \leqslant N$. Then the following conditions are equivalent:*

(a) *the set $u_0, \dots, u_l, x_1, \dots, x_N$ is distributive;*

(b) *all the sets $u_s \wedge x_1, \dots, u_s \wedge x_N$ are distributive and the equation*

$$u_s \wedge \left(\bigvee_{i\in I} x_i\right) = \bigvee_{i\in I} (u_s \wedge x_i)$$

holds for any subset $I \subset [1, N]$ and $1 \leqslant s \leqslant l$;

(b*) *all the sets* $u_s \vee x_1, \ldots, u_s \vee x_N$ *are distributive and the equation*

$$u_s \vee \big(\bigwedge\nolimits_{i\in I} x_i\big) = \bigwedge\nolimits_{i\in I}(u_s \vee x_i)$$

holds for any subset $I \subset [1, N]$ *and* $0 \leqslant s \leqslant l-1$;

(c) *all the sets* $u_s \wedge x_1 \vee u_{s-1}, \ldots, u_s \wedge x_N \vee u_{s-1}$ *are distributive and for every* $1 \leqslant i < j \leqslant N$ *one of the following two sets of equations holds:*

$$u_s \wedge (x_i \vee x_j) \vee u_{s-1} = (u_s \wedge x_i) \vee (u_s \wedge x_j) \vee u_{s-1},\ 1 \leqslant s \leqslant l;$$

$$u_s \wedge (x_i \wedge x_j) \vee u_{s-1} = u_s \wedge (x_i \vee u_{s-1}) \wedge (x_j \vee u_{s-1}),\ 1 \leqslant s \leqslant l$$

(the expressions without brackets make sense due to the modularity identity).

Proof: The equivalence of (a), (b), and (b*) follows immediately from Corollaries 6.4 and 6.5. Let us prove the only nontrivial part (c) $\Longrightarrow$ (a). Suppose the first set of equations holds for some i, j. Then we can prove by induction in s that the triple u_s, x_i, x_j is distributive for every s. Indeed, in the induction step one has to apply $\wedge(x_i \vee x_j)$ to the s-th equation and use modularity and the induction assumption. A dual argument shows distributivity of u_s, x_i, x_j under the assumption that the second set of equations holds. Next, we prove by induction in s that the subsets $u_s \wedge x_1, \ldots, u_s \wedge x_N$ are distributive. The induction step follows easily by applying Corollary 6.5 (c) $\Longrightarrow$ (a) to the elements $u = u_{s-1}, u_s \wedge x_1, \ldots, u_s \wedge x_N$. Dually, we can check that all the subsets $u_s \vee x_1, \ldots, u_s \vee x_N$ are distributive. Applying Corollary 6.5 (c) $\Longrightarrow$ (a) to the set $u_s, x_1, \ldots, x_N$ we derive its distributivity. Finally, by Corollary 6.4, this implies distributivity of the entire set $u_0, \ldots, u_l, x_1, \ldots, x_N$. □

7. Lattices of vector spaces

Let W be a vector space. The set Ω_W of all its linear subspaces is a lattice with respect to the operations of sum and intersection: $X \vee Y = X + Y$ and $X \wedge Y = X \cap Y$. This lattice is modular but not distributive: the equation $(X+Y) \cap Z = X \cap Z + Y \cap Z$ does not hold in general. In fact, it is not difficult to classify all triples of vector subspaces X, Y, $Z \subset W$ up to isomorphism. All indecomposable triples but one are distributive and have $\dim W = 1$. The only nondistributive indecomposable triple is that of three lines in a plane.

We will say that a collection of subspaces $X_1, \ldots, X_N \subset W$ is *distributive* if it generates a distributive lattice of subspaces of W.

The dual lattice Ω_W° is naturally identified with the lattice Ω_{W^*} of all vector subspaces of the dual vector space W^*: a subspace $X \subset W$ corresponds to its orthogonal complement $X^\perp \subset W^*$. It follows that the orthogonal complement collection $X_1^\perp, \ldots, X_N^\perp \subset W^*$ is distributive iff the collection $X_1, \ldots, X_N \subset W$ is.

It is easy to see that the direct sum collection $X_1' \oplus X_1'', \ldots, X_N' \oplus X_N'' \subset W' \oplus W''$ is distributive iff both collections $X_1', \ldots, X_N' \subset W'$ and $X_1'', \ldots, X_N'' \subset W''$ are distributive. If both spaces W' and W'' are nonzero then a similar statement holds with the direct sum collection replaced by the collection of $N' + N''$ subspaces $X_s' \otimes W''$, $W' \otimes X_t'' \subset W' \otimes W''$. Indeed, the "only if" part is trivial and the "if" part is a consequence of the following result.

PROPOSITION 7.1. *Let* W *be a vector space and* $X_1, \ldots, X_N \subset W$ *be a collection of its subspaces. Then the following conditions are equivalent:*

(a) *the collection* $X_1, \ldots, X_N$ *is distributive;*

(b) *there exists a direct sum decomposition* $W = \bigoplus_{\eta\in\mathcal{H}} W_\eta$ *of the vector space* W *such that each of the subspaces* X_i *is the sum of a set of subspaces* W_η.

(c) *there exists a basis* $\{w_\alpha : \alpha \in \mathcal{A}\}$ *of the vector space* W *such that each of the subspaces* X_i *is the linear span of a set of vectors* w_α.

Proof: (b) $\Longrightarrow$ (a) and (b) $\Longleftrightarrow$ (c) are clear. Let us prove (a) $\Longrightarrow$ (b). For every subset $\eta \subset [1, N]$ let us choose a subspace $W_\eta \subset \bigcap_{i\in\eta} X_i$ such that

$$\bigcap\nolimits_{i\in\eta} X_i = W_\eta \oplus \big(\big(\bigcap\nolimits_{i\in\eta} X_i\big) \cap \big(\sum\nolimits_{j\notin\eta} X_j\big)\big).$$

Then we claim that

$$X_i = \sum\nolimits_{\eta\ni i} W_\eta.$$

More generally, we can prove by descending induction in a subset $\xi \subset [1, N]$ that

$$\sum_{\eta\supset\xi} W_\eta = \bigcap\nolimits_{i\in\xi} X_i.$$

Indeed, suppose this is true for all strictly larger sets $\xi' \supset \xi$. Then using the definition of W_ξ and the distributivity identity one can easily derive the above equation for ξ. Note that in the case $\xi = \varnothing$ this equation states that the subspaces W_η generate W.

It remains to prove that the subspaces (W_η), $\eta \subset [1, N]$, are linearly independent. Assume that $\sum_s w_{\eta_s} = 0$ for a set of nonzero vectors $w_{\eta_s} \in W_{\eta_s}$, where all subsets η_s are distinct. Choose s_0 such that the subset $\eta_{s_0} \subset [1, N]$ does not contain any other subsets η_s. Then we have

$$\sum_{s\neq s_0} W_{\eta_s} \subset \sum_{s\neq s_0} \bigcap\nolimits_{j\in\eta_s} X_j \subset \sum_{j\notin\eta_{s_0}} X_j.$$

This is a contradiction since $W_{\eta_{s_0}}$ does not intersect $\sum_{j\notin\eta_{s_0}} X_j$ by the definition. ☐

We say that a basis of W satisfying property (c) above is *distributing* the collection $X_1, \ldots, X_N \subset W$.

Using the above proposition we immediately see that for any pair of distributive collections $X'_1, \ldots, X'_N \subset W'$ and $X''_1, \ldots, X''_N \subset W''$ the tensor product collection $X'_i \otimes X''_i \subset W' \otimes W''$ is also distributive. Note that the converse statement is not true: take X'_1, X'_2, X'_3 to be any nondistributive collection and $X''_2 = 0$.

PROPOSITION 7.2. *Let* W *be a vector space and* $X_1, \ldots, X_N \subset W$ *be a collection of subspaces such that any proper subset* $X_1, \ldots, \widehat{X}_k, \ldots, X_N$ *is distributive. Then the following conditions are equivalent:*

(a) *the collection* $X_1, \ldots, X_N$ *is distributive;*

(b) *the following complex of vector spaces* $K_\bullet(W; X_1, \ldots, X_N)$ *is exact*

$$0 \longrightarrow X_1 \cap \cdots \cap X_N \longrightarrow X_2 \cap \cdots \cap X_N \longrightarrow X_3 \cap \cdots \cap X_N / X_1 \longrightarrow$$
$$\cdots \longrightarrow \bigcap\nolimits_{s=i+1}^{N} X_s \Big/ \sum\nolimits_{t=1}^{i-1} X_t \longrightarrow \cdots \longrightarrow$$
$$X_N/(X_1 + \cdots + X_{N-2}) \longrightarrow W/(X_1 + \cdots + X_{N-1}) \longrightarrow W/(X_1 + \cdots + X_N) \longrightarrow 0$$

where we denote $Y/Z = Y/Y \cap Z$;

(c) *the following complex of vector spaces* $B_\bullet(W; X_1, \dots, X_N)$

$$W \longrightarrow \bigoplus_t W/X_t \longrightarrow \cdots \longrightarrow \bigoplus_{t_1<\cdots<t_{N-i}} W/\textstyle\sum_{s=1}^{N-i} X_{t_s} \longrightarrow \cdots \longrightarrow W/\sum_s X_s \longrightarrow 0$$

is exact everywhere except for the leftmost term;

(c*) *the following complex of vector spaces* $B^\bullet(W; X_1, \dots, X_N)$

$$0 \longrightarrow \textstyle\bigcap_s X_s \longrightarrow \cdots \longrightarrow \bigoplus_{t_1<\cdots<t_{N-i}} \bigcap_{s=1}^{N-i} X_{t_s} \longrightarrow \cdots \longrightarrow \bigoplus_t X_t \longrightarrow W$$

is exact everywhere except for the rightmost term.

Proof: It is immediate to verify that the conditions of exactness of the complex $K_\bullet(W; X_1, \dots, X_N)$ coincide with the equations of Theorem 6.3 characterizing distributive sets in modular lattices, so (a) $\iff$ (b) follows. Let us prove (a) $\iff$ (c). There is a natural exact sequence of complexes

$$0 \longrightarrow B_\bullet(W/X_1;\, (X_2 + X_1)/X_1,\, \dots\, ,\, (X_N + X_1)/X_1) \longrightarrow B_\bullet(W; X_1, \dots, X_N) \longrightarrow B_\bullet(W; X_2, \dots, X_N)[-1] \longrightarrow 0,$$

where $[-1]$ denotes the shift of homological degree. Since we assume that any proper subcollection is distributive, the third complex is exact in homological degree $\neq N$. Note that $H_N B_\bullet(W; X_1, \dots, X_N) = X_1 \cap \cdots \cap X_N$. It follows easily that the second complex is exact at the desired terms iff the first complex is exact in degree $\neq N-1$ and the connecting map

$$X_2 \cap \cdots \cap X_N \longrightarrow (X_2 + X_1) \cap \cdots \cap (X_N + X_1)/X_1$$

is surjective. Using induction in N we conclude that exactness of the first complex is equivalent to distributivity of the collection $X_1 + X_2, \dots, X_1 + X_N$. It remains to apply Corollary 6.5, (a) $\iff$ (b*), with u corresponding to X_1. □

In conclusion, let us reformulate the statement of Corollary 6.6, (a) $\iff$ (c) for the case of lattices of vector spaces. Assume we are given a vector space W equipped with a filtration

$$0 = F_0W \subset F_1W \subset \cdots \subset F_{l-1}W \subset F_lW = W$$

and a collection of subspaces $X_1, \dots, X_N \subset W$. Let $\mathrm{gr}^F W = \bigoplus_{s=1}^l F_sW/F_{s-1}W$ be the associated graded space. We can consider the associated collection of subspaces $\mathrm{gr}^F X_1, \dots, \mathrm{gr}^F X_N \subset \mathrm{gr}^F W$, where $\mathrm{gr}^F X = \bigoplus_s F_sW \cap X/F_{s-1}W \cap X$. The following result can be considered as a version of the PBW-theorem (discussed later in the book) for collections of subspaces.

COROLLARY 7.3. *Let* $0 = F_0W \subset F_1W \subset \cdots \subset F_{l-1}W \subset F_lW = W$ *be a filtered vector space and let* $X_1, \dots, X_N \subset W$ *be a collection of subspaces. Then the following two conditions are equivalent:*

(a) *the set of subspaces* $F_0W, \dots, F_lW, X_1, \dots, X_N \subset W$ *is distributive;*
(b) *the associated graded collection* $\mathrm{gr}^F X_1, \dots, \mathrm{gr}^F X_N$ *in the associated graded vector space* $\mathrm{gr}^F W$ *is distributive and for any* $1 \leqslant i < j \leqslant N$ *one of the two equivalent conditions holds:*

$$\mathrm{gr}^F(X_i + X_j) = \mathrm{gr}^F X_i + \mathrm{gr}^F X_j \quad \text{or} \quad \mathrm{gr}^F(X_i \cap X_j) = \mathrm{gr}^F X_i \cap \mathrm{gr}^F X_j.$$

Proof: The main statement follows from Corollary 6.6, (a) $\iff$ (c). The equivalence of two equations in (b) follows from the proof of that corollary. □

CHAPTER 2

Koszul algebras and modules

In this chapter we discuss the notion of Koszulness, central for this book. In section 1 we give definitions of a Koszul algebra and of a Koszul module. We should stress that our setting is not the most general since we assume that A_0 coincides with the ground field (except for section 9). An excellent introduction to Koszulness without this assumption can be found in sections 1 and 2 of [**24**]. In section 2 we consider a numerical identity between Hilbert series that gives an important necessary condition of Koszulness. In section 3 we define Koszul complexes and in section 4 we use them to prove Backelin's criterion of Koszulness in terms of distributivity of certain lattices. Then in sections 5 and 6 we explain how one can control Koszulness under certain types of homomorphisms between algebras. In section 7 we consider examples of projective varieties whose homogeneous coordinate algebras are Koszul. In section 8 we describe a construction of a graded infinitesimal Hopf algebra associated with a Koszul algebra that will be used later in chapter 7. In section 9 we consider the definition of a Koszul algebra in a more general context of monoidal abelian categories and prove that it amounts to having a monoidal functor from a certain fixed monoidal category (without unit). Finally, in section 10 we consider the notion of a relatively Koszul pair of modules over an algebra.

1. Koszulness

Recall that we always assume our graded algebras A (resp., A-modules) to be nonnegatively graded and locally finite-dimensional with $A_0 = \Bbbk$ (resp. locally finite-dimensional and bounded below).

Definition 1. A graded algebra A is called *Koszul* if the following equivalent conditions hold:

(a) $\mathrm{Ext}_A^{ij}(\Bbbk, \Bbbk) = 0$ for $i \neq j$;
(b) A is one-generated and the algebra $\mathrm{Ext}_A^*(\Bbbk, \Bbbk)$ is generated by $\mathrm{Ext}_A^1(\Bbbk, \Bbbk)$;
(c) A is quadratic and $\mathrm{Ext}_A^*(\Bbbk, \Bbbk) \simeq A^!$;
(d) the algebra $\mathrm{Ext}_A(\Bbbk, \Bbbk)$ equipped with internal grading is one-generated.

Definition 2(M). Let A be a Koszul algebra. Then a graded A-module M is called *Koszul* if the following equivalent conditions hold:

(a) $\mathrm{Ext}_A^{ij}(M, \Bbbk) = 0$ for $i \neq j$;
(b) M is generated by M_0 and the $\mathrm{Ext}_A^*(\Bbbk, \Bbbk)$-module $\mathrm{Ext}_A^*(M, \Bbbk)$ is generated by $\mathrm{Ext}_A^0(M, \Bbbk)$;
(c) M is quadratic and $\mathrm{Ext}_A^*(M, \Bbbk) \simeq M_A^!$;
(d) the $\mathrm{Ext}_A^*(\Bbbk, \Bbbk)$-module $\mathrm{Ext}_A^*(M, \Bbbk)$ equipped with internal grading is generated in degree zero.

The equivalence follows easily from the results of sections 3 and 5 of chapter 1. Additional equivalent conditions will be given in sections 3 and 4. We will see in section 3 that quadratic dual algebras or modules are Koszul simultaneously. Note also that algebras A and A^{op} with opposite multiplication are Koszul simultaneously, since their Ext-spaces are naturally isomorphic (see section 1 of chapter 1). We leave for the reader to define the notions of being quadratic (resp. Koszul) for a right A-module R and to check that this is equivalent to imposing the same conditions on the corresponding left A^{op}-module R^{op}.

The condition (a) in Definition 2 is equivalent to the condition that the module M has a *linear free resolution*, i.e., a resolution by free A-modules

$$\ldots \longrightarrow P_2 \longrightarrow P_1 \longrightarrow P_0 \longrightarrow M \longrightarrow 0$$

such that P_i is generated in degree i. The algebra A is Koszul iff $\Bbbk$ admits a linear free resolution.

Example. The symmetric algebra and the exterior algebra of a vector space V are Koszul. Indeed, consider the standard Koszul complex

$$\ldots \longrightarrow \bigwedge^3(V^*)\otimes \mathbb{S}(V) \longrightarrow \bigwedge^2(V^*)\otimes \mathbb{S}(V) \longrightarrow V^*\otimes \mathbb{S}(V) \longrightarrow \mathbb{S}(V)$$

with the differential $a^*\otimes a \longmapsto \sum_{i=1}^n (a^*\wedge e_i^*)\otimes(e_i a)$, where (e_i) and (e_i^*) are dual bases of V and V^*. It is easy to see that this complex is isomorphic to the tensor product of n complexes of the form $\Bbbk[x] \longrightarrow \Bbbk[x] : f \longmapsto xf$. Hence, it is a linear free resolution of the trivial $\mathbb{S}(V)$-module $\Bbbk$. The space $\bigwedge(V^*)\otimes\mathbb{S}(V)$ with the above differential can also be viewed as a linear free resolution of $\Bbbk$ as a $\bigwedge(V^*)$-module.

It is clear that for a given Koszul algebra A the category of Koszul modules over A is stable under extensions. We can also consider the truncation operations on a nonnegatively graded A-module M. Namely, for every $r \geqslant 0$ we define the *truncated module* $M^{[r]} = M_{\geqslant r}(r)$, so that $M_n^{[r]} = M_{r+n}$ for $n \geqslant 0$ and $M_n^{[r]} = 0$ for $n < 0$.

PROPOSITION 1.1. *Let A be a one-generated algebra. If an A-module M is generated in degree zero then the same is true for $M^{[r]}$ with $r \geqslant 0$. Assume in addition that A and M are quadratic or Koszul. Then the same is true for $M^{[r]}$. More precisely, one has $\mathrm{Tor}_{ij}^A(M^{[r]},\Bbbk) = 0$ if $\mathrm{Tor}_{i,j+r}^A(M,\Bbbk) = 0$ and $\mathrm{Tor}_{i+1,k}^A(\Bbbk,\Bbbk) = 0$ for all $j+1 \leqslant k \leqslant j+r$.*

Proof: Since the operation $M \longrightarrow M^{[r]}$ is the r-th iteration of the operation $M \longrightarrow M^{[1]}$, it suffices to consider the case $r = 1$. In this case the assertion about homology follows from the long exact sequence corresponding to the exact triple

$$0 \longrightarrow M^{[1]}(-1) \longrightarrow M \longrightarrow M_0\otimes\Bbbk \longrightarrow 0$$

□

Avramov and Eisenbud showed in [**14**] that for every finitely generated module M over a *commutative* Koszul algebra the truncated module $M^{[r]}$ is Koszul for $r \gg 0$ (we will reprove this in Corollary 7.2). This result was generalized to some noncommutative Koszul algebras in [**70**].

2. Hilbert series

Let $V = \bigoplus_{k\in\mathbb{Z}} V_k$ be a graded vector space. Then its *Hilbert series* is a formal power series defined by

$$h_V(z) = \sum_{k\in\mathbb{Z}} (\dim V_k) z^k.$$

For a graded module M over a graded algebra A we consider also the *double Poincaré series*

$$P_{A,M}(u,z) = \sum_{i,j\in\mathbb{Z}} (\dim \mathrm{Ext}^{ij}(M,\Bbbk)) u^i z^j.$$

Note that the spaces $\mathrm{Ext}^{ij}(M,\Bbbk)$ are finite-dimensional as cohomology of the locally finite-dimensional complex $\mathcal{C}ob^\bullet(A,M)$ (see section 1 in Chapter 1). For the trivial A-module $\Bbbk$ we denote $P_A(u,z) = P_{A,\Bbbk}(u,z)$.

PROPOSITION 2.1. *For any graded algebra A and graded A-module M, the following numerical relation holds:*

$$h_A(z)P_{A,M}(-1,z) = h_M(z).$$

In particular, we have

$$h_A(z)P_A(-1,z) = 1. \tag{2.1}$$

Proof: Let $F_\bullet$ be a minimal free graded A-module resolution of M. By the definition, we have $\mathrm{Ext}^i_A(M,\Bbbk) = (\Bbbk \otimes F_i)^*$ and therefore $h_{F_i}(z) = h_{\mathrm{Ext}^i_A(M,\Bbbk)}(z)h_A(z)$. The desired relation follows immediately by passing to the Euler characteristic of the complex $F_\bullet$. □

Remark 1. The above proposition implies that the dimensions $(\dim A_n)$ of grading components of A are determined by the dimensions of the homology spaces $\mathrm{Tor}^A_{ij}(\Bbbk,\Bbbk)$. Let us show what happens on the level of vector spaces for $n = 2$ and $n = 3$. In the case $n = 2$ consider the subspace $A_1A_1 \subset A_2$. Then we have an isomorphism

$$A_2/A_1A_1 \simeq \mathrm{Tor}^A_{1,2}(\Bbbk,\Bbbk).$$

and an exact sequence

$$0 \longrightarrow \mathrm{Tor}^A_{2,2}(\Bbbk,\Bbbk) \xrightarrow{\Delta'} \mathrm{Tor}^A_{1,1}(\Bbbk,\Bbbk) \otimes \mathrm{Tor}^A_{1,1}(\Bbbk,\Bbbk) \longrightarrow A_1A_1 \longrightarrow 0,$$

where Δ' is the comultiplication map (see section 3). In the case $n = 3$ we consider the natural filtration $A_1A_1A_1 \subset (A_1A_2 + A_2A_1) \subset A_3$. Then we have an isomorphism

$$\mathrm{Tor}^A_{1,3}(\Bbbk,\Bbbk) \simeq A_3/(A_1A_2 + A_2A_1),$$

and exact sequences

$$0 \to \ker(\Delta'') \to \mathrm{Tor}^A_{2,3} \xrightarrow{\Delta''} \mathrm{Tor}^A_{1,1}(\Bbbk,\Bbbk) \otimes \mathrm{Tor}^A_{1,2}(\Bbbk,\Bbbk) \oplus \mathrm{Tor}^A_{1,2}(\Bbbk,\Bbbk) \otimes \mathrm{Tor}^A_{1,1}(\Bbbk,\Bbbk)$$
$$\to (A_1A_2 + A_2A_1)/A_1A_1A_1 \to 0,$$

$$0 \to \ker(\Delta'') \xrightarrow{f} \mathrm{coker}(g) \to A_1A_1A_1 \to 0,$$

$$0 \to \mathrm{Tor}^A_{3,3}(\Bbbk,\Bbbk) \to \mathrm{Tor}^A_{1,1}(\Bbbk,\Bbbk) \otimes \mathrm{Tor}^A_{2,2}(\Bbbk,\Bbbk) \oplus \mathrm{Tor}^A_{2,2}(\Bbbk,\Bbbk) \otimes \mathrm{Tor}^A_{1,1}(\Bbbk,\Bbbk)$$
$$\xrightarrow{g} \mathrm{Tor}^A_{1,1}(\Bbbk,\Bbbk)^{\otimes 3} \to \mathrm{coker}(g) \to 0,$$

where $g = \Delta' \otimes \mathrm{id} + \mathrm{id} \otimes \Delta'$ and f is the map dual to the triple Massey product (see Appendix). Considering dimensions in these exact sequences one gets the same expression for $\dim A_3$ in terms of $\dim \mathrm{Tor}^A_{ij}(\Bbbk, \Bbbk)$ as from Proposition 2.1.

COROLLARY 2.2. *For quadratic dual Koszul algebras A and $A^!$ one has*

$$h_A(z)h_{A^!}(-z) = 1. \tag{2.2}$$

For a Koszul module M over a Koszul algebra A one has

$$h_{M^!_A}(z) = h_{A^!}(z)h_M(-z),$$

where $M^!_A$ is the quadratic dual module over $A^!$.

Proof: One just has to observe that a graded algebra A is Koszul iff $P_A(u, z) = h_{A^!}(uz)$, and an A-module M is Koszul iff $P_{A,M}(u, z) = h_{M^!_A}(uz)$. □

Example 1. The above corollary provides a way to check that certain algebras are not Koszul. For example, let A be the universal enveloping algebra of the Heisenberg Lie algebra associated with a symplectic vector (V, ω) of dimension $2n \geqslant 4$ (this algebra is quadratic, see Example 5 of section 2 of chapter 1). Then A has a central degree 2 element t such that $B = A/(t)$ is isomorphic to the symmetric algebra of V, and t is not a zero divisor by the PBW theorem. This implies the following equality for Hilbert series:

$$h_A(z) = \frac{h_B(z)}{1 - z^2} = \frac{1}{(1 - z)^{2n}(1 - z^2)}.$$

Hence, $h_A(-z)^{-1} = (1 + z)^{2n}(1 - z^2)$ has the negative top degree coefficient. By Corollary 2.2 this implies that A cannot be Koszul. A similar argument works for the universal enveloping algebra of the Heisenberg Lie algebra associated with a possibly degenerate skew-symmetric form ω provided that the rank of ω is $\geqslant 4$ (otherwise this algebra is not quadratic).

Example 2. More generally, let $A = U(\mathfrak{g})$ be the universal enveloping algebra of a graded Lie algebra $\mathfrak{g} = \bigoplus_{i \geqslant 1} \mathfrak{g}_i$. Then we claim that Koszulness of A implies that either $\mathfrak{g} = \mathfrak{g}_1$ (and hence, $A = \mathbb{S}(\mathfrak{g}_1)$) or the algebra $\mathfrak{g}$ has exponential growth. Indeed, by the PBW theorem the Hilbert series of A is equal to

$$h_A(z) = \prod_{i \geqslant 1} \frac{1}{(1 - z^i)^{a_i}},$$

where $a_i = \dim \mathfrak{g}_i$. If A is Koszul then by Corollary 2.2 one has

$$h_{A^!}(-z) = h_A(z)^{-1} = \prod_{i \geqslant 1} (1 - z^i)^{a_i}. \tag{2.3}$$

But $A^!$ is the quotient of the exterior algebra, so $h_{A^!}$ is a polynomial. Writing $h_{A^!}(-z) = (1 - \alpha_1 z) \dots (1 - \alpha_m z)$, where $\alpha_j \in \mathbb{C}$, and taking the logarithm of both sides of (2.3) we get the equality

$$\sum_{j=1}^{m} \log(1 - \alpha_j z) = \sum_{i \geqslant 1} a_i \log(1 - z^i).$$

Comparing the coefficients with z^n for $n \geqslant 1$ we get the identity

$$\sum_{j=1}^{m} \alpha_j^n = \sum_{d|n} d a_d.$$

This leads to inequalities

$$na_n \leqslant \sum_{j=1}^{m} \alpha_j^n \leqslant n \sum_{i=1}^{n} a_i.$$

If $|\alpha_j| > 1$ for some j then the second inequality shows that $\mathfrak{g}$ has exponential growth. On the other hand, if $|\alpha_j| \leqslant 1$ for all j then the first inequality shows that $a_n = 0$ for $n > m$. We claim that in this case $a_2 = 0$. Indeed, equation (2.3) shows that for $a_2 > 0$ one has $h_{A^!}(1) = 0$, which is a contradiction.

Example 3. Now let $A = U(L)$ be the universal enveloping algebra of a graded Lie superalgebra $L = \bigoplus_{i \geqslant 1} L_i$ (where the parity is induced by the $\mathbb{Z}$-grading). Then we claim that Koszulness of A implies that either L has exponential growth, or $L_3 = 0$ and $\dim L_2 \leqslant \dim L_1$. Indeed, in this case we have

$$h_A(z) = \prod_{i \geqslant 1} (1 - (-z)^i)^{(-1)^{i-1} a_i},$$

where $a_i = \dim L_i$. Since the algebra $A^!$ is commutative and has a_1 generators, we obtain an equality of the form

$$h_{A^!}(z) = \frac{(1 - \alpha_1 z) \dots (1 - \alpha_m z)}{(1 - z)^{a_1}} = \prod_{i \geqslant 1} (1 - z^i)^{(-1)^i a_i},$$

where $\alpha_j \in \mathbb{C}$, or equivalently,

$$(1 - \alpha_1 z) \dots (1 - \alpha_m z) = \prod_{i \geqslant 2} (1 - z^i)^{(-1)^i a_i}. \tag{2.4}$$

As in the previous example, taking the logarithm and comparing the coefficients with z^n we derive the identity

$$\sum_{j=1}^{m} \alpha_j^n = \sum_{d|n, d>1} (-1)^d d a_d.$$

This leads to inequalities

$$2^k a_{2^k} \leqslant \sum_{j=1}^{m} \alpha_j^{2^k}, \quad \sum_{j=1}^{m} \alpha_j^n \leqslant n \sum_{i=1}^{n} a_i.$$

As before, these inequalities imply that either L has exponential growth or it is finite-dimensional. Let us show that in the latter case one has $a_3 = 0$. Assume that $a_3 \neq 0$ and let n be the maximal odd number such that $a_n \neq 0$. Then the right-hand side of (2.4) has a pole at a primitive n-th root of unity, which is a contradiction. Finally, the condition that $(1 - z^2)^{a_2}/(1 - z)^{a_1}$ has positive coefficients implies that $a_2 \leqslant a_1$.

It was unknown for some time whether any quadratic algebra A satisfying numerical relation (2.2) is Koszul. Counterexamples were found independently in [**106**] and [**101**]. In section 5 of chapter 3 we will present an even more remarkable counterexample due to Piontkovskii [**92**] showing that it is impossible to decide whether a quadratic algebra A is Koszul knowing only Hilbert series of A and $A^!$.

We conclude this section with two results that show that in certain special situations relation (2.2) does imply Koszulness. The first of these results is related to the *Golod-Shafarevich inequality* (see [**63**]). For a power series $f(z) \in \mathbb{R}[[z]]$ we denote by $|f(z)|$ the series obtained by deleting all terms starting from the first negative term. Also, for two power series we write $f(z) \geqslant g(z)$ if this inequality holds coefficient-wise.

PROPOSITION 2.3. *For any quadratic algebra A with m generators and r relations one has*

$$h_A(z) \geqslant |(1 - mz + rz^2)^{-1}|. \tag{2.5}$$

Also, if

$$h_A(z) = (1 - mz + rz^2)^{-1}$$

then A is Koszul.

Proof: For any quadratic algebra $A = \{V, I\}$ we have an exact sequence

$$I \otimes A(-2) \longrightarrow V \otimes A(-1) \longrightarrow A \longrightarrow \Bbbk \longrightarrow 0.$$

Therefore, we get the inequality $(1 - mz + rz^2)h_A(z) - 1 \geqslant 0$. If the equality holds, then the above sequence is also exact at the term $I \otimes A(-2)$; hence, we get a linear free resolution for $\Bbbk$. We claim that in any case we get the required inequality. Indeed, it suffices to check that for a pair of power series $f(z), g(z) \in \mathbb{R}[[x]]$, where $g(0) > 0$, the inequalities $f(z) \geqslant 0$ and $f(z)g(z) \geqslant 1$ imply that $f(z) \geqslant |g(z)^{-1}|$. Indeed, assume that the first negative coefficient of $g(z)^{-1}$ appears with z^N. Then we just have to observe that $|g(z)^{-1}| = g(z)^{-1} \mod (z^N)$ and that $f(z) \geqslant g(z)^{-1} \mod (z^N)$. □

COROLLARY 2.4. *Let A be a quadratic algebra such that either $A_3 = 0$ or $A^!_3 = 0$. Assume that relation* (2.2) *holds. Then A is Koszul.*

The following theorem is a slight generalization of Theorem 2.2 of [**112**] (however, the proof is the same). Recall that a graded algebra A is called *Gorenstein* if $\mathrm{Ext}^*_A(\Bbbk, A)$ is one-dimensional.

THEOREM 2.5. *Let A be a graded Gorenstein algebra of finite cohomological dimension d. Then $h_A(z)$ is a rational function. Furthermore, the following conditions are equivalent:*
(i) A is Koszul;
(ii) $\mathrm{Ext}^d_A(\Bbbk, A)$ *has internal degree d;*
(iii) h_A has degree $-d$.

Proof: Let

$$0 \longrightarrow P_d \longrightarrow \ldots \longrightarrow P_2 \longrightarrow P_1 \longrightarrow P_0 \longrightarrow \Bbbk \longrightarrow 0$$

be a minimal free resolution of $\Bbbk$ as a left A-module. The Gorenstein condition implies that the dual complex of right A-modules

$$P_0^\vee \longrightarrow P_1^\vee \longrightarrow \ldots \longrightarrow P_{d-1}^\vee \longrightarrow P_d^\vee \tag{2.6}$$

(where $P_i^\vee = \mathrm{Hom}_A(P_i, A)$) has cohomology $\Bbbk$. Note that the complex (2.6) is also minimal. In particular, the map $P_{d-1}^\vee \longrightarrow P_d^\vee$ cannot be surjective. Hence, the complex (2.6) is a minimal resolution of $\Bbbk(n)$ as a right A-module for some $s \in \mathbb{Z}$. This implies that $P_d^\vee \simeq A(n)$ and generators of $P_{d-i}^\vee$ live in degrees $\geqslant -n + i$. In other words, $P_d \simeq A(-n)$ and P_{d-i} has generators in degrees $\leqslant n - i$. It follows that $h_{P_d}(z) = z^n h_A(z)$ while $h_{P_{d-i}}(z) = p_{d-i}(z) h_A(z)$ for some polynomials $p_{d-i}(z)$ such that $\deg p_{d-i} \leqslant n - i$. Since $P_\bullet$ is a resolution of $\Bbbk$ we obtain the identity

$$1 = h_A(z) \cdot ((-1)^d z^n + \sum_{i=1}^{d} (-1)^{d-i} p_{d-i}(z)).$$

Therefore, $h_A(z)$ is rational and has degree $-n$. From this argument one can immediately deduce the equivalence of (ii) and (iii). If A is Koszul then P_d is generated in degree d, hence $n = d$. This proves the implication (i) $\Longrightarrow$ (ii). Conversely, if $n = d$ then P_{d-i} has generators in degrees $d-i$, so $\Bbbk$ has a linear free resolution, i.e., A is Koszul. □

COROLLARY 2.6. *Let A be a quadratic Gorenstein algebra of finite cohomological dimension satisfying relation* (2.2). *Then A is Koszul.*

Remark 2. One can show that a graded algebra A is Gorenstein of finite homological dimension iff its cohomology algebra $\mathrm{Ext}_A^*(\Bbbk,\Bbbk)$ is *Frobenius*, i.e., $\dim \mathrm{Ext}_A^i(\Bbbk,\Bbbk) = 0$ for $i > d$, $\dim \mathrm{Ext}_A^d(\Bbbk,\Bbbk) = 1$, and the multiplication maps

$$\mathrm{Ext}_A^i(\Bbbk,\Bbbk) \otimes \mathrm{Ext}_A^{d-i}(\Bbbk,\Bbbk) \longrightarrow \mathrm{Ext}_A^d(\Bbbk,\Bbbk)$$

are nondegenerate pairings. Hence, a Koszul algebra is Gorenstein of finite homological dimension iff its quadratic dual algebra is Frobenius.

Remark 3. Another situation when Koszulness can be derived from other homological properties together with information about the Hilbert series is considered in Theorem 2 of [**53**] that implies that every commutative quadratic extremal Gorenstein algebra is Koszul (here *extremality* means that $\deg h_A = -\dim A + 2$). For example, this gives Koszulness of the following algebras: (i) $\Bbbk[x_{ij}, 1 \leqslant i,j \leqslant 3]/I$, where I is generated by all 2×2-minors; (ii) $\Bbbk[x_{ij}, 1 \leqslant i,j \leqslant 5]/I$, where $x_{ij} = -x_{ji}$ and I is generated by the Pfaffians of all principal 4×4-submatrices (see [**109**]).

3. Koszul complexes

The following construction generalizing the standard Koszul complex (see Example in section 1) is due to S. Priddy [**104**].

Let A and $A^!$ be dual quadratic algebras. Let us equip the tensor product $A\otimes A^!$ with the standard algebra structure (so that subalgebras A and $A^!$ commute) and consider the identity element $e_A \in A_1 \otimes A_1^!$. One can check easily that $e_A^2 = 0$ in the algebra $A \otimes A^!$. Indeed, let $A = \{V, I\}$. Then the multiplication map $(A_1 \otimes A_1^!)^{\otimes 2} \to A_2 \otimes A_2^!$ can be identified with the natural map

$$\mathrm{Hom}(V^{\otimes 2}, V^{\otimes 2}) \to \mathrm{Hom}(I, V^{\otimes 2}/I).$$

Under this identification the element $e_A \otimes e_A$ corresponds to the identity element in $\mathrm{Hom}(V^{\otimes 2}, V^{\otimes 2})$; hence, the above map sends it to zero.

More generally, for any graded algebra A consider the algebra $B = (\mathrm{q}A)^!$, dual to the quadratic part of A. Then the image of the element $e_{\mathrm{q}A}$ under the natural homomorphism $\mathrm{q}A \longrightarrow A$ defines an element $e_A \in A_1 \otimes B_1$ with the same property $e_A^2 = 0$ in $A \otimes B$. Let R (resp., L) be a graded right (resp., left) A-module (resp., B-module). Note that the graded dual space L^* has a natural structure of a right B-module. Hence, the tensor product $R \otimes L^*$ is a right $A \otimes B$-module. Now the action of $e_A \in A \otimes B$ equips $R \otimes L^*$ with a differential ∂. This complex of vector spaces is called the *Koszul complex* of L and R and is denoted $K^A(R, L)$.

It will be convenient for us to use two different notations for the natural bigrading on $K^A(R,L)$. Namely, we set

$$K_{ij}^A(R,L) = {}_{pq}K^A(R,L) = R_q \otimes L_p^*, \qquad \text{for } i = p,\ j = p+q.$$

The differential ∂ maps ${}_{pq}K$ to ${}_{p-1,q+1}K$ and K_{ij} to $K_{i-1,j}$. Note that for the subscript (p,q) the roles of R and L are symmetric. On the other hand, using (i,j)

we can consider $K^A(R, L)$ as a complex of graded vector spaces, with i being the homological grading and j the internal one.

For a left nonnegatively graded A-module M denote by $N = (\mathrm{q}_A M)^!_{(\mathrm{q}A)}$ the quadratic dual left B-module and set

$$'K_\bullet(A, M) = K^A_\bullet(A, N),$$

where A is viewed as a right A-module. Then $'K_\bullet(A, M)$ is a complex of free graded left A-modules

$$\cdots \longrightarrow A \otimes N_3^* \longrightarrow A \otimes N_2^* \longrightarrow A \otimes N_1^* \longrightarrow A \otimes N_0^* \longrightarrow 0,$$

where N_i^* is equipped with internal degree i. Note that $N_0^* = M_0$ is the space of degree 0 generators of M and $N_1^* = K_M$ is the space of quadratic relations in M.

It is easy to see that the natural A-module morphism $A \otimes M_0 \longrightarrow M$ factors through $H_0\, 'K_\bullet(A, M)$. Hence, we can extend the above complex by adding M on the right. Let $'\overline{K}_\bullet(A, M)$ denote the obtained complex, so that $'\overline{K}_i(A, M) = {}'K_i(A, M)$ for $i \neq -1$ and $'\overline{K}_{-1}(A, M) = M$.

PROPOSITION 3.1. *Let M be a nonnegatively graded module over a graded algebra A. Then M admits a linear free resolution iff the complex $'K_\bullet(A, M)$ is a resolution of M, i.e., the complex $'\overline{K}_\bullet(A, M)$ is exact. More precisely, for any $a, b \geqslant 0$ the following two conditions are equivalent:*

(i) $\mathrm{Tor}^A_{ij}(\Bbbk, M) = 0$ *for all* $i \neq j$, $i \leqslant a$, $j - i \leqslant b$;

(ii) $H_{i-1,j}\, '\overline{K}_\bullet(A, M) = 0$ *for all* $i \leqslant a$, $j - i \leqslant b$.

Furthermore, there are natural morphisms $\psi^{A,M}_{i,j} : H_{i-1,j}\, '\overline{K}_\bullet(A, M) \longrightarrow \mathrm{Tor}^A_{ij}(\Bbbk, M)$ for all $i < j$. If the above equivalent conditions hold then $\psi_{a,a+b+1}$ is injective and $\psi_{a+1,a+b+1}$ is surjective.

Proof: There is a natural morphism of complexes of free A-modules

$$(3.1) \qquad 'K_\bullet(A, M) \longrightarrow \widetilde{\mathcal{B}ar}_\bullet(A, M), \qquad a \otimes x \longmapsto a \otimes \rho(x),$$

where $\rho \colon N_i^* \longrightarrow A_1^{\otimes i} \otimes M_0$ is the embedding dual to the action map $B_1^{\otimes i} \otimes N_0 \longrightarrow N_i$. Let $C_\bullet$ be the cone of this morphism. Then $C_\bullet$ is also a complex of free A-modules. Also, since $\widetilde{\mathcal{B}ar}_\bullet(A, M)$ is a resolution of M, we have $H_{ij}(C_\bullet) = H_{i-1,j}\, '\overline{K}_\bullet(A, M)$. On the other hand, the complex $\Bbbk \otimes_A {}'K(A, M)$ has a zero differential and the morphism (3.1) tensored with $\Bbbk$ induces an isomorphism of $H_i(\Bbbk \otimes_A {}'K(A, M)$ with $H_{ii}(\Bbbk \otimes_A \widetilde{\mathcal{B}ar}(A, M)) = \mathrm{Tor}^A_{ii}(\Bbbk, M)$ (see section 3 of chapter 1). Hence, we have $H_{ij}(\Bbbk \otimes_A C_\bullet) = \mathrm{Tor}^A_{ij}(\Bbbk, M)$ for $i \neq j$ and $H_{ii}(\Bbbk \otimes_A C_\bullet) = 0$. Now we define $\psi^{A,M}_{i,j}$ as the maps $\psi^{C_\bullet}_{i,j} : H_{i,j}(C_\bullet) \longrightarrow H_{i,j}(\Bbbk \otimes_A C_\bullet)$ induced by the natural morphism of complexes of graded vector spaces $C_\bullet \longrightarrow \Bbbk \otimes_A C_\bullet$.

It remains to prove that for an arbitrary bounded above complex $C_\bullet$ of free graded A-modules and a pair of integers (a, b), the vanishing of $H_{ij}(C_\bullet)$ for $i \leqslant a$ and $j - i \leqslant b$ is equivalent to the vanishing of $H_{ij}(\Bbbk \otimes_A C_\bullet)$ in the same region, and that in this case $\psi_{a,a+b+1}$ (resp., $\psi_{a+1,a+b+1}$) is injective (resp., surjective). According to Proposition 4.2 we can assume the complex $C_\bullet$ to be minimal (since an acyclic complex of free A-modules remains acyclic after tensoring with $\Bbbk$). It is easy to deduce from minimality that the morphism $\psi^{C_\bullet}_{i_0,j_0}$ is injective (resp., surjective) if $H_{ij}(\Bbbk \otimes_A C_\bullet) = 0$ for $i = i_0$ and $j < j_0$ (resp., for $i = i_0 - 1$ and $j < j_0$). Using this observation about injectivity one can immediately check that the vanishing of $H_{ij}(\Bbbk \otimes_A C_\bullet)$ for $i \leqslant a$, $j - i \leqslant b$ implies the vanishing of $H_{ij}(C_\bullet)$ in the same region. To prove the converse we can use induction in a. Note that since $C_\bullet$ is bounded above, the assertion is trivial for $a \ll 0$. Assume that $H_{ij}(C_\bullet) = 0$ for $i \leqslant a$ and $j-i \leqslant b$. By the induction assumption this implies that $H^{ij}(\Bbbk \otimes_A C_\bullet) = 0$ for $i \leqslant a-1$ and $j-i \leqslant b$. Using the above observation we derive that $\psi_{i,j}^{C_\bullet}$ is surjective for $i \leqslant a$ and $j - i \leqslant b$, and the assertion follows. □

The *Koszul complex* of a graded algebra A is defined as $K_\bullet(A) = {}'K_\bullet(A, \Bbbk)$. Since $N = \Bbbk^!_{qA} = (qA)^!$, it has the following form:

$$\cdots \longrightarrow A \otimes A_3^{!*} \longrightarrow A \otimes A_2^{!*} \longrightarrow A \otimes A_1^{!*} \longrightarrow A \longrightarrow 0.$$

Here $A_1^{!*} = A_1$ is the space of degree 1 generators of A and $A_2^{!*} = I_A$ is the space of quadratic relations of A.

COROLLARY 3.2. *Let A be a graded algebra, M be a graded A-module.*
(i) A is quadratic iff $H_{ij}K_\bullet(A) = 0$ for $j > 0$ and $i = 0, 1$.
(iM) Assume that A is quadratic. Then M is quadratic iff $H_i\, {}'\overline{K}_\bullet(A, M) = 0$ for $i = -1$ and 0.
(ii) A is Koszul iff the Koszul complex $K_\bullet(A)$ is acyclic in all positive internal degrees.
(iiM) Assume that A is Koszul. Then M is Koszul iff the Koszul complex ${}'K_\bullet(A, M)$ is a resolution of M.
(iii) One has $\mathrm{Ext}_A^{ij}(\Bbbk, \Bbbk) = 0$ for $i \neq j$ and $j \leqslant n$ iff $K_\bullet(A)$ is acyclic in all positive internal degrees $\leqslant n$.
(iiiM) One has $\mathrm{Ext}_A^{ij}(M, k) = 0$ for $i \neq j$ and $j \leqslant n$ iff the complex ${}'\overline{K}_\bullet(A, M)$ is acyclic in all internal degrees $\leqslant n$. □

Let A be a quadratic algebra A. Then for a left $A^!$-module L and a right A-module R we have a natural isomorphism

$${}_{pq}K^A(R, L) = {}_{qp}K^{A^{!\mathrm{op}}}(L, R)^*$$

and the corresponding differentials are dual to each other. Therefore, for any p and q one has ${}_{pq}HK^A(R, L) = 0$ iff ${}_{qp}HK^{A^{!\mathrm{op}}}(L, R)^* = 0$.

COROLLARY 3.3. *Quadratic dual algebras A and $A^!$ are Koszul simultaneously. More precisely, for fixed a and b one has $\mathrm{Ext}_A^{ij}(\Bbbk, \Bbbk) = 0$ for all $i - 1 \leqslant a$, $1 < j - i + 1 \leqslant b$ iff $\mathrm{Ext}_{A^!}^{ij}(\Bbbk, \Bbbk) = 0$ for all $i - 1 \leqslant b$, $1 < j - i + 1 \leqslant a$. In addition, there is a natural pairing*

$$\langle \cdot, \cdot \rangle : \mathrm{Ext}_A^{c+1, c+d}(\Bbbk, \Bbbk) \otimes \mathrm{Ext}_{A^!}^{d+1, c+d}(\Bbbk, \Bbbk) \longrightarrow \Bbbk$$

for any c, $d \geqslant 2$. It is nondegenerate on the left (i.e., $\langle x, y \rangle = 0$ for all y implies $x = 0$) if the above condition holds for $a = c - 1$, $b = d$. The pairing is nondegenerate on the right if the above condition holds for $a = c$, $b = d - 1$.

Proof: As we have observed above, the complexes $K(A) = K^A(A, A^!)$ and $K(A^{!\mathrm{op}}) = K^{A^{!\mathrm{op}}}(A^!, A)$ are dual to each other. The pairing $\langle \cdot, \cdot \rangle$ corresponds to the map

$$\psi^{A^{!\mathrm{op}}, \Bbbk}_{d+1, c+d} \circ (\psi^{A, \Bbbk}_{c+1, c+d})^* \colon \mathrm{Ext}_A^{c+1, c+d}(\Bbbk, \Bbbk) \simeq \mathrm{Tor}^A_{c+1, c+d}(\Bbbk, \Bbbk)^* \longrightarrow H_{c, c+d}K_\bullet(A)^*$$
$$\simeq H_{d, c+d}K_\bullet(A^{!\mathrm{op}}) \longrightarrow \mathrm{Tor}^{A^{!\mathrm{op}}}_{d+1, c+d}(\Bbbk, \Bbbk) \simeq \mathrm{Tor}^{A^!}_{d+1, c+d}(\Bbbk, \Bbbk) \simeq \mathrm{Ext}_{A^!}^{d+1, c+d}(\Bbbk, \Bbbk)^*,$$

where the morphisms $\psi_{i,j}$ were defined in Proposition 3.1. All the statements follow immediately from this proposition. □

Assume for a moment that an algebra A is Koszul. Then the complex $K_\bullet(A) = K_\bullet^A(A, A^!)$ is a free graded resolution of the left A-module $\Bbbk$. Similarly, the complex $K_\bullet(A^{\mathrm{op}}) = K_\bullet^{A^{\mathrm{op}}}(A, A^!)$ is a resolution of the right A-module $\Bbbk$. Therefore, in this case for every A-module M the homology of the complex

$$''K_\bullet(A, M) := K_\bullet^{A^{\mathrm{op}}}(M, A^!) = K_\bullet^{A^{\mathrm{op}}}(A, A^!) \otimes_A M :$$

$$\cdots \longrightarrow A_3^{!*} \otimes M \longrightarrow A_2^{!*} \otimes M \longrightarrow A_1^{!*} \otimes M \longrightarrow M \longrightarrow 0$$

is isomorphic to $\mathrm{Tor}_{ij}^A(\Bbbk, M)$. Recall that M is a Koszul module iff this homology is concentrated on the diagonal. Moreover, the diagonal homology is isomorphic to $M_A^{!*}$ (see Proposition 3.1 of chapter 1). Let $''\overline{K}_\bullet(A, M)$ denote the cone of the natural embedding $M_A^{!*} \longrightarrow {''K}_\bullet(A, M)$, where $M_A^{!*}$ is considered as a complex with zero differential. It follows that a module M is Koszul iff the complex $''\overline{K}_\bullet(A, M)$ is acyclic.

PROPOSITION 3.4. *For any graded algebra A and graded A-module M, there are natural morphisms $\varphi_{i,j}^{A,M} : H_{ij}\, ''K_\bullet(A, M) \longrightarrow \mathrm{Tor}_{ij}^A(\Bbbk, M)$. If for some a and b one has $\mathrm{Tor}_{ij}^A(\Bbbk, \Bbbk) = 0$ for all $i-1 \leqslant a,\ 1 < j-i+1 \leqslant b$, then these morphisms are injective for all $i \leqslant a,\ j-i \leqslant b$ and surjective for all $i-1 \leqslant a,\ 1 < j-i+1 \leqslant b$.*

Proof: The morphisms $\varphi_{i,j}$ are obtained by applying the hyper-Tor functors $\mathrm{Tor}_{ij}^A(\cdot, M)$ to the morphism of complexes of right A-modules $K_\bullet(A^{\mathrm{op}}) \longrightarrow \Bbbk$. Note that the cocone of this morphism coincides with the complex $'\overline{K}_\bullet(A^{\mathrm{op}}, \Bbbk) = \overline{K}_\bullet(A^{\mathrm{op}})$. Hence, it suffices to show the vanishing of $\mathrm{Tor}_{ij}^A(\overline{K}_\bullet(A^{\mathrm{op}}),\, M)$ for all $i \leqslant a$ and $j-i \leqslant b$ under our assumptions on $\mathrm{Tor}_{ij}^A(\Bbbk, \Bbbk)$. But this follows immediately from the hyperhomology spectral sequence

$$E_{pq}^2 = \mathrm{Tor}_p^A(H_q\overline{K}_\bullet(A^{\mathrm{op}}),\, M) \Longrightarrow \mathrm{Tor}_{p+q}^A(\overline{K}_\bullet(A^{\mathrm{op}}),\, M)$$

and Proposition 3.1 applied to $M = \Bbbk$. □

COROLLARY 3.5.(M). *Quadratic dual modules M and $M_A^!$ over quadratic dual Koszul algebras A and $A^!$ are Koszul simultaneously. More precisely, if A is Koszul, then one has $\mathrm{Ext}_A^{ij}(M, \Bbbk) = 0$ for all $i-1 \leqslant a,\ 0 < j-i \leqslant b$ iff $\mathrm{Ext}_{A^!}^{ij}(M^!, \Bbbk) = 0$ for all $i-1 \leqslant b,\ 0 < j-i \leqslant a$. More generally, for a quadratic algebra A and a quadratic A-module M there is a natural pairing*

$$\mathrm{Ext}_A^{c+1,c+d+1}(M, \Bbbk) \otimes \mathrm{Ext}_{A^!}^{d+1,c+d+1}(M^!, \Bbbk) \longrightarrow \Bbbk$$

for any $c,\ d \geqslant 1$. It is nondegenerate on the left (resp., right) if $\mathrm{Ext}_A^{ij}(\Bbbk, \Bbbk) = 0$ for $i-1 \leqslant c,\ 1 < j-i+1 \leqslant d+1$ (resp., $i-1 \leqslant c+1,\ 1 < j-i+1 \leqslant d$) and the above conditions hold for $a = c-1,\ b = d$. (resp., $a = c,\ b = d-1$).

Proof: Note that the complexes $'K(A, M) = K^A(A, M^!)$ and $''K(A^!, M^!) = K^{A^{!\mathrm{op}}}(M^!, A)$ are dual to each other. The desired pairing corresponds to the map

$$\varphi_{d+1,c+d}^{A^!,M^!} \circ (\psi_{c+1,c+d}^{A,M})^* \colon \mathrm{Ext}_A^{c+1,c+d+1}(M, \Bbbk) \longrightarrow H_{c,c+d+1}\, 'K_\bullet(A, M)^*$$
$$\simeq H_{d+1,c+d+1}\, ''K_\bullet(A^!, M^!) \longrightarrow \mathrm{Ext}_{A^!}^{d+1,c+d}(M^!, \Bbbk)^*,$$

where the morphisms ψ and φ were defined in Propositions 3.1 and 3.4. All the assertions follow from these propositions. □

Remark 1. In the simplest nontrivial cases the pairing from Corollary 3.3 provides an isomorphism $\mathrm{Ext}_A^{3,4}(\Bbbk, \Bbbk) \simeq \mathrm{Ext}_{A^!}^{3,4}(\Bbbk, \Bbbk)^*$ and a surjective morphism

$\mathrm{Ext}_A^{4,5}(\Bbbk, \Bbbk) \longrightarrow \mathrm{Ext}_{A^!}^{3,5}(\Bbbk, \Bbbk)^*$ for any quadratic algebra A. One can check that the kernel of the latter morphism coincides with the image of the Yoneda multiplication map

$$\mathrm{Ext}_A^{1,1}(\Bbbk, \Bbbk) \otimes \mathrm{Ext}_A^{3,4}(\Bbbk, \Bbbk) \oplus \mathrm{Ext}_A^{3,4}(\Bbbk, \Bbbk) \otimes \mathrm{Ext}_A^{1,1}(\Bbbk, \Bbbk) \longrightarrow \mathrm{Ext}_A^{4,5}(\Bbbk, \Bbbk).$$

The map from each of the direct summands can have a kernel and their images can intersect each other. For this reason it is impossible in general to recover the dimension of $\mathrm{Ext}_{A^!}^{3,5}(\Bbbk, \Bbbk)$ from the dimensions of $\mathrm{Ext}_A^{ij}(\Bbbk, \Bbbk)$ [**101**]. Similarly, from Corollary 3.5 for a quadratic M over a quadratic algebra A we get an isomorphism $\mathrm{Ext}_A^{2,3}(M, \Bbbk) \simeq \mathrm{Ext}_{A^!}^{2,3}(M^!_A, \Bbbk)^*$ and a surjective morphism $\mathrm{Ext}_A^{3,4}(M, \Bbbk) \longrightarrow \mathrm{Ext}_{A^!}^{2,4}(M^!_A, \Bbbk)^*$ whose kernel coincides with the image of the multiplication

$$\mathrm{Ext}_A^{1,1}(\Bbbk, \Bbbk) \otimes \mathrm{Ext}_A^{2,3}(M, \Bbbk) \oplus \mathrm{Ext}_A^{3,4}(\Bbbk, \Bbbk) \otimes \mathrm{Ext}_A^{0,0}(M, \Bbbk) \longrightarrow \mathrm{Ext}_A^{3,4}(M, \Bbbk).$$

For a Koszul algebra A the second summand vanishes. More generally, one can show that the images of the Yoneda multiplication are always contained in the kernels of the pairings from Corollaries 3.3 and 3.5.

Remark 2. There exist also natural complexes computing Hochschild and cyclic homology of a Koszul algebra (see [**49**], [**123**]). Furthermore, it is shown in [**124**] that for a Koszul Gorenstein algebra of finite global dimension there is a duality between Hochschild cohomology and Hochschild homology (with some twist).

4. Distributivity and n-Koszulness

The following criterion is due to J. Backelin [**15**]. The generalization for modules was proved in [**24**].

THEOREM 4.1. *A quadratic algebra $A = \{V, I\}$ is Koszul iff for all $n \geqslant 0$ the collection of subspaces*

$$X_i = V^{\otimes i-1} \otimes I \otimes V^{\otimes n-i-1} \subset V^{\otimes n}, \qquad i = 1, \dots, n-1,$$

is distributive. More precisely, the following conditions are equivalent:

(a) $\mathrm{Ext}_A^{ij}(\Bbbk, \Bbbk) = 0$ *for all* $i < j \leqslant n$;
(b) *the Koszul complex* $K_\bullet(A)$ *is acyclic in positive internal degrees* $\leqslant n$;
(c) *the collection* $(X_1, \dots, X_{n-1})$ *in* $V^{\otimes n}$ *is distributive.*

Proof:
(b) $\iff$ (c). One can immediately check that the complex $K_\bullet(V^{\otimes n}; X_1, \dots, X_{n-1})$ from Proposition 7.2 of chapter 1 coincides with the (internal) degree-n component of the Koszul complex $K_\bullet(A)$. On the other hand, it is easy to see that any proper subcollection of the collection of subspaces $V^{\otimes i-1} \otimes I \otimes V^{\otimes n-i-1}$ in $V^{\otimes n}$ is distributive iff similar collections of subspaces in $V^{\otimes j}$ are distributive for all $j < n$. Now the assertion follows by induction in n.
(a) $\iff$ (c). This immediately follows from Proposition 7.2 of chapter 1 by observing that the complex $B_\bullet(V^{\otimes n}; X_1, \dots, X_{n-1})$ can be identified with the degree-n component of the bar-complex $\mathcal{B}ar_\bullet(A)$. $\square$

Note that the equivalence of (a) and (b) holds for not necessarily quadratic algebras A (by Corollary 3.2).

Definition 1. A graded algebra A is called *n-Koszul* if it satisfies the equivalent conditions (a) and (b) of Theorem 4.1. This is also equivalent to the condition that the algebra

$$\operatorname{Ext}_A(\Bbbk,\Bbbk)/(\bigoplus_{i\geqslant 1, j>n}\operatorname{Ext}_A^{ij}(\Bbbk,\Bbbk))$$

equipped with internal grading is one-generated (the equivalence follows from Proposition 3.1).

For example, any graded algebra is 1-Koszul, any one-generated algebra is 2-Koszul, and any quadratic algebra is 3-Koszul. An algebra is Koszul iff it is n-Koszul for every $n \geqslant 1$.

Since the subspaces $X_1^!,\dots,X_{n-1}^! \subset A_1^{!\otimes n} = V^{*\otimes n}$ corresponding to the quadratic dual algebra $A^!$ are the orthogonal complements to the subspaces $X_1,\dots,X_{n-1} \subset V^{\otimes n}$, it follows that A and $A^!$ are n-Koszul simultaneously (this can also be deduced from Corollary 3.3). Note also that the complex $B^\bullet(V^{\otimes n},X_1,\dots,X_{n-1})$ from Proposition 7.2 of chapter 1 can be identified with the degree-n part of the cobar-complex $\mathcal{C}ob(A^!)$.

THEOREM 4.2.(M). *A quadratic module $M = \langle H,K\rangle_A$ over a Koszul algebra $A = \{V,I\}$ is Koszul iff for any $n \geqslant 0$ the collection of subspaces*

$$Y_i = \begin{cases} V^{\otimes i-1}\otimes I\otimes V^{\otimes n-i-1}\otimes H, & i=1,\dots,n-1\\ V^{\otimes n-1}\otimes K, & i=n\end{cases}$$

in $V^{\otimes n}\otimes H$ is distributive. More generally, assume only that A is n-Koszul. Then the following conditions are equivalent:

(a) $\operatorname{Ext}_A^{ij}(M,\Bbbk)=0$ *for all $i<j\leqslant n$;*
(b′) *the Koszul complex $'\overline{K}_\bullet(A,M)$ is acyclic in internal degrees $\leqslant n$;*
(b″) *the Koszul complex $''\overline{K}_\bullet(A,M)$ is acyclic in internal degrees $\leqslant n$;*
(c) *the collection $(Y_1,\dots,Y_n)$ in $V^{\otimes n}\otimes H$ is distributive.*

Proof: Set $W = V^{\otimes n}\otimes H$. We are going to use Proposition 7.2 of chapter 1 for the collection $(Y_1,\dots,Y_n)$ of subspaces in W. One can easily see that the complex $B_\bullet(W;Y_1,\dots,Y_n)$ coincides with the degree-n part of the bar-complex $\mathcal{B}ar_\bullet(A,M)$, and that the complex $K_\bullet(W;Y_1,\dots,Y_n)$ (resp., $K_\bullet(W;Y_n,\dots,Y_1)$) coincides with the degree-n part of the complex $'\overline{K}_\bullet(A,M)$ (resp., $''\overline{K}_\bullet(A,M)$). Using n-Koszulness of A and Theorem 4.1 one can easily show that any proper subcollection of $X_1,\dots,X_n$ is distributive iff similar collections of subspaces in $V^{\otimes j}\otimes H$ are distributive for all $j<n$. It remains to apply Proposition 7.2 of chapter 1 together with induction in n. □

An important class of quadratic algebras is formed by *monomial* quadratic algebras. By definition these are algebras of the form

$$A = \Bbbk\{x_1,\dots,x_n\}/(x_ix_j,\ (i,j)\in S),$$

where $S\subset[1,n]\times[1,n]$ is any subset of pairs of indices. Quadratic *monomial modules* over such an algebra A are modules of the form

$$M = (Ae_1\oplus\dots\oplus Ae_m)/\sum_{(i,j)\in T} Ax_ie_j$$

for some $T\subset[1,n]\times[1,m]$.

COROLLARY 4.3. *A monomial quadratic algebra is Koszul. A monomial quadratic module over a monomial quadratic algebra is Koszul.*

Proof: In this case the natural monomial bases of the relevant spaces are distributive. □

Note that equivalence of (a), (b′) and (b″) in Theorem 4.2 holds for a not necessarily quadratic module M (by Corollary 3.2 and Proposition 3.4).

Definition 2(M). A graded module M over an n-Koszul algebra A is called *n-Koszul* if the equivalent conditions (a), (b′), or (b″) of Theorem 4.2 are satisfied (see Corollary 3.2). This is also equivalent to the condition that the $\mathrm{Ext}_A(\Bbbk,\Bbbk)$-module

$$\mathrm{Ext}_A(M,\Bbbk)/(\bigoplus_{i\geqslant 0, j>n} \mathrm{Ext}^{ij}(M,\Bbbk))$$

is generated in degree zero.

For example, any module generated in degree 0 is 1-Koszul and any quadratic module is 2-Koszul. As before, using lattices one can immediately see that quadratic dual modules (over dual n-Koszul algebras) are n-Koszul simultaneously.

Clearly, the distributivity condition (c) of Theorem 4.2 implies n-Koszulness of the algebra A itself, provided $H \neq 0$. The following result shows how one can deduce n-Koszulness of A imposing homological conditions on *both* quadratic dual modules M and $M^!$.

PROPOSITION 4.4. *Let $A = \{V, I\}$ be a quadratic algebra, $M = \langle H, K\rangle_A$ a nonzero quadratic A-module, and $M^!$ the dual quadratic $A^!$-module. Then for any $n \geqslant 0$ the following conditions are equivalent:*

(a) $\mathrm{Ext}^{ij}_A(M,\Bbbk) = 0$ *and* $\mathrm{Ext}^{ij}_{A^!}(M^!,\Bbbk) = 0$ *for all* $i < j \leqslant n$;

(b) *both Koszul complexes* $'\overline{K}_\bullet(A,M)$ *and* $''\overline{K}_\bullet(A,M)$ *are acyclic in internal degrees* $\leqslant n$;

(c) *the algebra A is n-Koszul and the A-module M is n-Koszul.*

Proof: According to Proposition 3.1, our vanishing condition for $\mathrm{Ext}^{ij}_A(M,\Bbbk)$ is equivalent to exactness of the complex of $'\overline{K}_\bullet(A,M)$. Since the complex $''\overline{K}_\bullet(A,M)$ is dual to $'\overline{K}_\bullet(A^!,M^!)$, its exactness is equivalent to our vanishing condition on $\mathrm{Ext}_{A^!}(M^!,\Bbbk)$. This proves (a) $\Longleftrightarrow$ (b). The implication (c) $\Longrightarrow$ (a) follows from the observation that the quadratic dual module to an n-Koszul module is n-Koszul. It remains to prove that (b) $\Longrightarrow$ (c). We are going to use notation from the proof of Theorem 4.2. Recall that the complex $K_\bullet(W;Y_1,\dots,Y_n)$ (resp., $K_\bullet(W;Y_n,\dots,Y_1)$) can be identified with the degree-n part of $'\overline{K}_\bullet(A,M)$ (resp., $''\overline{K}_\bullet(A,M)$). Using induction in n we can assume that any proper subcollection of $Y_1,\dots,Y_{n-1}$ is distributive together with Y_n. As was noticed in the proof of Proposition 7.2, the conditions of exactness of $K_\bullet(W;Y_1,\dots,Y_n)$ (resp., $K_\bullet(W;Y_n,\dots,Y_1)$) coincide with the lattice equations on $Y_1,\dots,Y_n$ (resp., $Y_n,\dots,Y_1$) from Theorem 6.3. Now we can apply Corollary 6.5, (c) $\Longrightarrow$ (a) for $N = n-1$ and u corresponding to Y_n. □

Let us conclude with one more result relating $(n-1)$-Koszulness, n-Koszulness and quadratic duality.

PROPOSITION 4.5. *Let A be an $(n-1)$-Koszul quadratic algebra. Then for every $2 < i < n$, there is a natural perfect pairing*

$$\mathrm{Ext}^{i,n}_A(\Bbbk,\Bbbk) \otimes \mathrm{Ext}^{n-i+2,n}_{A^!}(\Bbbk,\Bbbk) \longrightarrow \Bbbk.$$

If A is an n-Koszul quadratic algebra and M is an $(n-1)$-Koszul quadratic module over A then for every $1 < i < n$ there is a natural perfect pairing

$$\operatorname{Ext}_A^{i,n}(M, \Bbbk) \otimes \operatorname{Ext}_{A^!}^{n-i+1,n}(M^!, \Bbbk) \longrightarrow \Bbbk.$$

Proof: This follows immediately from Corollaries 3.3 and 3.5. □

5. Homomorphisms of algebras and Koszulness. I

In this section we start to discuss the relation between Koszulness of graded algebras connected by certain special kinds of homomorphisms $A \longrightarrow B$. The motivation for this discussion comes mainly from two situations studied in [**20**] when B is a quotient of A by an ideal generated by a normal element of degree 1 or of degree 2. Below we present some generalizations of the results obtained in [**20**].

It is convenient to consider the following relative version of Koszulness.

Definition. A homomorphism $A \longrightarrow B$ of graded algebras is called *left Koszul* (resp., *right Koszul*) if B has a linear free resolution as a left (resp., right) module over A.

Note that a Koszul homomorphism is necessarily surjective (since B is generated by $B_0 = \Bbbk$ as an A-module). Also, a graded algebra A is Koszul iff the augmentation homomorphism $A \rightarrow \Bbbk$ is Koszul.

PROPOSITION 5.1. *Composition of left (resp., right) Koszul homomorphisms is again a left (resp., right) Koszul homomorphism.*

Proof: Let $A \longrightarrow B \longrightarrow C$ be a pair of left Koszul homomorphisms. Consider the standard spectral sequence

$$E_{pq}^2 = \operatorname{Tor}_p^B(\operatorname{Tor}_q^A(\Bbbk, B), C) \Longrightarrow \operatorname{Tor}_{p+q}^A(\Bbbk, C). \tag{5.1}$$

Since $A \longrightarrow B$ is Koszul, the space $\operatorname{Tor}_q^A(\Bbbk, B)$ is concentrated in internal degree q. In particular, the right action of B on it is trivial. Hence,

$$E_{pq}^2 \simeq \operatorname{Tor}_q^A(\Bbbk, B) \otimes \operatorname{Tor}_p^B(\Bbbk, C).$$

Since $B \longrightarrow C$ is Koszul, this space is concentrated in internal degree $p+q$; hence, $A \longrightarrow C$ is Koszul. □

Our next result involves a slightly more general class of homomorphisms (cf. Thm. 5 of [**102**]).

THEOREM 5.2. *Let $A \longrightarrow B$ be a homomorphism of graded algebras such that $\operatorname{Tor}_{i,j}^A(\Bbbk, B) = 0$ for $j > i+1$, and let $B \longrightarrow C$ be an arbitrary homomorphism of graded algebras such that the composed homomorphism $A \longrightarrow C$ is left Koszul. Then $B \longrightarrow C$ is left Koszul.*

Proof: Below by *degree* we always mean the *internal* degree. Consider again the spectral sequence (5.1). Let us prove by induction in n that $\operatorname{Tor}_n^B(\Bbbk, C)$ has internal degree n. This is true for $n = 0$ since surjectivity of $A \longrightarrow C$ implies surjectivity of $B \longrightarrow C$. Assume now that the assertion is true for all $n' < n$. Note that the graded right B-module $\operatorname{Tor}_q^A(\Bbbk, B)$ is an extension of its grading components of degrees q and $q+1$ (endowed with the trivial B-module structures). It follows that for every $p < n$ the space E_{pq}^2 is concentrated in degrees $p+q$ and $p+q+1$.

Next, we claim that the term $E_{n,0}^2 = \operatorname{Tor}_n^B(\operatorname{Tor}_0^A(\Bbbk, B), C)$ is concentrated in degree $\leqslant n$. Indeed, the differential d_N in our spectral sequence has the form

$$0 = E_{n+N,-N+1}^N \longrightarrow E_{n,0}^N \longrightarrow E_{n-N,N-1}^N.$$

Since $E^N_{n-N,N-1}$ is a subquotient of $E^2_{n-N,N-1}$, it is concentrated in degrees $n-1$ and n. On the other hand, the limit term is concentrated in degree n (by Koszulness of $A \longrightarrow C$), so it follows that $E^2_{n,0}$ is concentrated in degrees $n-1$ and n, which proves our claim.

By our assumption the right B-module $\mathrm{Tor}_0^A(\Bbbk, B)$ is concentrated in degrees 0 and 1, so we have an exact triple of right B-modules

$$0 \longrightarrow \mathrm{Tor}^A_{0,1}(\Bbbk, B) \otimes \Bbbk(-1) \longrightarrow \mathrm{Tor}^A_0(\Bbbk, B) \longrightarrow \Bbbk \longrightarrow 0.$$

Let us consider the fragment of the associated long exact sequence:

$$\ldots \longrightarrow \mathrm{Tor}^B_n(\mathrm{Tor}^A_0(\Bbbk, B), C) \longrightarrow \mathrm{Tor}^B_n(\Bbbk, C) \longrightarrow$$
$$\mathrm{Tor}^A_{0,1}(\Bbbk, B) \otimes \mathrm{Tor}^B_{n-1}(\Bbbk(-1), C) \longrightarrow \ldots$$

By the induction assumption the space $\mathrm{Tor}^B_{n-1}(\Bbbk(-1), C)$ has internal degree n. Hence, from this sequence we deduce that $\mathrm{Tor}^B_n(\Bbbk, C)$ lives in degrees $\leqslant n$. But it also lives in degrees $\geqslant n$ for trivial reasons. Therefore, it has to be concentrated in degree n. □

COROLLARY 5.3. *Let $A \longrightarrow B$ be a left Koszul homomorphism of graded algebras and let $B \longrightarrow C$ be an arbitrary homomorphism. Then $B \longrightarrow C$ is left Koszul iff the composed homomorphism $A \longrightarrow C$ is left Koszul.*

Proof: The "only if" follows from Proposition 5.1 and the "if" follows from Theorem 5.2. □

Taking $C = \Bbbk$ in the above corollary we obtain the following well-known result.

COROLLARY 5.4. *Let $A \longrightarrow B$ be a left Koszul homomorphism of graded algebras. Then A is Koszul iff B is Koszul.*

COROLLARY 5.5. *Let M be a Koszul left module over a Koszul algebra A. Let us equip the graded space $A_M = A \oplus M(-1)$ with the algebra structure by setting $(a, m) \cdot (a', m') = (aa', am')$. Then the algebra A_M is Koszul.*

Proof: One can check easily that the conditions of Theorem 5.2 are satisfied for the embedding $A \longrightarrow A_M$ (we take $B = A_M$ and $C = \Bbbk$). □

Example 1. Recall that an element $t \in A$ is called *normal* if $At = tA$. Let us say that t is *left (resp., right) normal* if $tA \subset At$ (resp., $At \subset tA$), i.e., At (resp., tA) is a two-sided ideal in A. Assume that t is a left normal element in A_1 that is not a right zero divisor (i.e., $at \neq 0$ for $a \neq 0$). Then the algebra $B = A/At$ has a very simple linear free resolution as a left module over A:

$$0 \longrightarrow At \longrightarrow A \longrightarrow B \longrightarrow 0.$$

Hence, in this case the homomorphism $A \longrightarrow B$ is Koszul. From the above corollary we get that in this situation A is Koszul iff B is Koszul (cf. [**20**] Thm. 4 (e); [**112**] Thm. 1.5). This statement is often used in the commutative situation (see Lemma 7.3).

Example 2. Let V be a vector space and let $q \in \mathbb{S}^2(V)$ be a nonzero element. Then the algebra $\mathbb{S}(V)/(q)$ is Koszul (see Example 1 of the next section for a more general statement). Indeed, this can be proved by induction in $\dim V$. If $\dim V = 1$, then this is clear. The induction step follows easily from the previous example: pick $v \in V$ such that q is not divisible by v in $\mathbb{S}(V)$. Then v is not a zero divisor in $\mathbb{S}(V)$, so the homomorphism $\mathbb{S}(V)/(q) \to \mathbb{S}(V)/(v, q)$ is Koszul. But

$\mathbb{S}(V)/(v,q) \simeq \mathbb{S}(V')/(q')$, where $V' = V/\Bbbk v$ and $q' \in \mathbb{S}^2(V')$ is the image of q, so this algebra is Koszul by the induction assumption.

Example 3. Assume that an algebra A contains a pair of elements t_1 and t_2 of degree 1 such that $At_1 + At_2$ is a two-sided ideal and that a relation $s_1t_1 = s_2t_2$ holds for some $s_1, s_2 \in A_1$. Assume in addition that $x = s_1t_1 = s_2t_2$ is not a right zero divisor and that $Ax = At_1 \cap At_2$. Then the algebra $B = A/(At_1 + At_2)$ has a linear free resolution as a left A-module of the form:

$$0 \longrightarrow Ax \longrightarrow At_1 \oplus At_2 \longrightarrow A \longrightarrow B \longrightarrow 0.$$

Hence, in this case the homomorphism $A \to B$ is Koszul. Some homomorphisms of this type can be obtained as compositions of two homomorphisms as in Example 1. However, not all of them can be represented in this form. For example, consider the quadratic algebra A generated by degree one elements s_1, s_2, t_1, t_2 with defining relations $s_1t_1 = s_2t_2$ and $t_is_j = t_it_j$ for all $i = 1, 2$ and $j = 1, 2$. Then the above conditions are satisfied. In this case B is the free algebra in s_1 and s_2.

The following result gives a useful criterion of Koszulness of a homomorphism (see sections 7 and 11 for applications).

THEOREM 5.6. *Let $f : A \longrightarrow B$ be a surjective homomorphism of graded algebras. Assume that B is Koszul and there exists a complex $K_\bullet$ of left (resp., right) free A-modules of the form*

$$\ldots \longrightarrow V_2 \otimes A(-2) \longrightarrow V_1 \otimes A(-1) \xrightarrow{\delta} A$$

(so that $K_i = V_i \otimes A(-i)$), where V_i are finite-dimensional vector spaces, such that $\mathrm{im}(\delta) = \ker(f)$ *and $H_i(K_\bullet)_j = 0$ for $i \geqslant 1$ and $j > i+1$. Then f is left (resp., right) Koszul.*

Proof: Assume that we are dealing with left modules. First, let us consider the spectral sequence

$$E^1_{p,q} = \mathrm{Tor}^A_p(\Bbbk, K_q) \implies \mathrm{Tor}^A_{p+q}(\Bbbk, K_\bullet).$$

Note that $E^1_{p,q} = 0$ for $p > 0$ and the differential d^1 on $E^1_{0,*}$ is zero due to the form of $K_\bullet$. Hence, the sequence degenerates at E^1 and $\mathrm{Tor}^A_n(\Bbbk, K_\bullet) \simeq \Bbbk \otimes_A K_n$ is concentrated in internal degree n. Next, let us prove by induction in n that $\mathrm{Tor}^A_i(\Bbbk, B)_j = 0$ for $i \leqslant n$ and $j > i$. This is true for $n = 0$ since f is surjective. Assume that the assertion is true for $n - 1$. Then the spectral sequence (5.1) with $C = \Bbbk$ shows that $\mathrm{Tor}^A_i(\Bbbk, \Bbbk)$ is concentrated in degree i for all $i < n$. Now let us consider another spectral sequence

$$E^2_{p,q} = \mathrm{Tor}^A_p(\Bbbk, H_q(K_\bullet)) \implies \mathrm{Tor}^A_{p+q}(\Bbbk, K_\bullet).$$

Since for $q \geqslant 1$ the A-module $H_q(K_\bullet)$ is an extension of $\Bbbk(-q)$ and $\Bbbk(-q-1)$, we obtain that for $p < n$ and $q \geqslant 1$ the term $E^2_{p,q}$ is concentrated in internal degrees $p+q$ and $p+q+1$. This implies that for $r \geqslant 2$ the differential $d_r : E^r_{n-r,r-1} \to E^r_{n,0}$ has zero components of internal degree $\geqslant n+1$. Hence, all components of internal degree $\geqslant n+1$ in $E^2_{n,0} = \mathrm{Tor}^A_n(\Bbbk, B)$ "survive" in the spectral sequence and give a contribution to $E^\infty_n = \mathrm{Tor}^A_n(\Bbbk, K_\bullet)$. But the latter space is concentrated in internal degree n, so we obtain that $\mathrm{Tor}^A_n(\Bbbk, B)_j = 0$ for $j > n$. □

COROLLARY 5.7 (cf. [**96**], Thm. 4). *Let A be a graded algebra such that there exists a complex $K_\bullet$ of right free A-modules of the form*

$$\cdot \longrightarrow V_2 \otimes A(-2) \longrightarrow V_1 \otimes A(-1) \longrightarrow A$$

where V_i are finite-dimensional vector spaces, such that $H_i(K_\bullet)_j = 0$ for $j \geqslant i \geqslant 2$ and also for $(i,j) \in \{(0,1),(1,1),(1,2)\}$. Then the quadratic part of A is Koszul.

Proof: The condition $H_i(K_\bullet)_j = 0$ for $i \geqslant 1$ and $j = i$ or $j = i+1$ implies that we have isomorphisms $V_i \simeq A_i^{!*}$, so that the complex $K_\bullet$ is identified with the Koszul complex of A. It is easy to see that the dual complex $A^! \otimes A^*$ considered as a complex of $A^!$-modules satisfies the assumptions of Theorem 5.6, where as homomorphism f we take the augmentation homomorphsim $A^! \longrightarrow \Bbbk$. This implies that $A^!$ is Koszul. □

The following simple result is very useful in relating homological properties of a homomorphism of Koszul algebras with those of the induced homomorphism of dual Koszul algebras.

THEOREM 5.8 ([**102**], Thm. 6). *Let $f: A \longrightarrow B$ be a homomorphism of Koszul algebras and let $f^!: B^! \longrightarrow A^!$ be the dual homomorphism. Then there are natural isomorphisms*

$$\mathrm{Tor}^A_{i,j}(\Bbbk, B)^* \simeq \mathrm{Tor}^{B^!}_{j-i,j}(A^!, \Bbbk),$$

where B is viewed as a left A-module and $A^!$ as a right $B^!$-module. Furthermore, the direct sum of these isomorphisms over all (i,j) is compatible with the natural left $A^!$-action (resp. right B-action).

Proof: This follows easily from the fact that both spaces are appropriate bigrading components of the cohomology of the complex $A^! \otimes B$ (see section 3). □

As an application of the above theorem, we can interpret Koszulness of a homomorphism between Koszul algebras in terms of the dual Koszul algebras.

COROLLARY 5.9. *Let $f: A \longrightarrow B$ be a homomorphism of Koszul algebras. Then f is left Koszul iff $A^!$ is a free right $B^!$-module. If this is the case then the dual morphism $B^! \longrightarrow A^!$ is injective, and B and $A^!/A^!B^!_+$ are dual Koszul modules over A and $A^!$.*

Proof: Both assertions follow easily from the theorem. For the first assertion one has also to use the fact that $A^!$ is a free $B^!$-module iff $\mathrm{Tor}^{B^!}_{i,j}(A^!, \Bbbk) = 0$ for $i \neq 0$. For the second assertion we apply the isomorphisms of Theorem 5.8 for $i = j$. □

COROLLARY 5.10. *Let $f: A \longrightarrow B$ be a homomorphism of commutative Koszul algebras. Then f is Koszul iff the dual homomorphism $f^!: B^! \longrightarrow A^!$ is injective.*

Proof: The "only if" part is immediate from the above theorem. For the converse, we observe that the quadrataic dual algebras to A and B are universal enveloping algebras of graded Lie superalgebras: $A^! = U(\mathfrak{g})$ and $B^! = U(\mathfrak{h})$ (see Example 4 in section 2 of chapter 1). Furthermore, the homomorphism $f^!: B^! \to A^!$ is induced by a Lie algebra homomorphism $\mathfrak{h} \to \mathfrak{g}$. By the PBW theorem if $f^!$ is injective then $A^!$ is a free right $B^!$-module. It remains to apply the above theorem. □

Here is another useful criterion of Koszulness of the quotient of a Koszul algebra by a number of linear relations. In the commutative case it does not give anything new, since in this case the homomorphism $A \longrightarrow B$ is Koszul. However, in the noncommutative case it deals with a new type of homomorphisms.

PROPOSITION 5.11. *Let $B = A/H$ where H is two-sided ideal generated in degree 1, free as a right A-module. Assume also that $H/HA_+(1)$ has a free linear resolution as a left B-module. Then A is Koszul iff B is Koszul. In this case the homomorphism $B^! \longrightarrow A^!$ is injective.*

Proof: Consider the spectral sequence

$$E^2_{pq} = \mathrm{Tor}^B_p(\Bbbk, \mathrm{Tor}^A_q(B, \Bbbk)) \implies \mathrm{Tor}^A_{p+q}(\Bbbk, \Bbbk).$$

From the exact sequence $0 \longrightarrow H \longrightarrow A \longrightarrow B \longrightarrow 0$ we get an isomorphism

$$\mathrm{Tor}^A_q(B, \Bbbk) \simeq \mathrm{Tor}^A_{q-1}(H, \Bbbk)$$

for $q \geqslant 1$. But H is a free right A-module, so the latter space is zero unless $q = 1$. Hence the term E^2 of our spectral sequence is concentrated on two lines: for $q = 0$ we get $E^2_{p0} = \mathrm{Tor}^B_p(\Bbbk, \Bbbk)$ and for $q = 1$ we get $E^2_{p1} = \mathrm{Tor}^B_p(\Bbbk, H/HA_+)$. Our assumption on H/HA_+ is equivalent to the condition that the space E^2_{p1} is concentrated in the internal degree $p + 1$. It follows that all the differentials

$$d_2 : \mathrm{Tor}_B^{p+2}(\Bbbk, \Bbbk) = E^2_{p+2,0} \longrightarrow E^2_{p,1}$$

vanish. Therefore, the spectral sequence degenerates at the term E^2 and gives rise to exact sequences

$$0 \longrightarrow E^2_{p-1,1} \longrightarrow \mathrm{Tor}^A_p(\Bbbk, \Bbbk) \longrightarrow \mathrm{Tor}^B_p(\Bbbk, \Bbbk) \longrightarrow 0.$$

Since $E^2_{p-1,1}$ has internal degree p this shows that A is Koszul iff B is Koszul. We also see that the map $B^!_p \longrightarrow A^!_p$ is injective being dual to the map $\mathrm{Tor}^A_p(\Bbbk, \Bbbk) \longrightarrow \mathrm{Tor}^B_p(\Bbbk, \Bbbk)$. □

Following [**7**] we call a homogeneous element $t \in A_d$ *strongly free* if A has the same Hilbert series as the free product of $A/(t)$ and $k[t]$ (where $\deg t = d$), i.e.,

$$h_{A/(t)}(z)^{-1} = h_A(z)^{-1} + z^d$$

(see equation (1.2) of chapter 3). Let $W \subset A$ be a graded complement to $(t) \subset A$ (so W is just a graded subspace of A). It is easy to see that the sequence

$$W \otimes_{\Bbbk} A(-1) \longrightarrow A \longrightarrow A/(t) \longrightarrow 0$$

is exact, where the left arrow is $w \otimes a \longmapsto wta$. Furthermore, the comparison of Hilbert series shows that t is strongly free iff the map $W \otimes_{\Bbbk} A(-1) \longrightarrow (t)$ above is an isomorphism. Now we can deduce the following result due to Backelin and Fröberg (see [**20**] Thm. 4(e)(i)).

COROLLARY 5.12. *Let $B = A/(t)$, where $t \in A_1$ is strongly free. Then A is Koszul iff B is Koszul.*

Proof: We apply the above proposition to $H = (t)$. As we have seen above, the assumption that t is strongly free implies that H is free as a right A-module and that one has an isomorphism $H/HA_+ \simeq A/(t)(-1)$. □

For example, if B is any Koszul algebra then the free product of B with $\Bbbk[t]$ (where $\deg t = 1$) is also Koszul. In section 1 of chapter 3 we will show that the free product of any two Koszul algebras is Koszul.

6. Homomorphisms of algebras and Koszulness. II

In this section we prove several results about Koszulness in the case when B is a quotient of A by a number of quadratic relations.

PROPOSITION 6.1 (cf. [**20**], Thm. 7(a)). *Let A be a Koszul algebra and $I \subset A$ be a two-sided ideal such that $I(2)$ is a Koszul left A-module. Then the algebra B is also Koszul.*

Proof: This follows immediately from Theorem 5.2. We only have to observe that since $I(2)$ has a linear free resolution as an A-module, it follows that B has a resolution of the form

$$\ldots \longrightarrow V_2 \otimes A(-3) \longrightarrow V_1 \otimes A(-2) \longrightarrow A \longrightarrow B \longrightarrow 0$$

where V_i are vector spaces (of degree 0). Therefore, $\mathrm{Tor}^A_{i,j}(\Bbbk, B) = 0$ for $i \geqslant 1$ and $j \neq i+1$. □

Example 1. Let us consider the case $A = \Bbbk[x_1, \ldots, x_n]$. Then according to the work of P. Schenzel [**109**] the assumption of the above proposition holds when A/I is a quadratic *extremal* Cohen-Macaulay (CM) ring. Note that A/I is a CM-ring iff its projective dimension over A is equal to $n - \dim A/I$. *Extremality* here means that $\deg h_{A/I}(z) = -\dim A/I + 1$. For example, A/I is a quadratic extremal CM-ring in the following examples: (i) I is generated by all maximal minors in the $2 \times n$-matrix of indeterminates; (ii) I is generated by all 2×2-minors in the symmetric 3×3-matrix of indeterminates.

From the above proposition we immediately recover the following results of [**20**] (Thm. 4(e)(iii),(iv)).

COROLLARY 6.2. *Let A be a Koszul algebra and let $B = A/tA$ where $t \in A_2$ satisfies $tA_+ = A_+t = 0$. Then B is also Koszul.*

COROLLARY 6.3. *Let A be a Koszul algebra and let $B = A/At$ where $t \in A_2$ is a right-normal element which is not a right zero-divisor. Then B is also Koszul.*

Example 2. The above corollary shows that if A is a Koszul algebra then $B = A/(t)$ is also Koszul, provided that $t \in A_2$ is a central element that is not a zero-divisor. Starting with the symmetric algebra and iterating this assertion we derive Koszulness of $\Bbbk[x_1, \ldots, x_n]/(q_1, \ldots, q_r)$ for any regular sequence of quadrics $(q_1, \ldots, q_r)$. Note that the converse to the above statement is not true: Koszulness of B does not necessarily imply Koszulness of A. Indeed, let A be the universal enveloping algebra of the Heisenberg Lie algebra associated with the symplectic vector (V, ω) of dimension $2n \geqslant 4$. This algebra is quadratic (see Example 5 of section 2 of chapter 1) but not Koszul (see Example 1 of section 2). By definition, A has a central degree 2 element t such that $B = A/(t)$ is isomorphic to the symmetric algebra of V, and t is not a zero divisor by the usual PBW theorem.

On the other hand, as was proved in [**20**] (Thm. 4(e)(iv)), if A is commutative and $B = A/(t)$, where $t \in A_2$ is not a zero divisor, then Koszulness of B does imply Koszulness of A. We will prove a slightly stronger assertion in Proposition 6.5 below.

Example 3. In the skew-commutative case quadrics are usually not Koszul. Namely, if V is a vector space and $\omega \in \bigwedge^2(V)$ is a nonzero element then the algebra $A = \bigwedge(V)/(\omega)$ is Koszul iff ω is decomposable, i.e., $\omega = v_1 \wedge v_2$ for some $v_1, v_2 \in V$.

Indeed, if ω is not decomposable then $A^!$ is isomorphic to the enveloping algebra of the (possibly degenerate) Heisenberg Lie algebra associated with ω, that has a non-Koszul Hilbert series as we have seen in Example 1 of section 2. Koszulness of A in the case when ω is decomposable can be checked in many ways. For example, it follows from Theorem 8.1 of chapter 4.

LEMMA 6.4. *Let A be a quadratic algebra, $t \in A_2$ be a nonzero element, $B = A/(t)$. Then there is a natural line $\Bbbk t^! \in B_2^!$ such that $A^! = B^!/(t^!)$. Assume in addition that t is central in A. Then the element $t^!$ is central in $B^!$ iff the equality $xt = 0$ in A for $x \in A_1$ implies that $x = 0$.*

Proof: Let $A = \{V, I_A\}$, $B = \{V, I_B\}$. Then $I_A \subset I_B$ and the element t is a generator of the 1-dimensional space $I_B/I_A \subset A_2$. We define $t^! \in B_2^!$ to be the dual of the functional

$$\phi : I_B \longrightarrow I_B/I_A \simeq \Bbbk.$$

Then $A_2^! = B_2^!/\Bbbk t^!$. The element $t^!$ is central in $B^!$ iff for every $r \in I_B \otimes V \cap V \otimes I_B$ one has

$$\phi \otimes \mathrm{id}(r) = \mathrm{id} \otimes \phi(r)$$

in V. Let $\widetilde{t} \in I_B$ be any element projecting to $t \in I_B/I_A$. Then $I_B = I_A + \Bbbk\widetilde{t}$ and for every $v \in V$ one has

$$\widetilde{t} \otimes v - v \otimes \widetilde{t} \in I_A \otimes V + V \otimes I_A$$

since t is central in A. Now let $r \in I_B \otimes V \cap V \otimes I_B$ be an arbitrary element. Then we can write

$$r = \widetilde{t} \otimes v_1 + x_1 = v_2 \otimes \widetilde{t} + x_2$$

where $v_1, v_2 \in V$, $x_1 \in I_A \otimes V$, $x_2 \in V \otimes I_A$. Furthermore, one has

$$\phi \otimes \mathrm{id}(r) = v_1, \ \mathrm{id} \otimes \phi(r) = v_2.$$

Thus, $t^!$ is central in $B^!$ iff the condition

$$\widetilde{t} \otimes v_1 - v_2 \otimes \widetilde{t} \in I_A \otimes V + V \otimes I_A$$

for $v_1, v_2 \in V$ implies $v_1 = v_2$. But

$$\widetilde{t} \otimes v_1 - v_2 \otimes \widetilde{t} \equiv \widetilde{t} \otimes (v_1 - v_2) \mod I_A \otimes V + V \otimes I_A.$$

Hence, $t^!$ is central iff the condition

$$\widetilde{t} \otimes v \in I_A \otimes V + V \otimes I_A$$

for $v \in V$ implies $v = 0$. □

Example 4. Let V be a vector space, $q \in \mathbb{S}^2(V)$ a nonzero element. Applying the above lemma to $A = \mathbb{S}(V)$ and $t = q$ we see that the dual quadratic algebra to $B = \mathbb{S}(V)/(q)$ has a central element $t^!$ of degree 2 such that the quotient $B^!/(t^!)$ is the exterior algebra of V^*. In fact, $B^!$ is the graded version of the Clifford algebra associated with q. Namely, it is generated by the degree-1 space V^* along with a central degree-2 element $t^!$ and has defining relations

$$\xi_1\xi_2 + \xi_2\xi_1 + (\xi_1 \cdot \xi_2)t^! = 0$$

for all $\xi_1, \xi_2 \in V^*$, where $\xi_1 \cdot \xi_2$ is the symmetric pairing on V^* associated with q.

PROPOSITION 6.5. *Let A be a commutative quadratic algebra, $t \in A_2$ be an element such that the equality $xt = 0$ in A for $x \in A_1$ implies $x = 0$. Assume that the algebra $B = A/(t)$ is Koszul. Then A is Koszul.*

Proof: According to the above lemma we have $A^! = B^!/(t^!)$ where $t^! \in B^!_2$ is a central element. Note that $B^!$ is the universal enveloping algebra $U(\mathfrak{g})$ of a graded Lie superalgebra $\mathfrak{g}$ and $t^!$ is a nonzero element of $\mathfrak{g}_2 \subset U(\mathfrak{g})$. Therefore, by the PBW-theorem $t^!$ is not a zero-divisor. Applying Corollary 6.3 we deduce that $A^!$ is Koszul and hence A is Koszul. □

Next, we have the following analogue of Corollary 5.9.

THEOREM 6.6. *Let $f : A \longrightarrow B$ be a homomorphism of Koszul algebras such that f induces an isomorphism $A_1 \widetilde{\longrightarrow} B_1$. Let $f^! : B^! \longrightarrow A^!$ be the dual homomorphism. Then* $\ker(f)(2)$ *is a Koszul left A-module iff $H := \ker(f^!)(2)$ is a free right $B^!$-module. In this case* $\ker(f)(2)$ *and $H/HB^!_+$ are dual Koszul modules over A and $A^!$.*

Proof: **"only if"**. Assume that $\ker(f)(2)$ is a Koszul left A-module. Let us consider the spectral sequence

$$E^2_{pq} = \mathrm{Ext}^p_{B^{\mathrm{op}}}(\mathrm{Tor}^{A^{\mathrm{op}}}_q(B, \Bbbk), \Bbbk) \implies \mathrm{Ext}^{p+q}_{A^{\mathrm{op}}}(\Bbbk, \Bbbk).$$

We have $E^2_{p0} = \mathrm{Ext}^p_{B^{\mathrm{op}}}(\Bbbk, \Bbbk) \simeq B^{!\mathrm{op}}$ and

$$E^2_{pq} \simeq \mathrm{Ext}^p_{B^{\mathrm{op}}}(\mathrm{Tor}^{A^{\mathrm{op}}}_{q-1}(\ker(f), \Bbbk), \Bbbk) \simeq \mathrm{Tor}^{A^{\mathrm{op}}}_{q-1}(\ker(f), \Bbbk) \otimes B^{!\mathrm{op}}_p$$

for $q \geqslant 1$. It follows that all the terms E^*_{pq} with $q \geqslant 1$ have internal degree $p+q+1$ and hence get killed in the limit. Furthermore, the only nonzero differentials in the spectral sequence have E^*_{p0} as a target. On the other hand, the natural morphism $\bigoplus_p E^2_{p0} \simeq B^{!\mathrm{op}} \longrightarrow A^{!\mathrm{op}}$ coincides with $f^!$. Hence, we get a filtration on $\ker(f^!)$ with the associated graded object

$$\bigoplus_{p\geqslant 0, q\geqslant 1} E^2_{pq} \simeq \mathrm{Tor}^{A^{\mathrm{op}}}_*(\ker(f), \Bbbk) \otimes B^{!\mathrm{op}}.$$

"if". Let us rename $A^!$, $B^!$ and $f^!$ to B, A and f, respectively. We know that $H = \ker(f)(2)$ is a free right A-module and we want to deduce that $\ker(f^!)(2)$ is a Koszul left $B^!$-module. Let us consider the spectral sequence

$$E^2_{pq} = \mathrm{Ext}^p_B(\mathrm{Tor}^A_q(B, \Bbbk), \Bbbk) \implies \mathrm{Ext}^{p+q}_A(\Bbbk, \Bbbk).$$

Since $H(-2) \longrightarrow A$ is a minimal free resolution of B as a right A-module, we have $\mathrm{Tor}^A_q(B, \Bbbk) = 0$ for $q > 1$ and $\mathrm{Tor}^A_1(B, \Bbbk) \simeq H/HA_+(-2)$. Hence, the spectral sequence degenerates in the term E^3 and we get an exact sequence

$$0 \longrightarrow \mathrm{Ext}^*_B(H/HA_+, \Bbbk)(-2) \longrightarrow B^! \longrightarrow A^! \longrightarrow 0.$$

It follows that for every $p \geqslant 0$ the space $\mathrm{Ext}^p_B(H/HA_+, \Bbbk)$ is concentrated in the internal degree p. Hence H/HA_+ is a Koszul B-module and $\ker(f^!)(2)$ is the Koszul dual $B^!$-module. □

Here is an analogue of Proposition 5.11 for degree-2 quotients.

PROPOSITION 6.7. *Let $f : A \longrightarrow B$ be a homomorphism of graded algebras such that $A_1 = B_1$ and let $H = \ker(f)(2)$. Assume that*
(i) A is Koszul;
(ii) H is free as a right A-module;
(iii) H/HA_+ has a linear free resolution as a left B-module.
Then B is Koszul.

Proof: Consider again the spectral sequence

$$E^2_{pq} = \mathrm{Tor}^B_p(\Bbbk, \mathrm{Tor}^A_q(B, \Bbbk)) \Longrightarrow \mathrm{Tor}^A_{p+q}(\Bbbk, \Bbbk).$$

Since $H(-2) \longrightarrow A$ is a minimal free resolution for B as a right A-module, we have $\mathrm{Tor}^A_q(B, \Bbbk) = 0$ for $q > 1$ and $\mathrm{Tor}^A_1(B, \Bbbk) \simeq H/HA_+(-2)$. Therefore, the sequence degenerates in the term E_3 and the only nonzero differentials d_2 have the form

$$d_2 : E^2_{p1} = \mathrm{Tor}^B_p(\Bbbk, H/HA_+)(-2) \longrightarrow E^2_{p+2,0} = \mathrm{Tor}^B_{p+2}(\Bbbk, \Bbbk).$$

Since H/HA_+ has a linear free resolution over B we see that E^2_{p1} has internal degree $p+2$. Since A is Koszul this implies that $E^2_{p+2,0}$ has internal degree $p+2$, so B is also Koszul. □

Finally, we have the following degree-2 analogue of Corollary 5.12.

COROLLARY 6.8. *Let A be a quadratic algebra and let $B = A/(t)$, where $t \in A_2$ is strongly free. Then A is Koszul iff B is Koszul.*

Proof: The conditions of the above proposition are easily checked, so if A is Koszul then B is also Koszul. To prove the converse let us consider the dual quadratic algebras. We have $A^! = B^!/(t^!)$. We claim that $t^! B^!_1 = B^!_1 t^! = 0$ in $B^!$. Indeed, in terms of the original algebras this means that $I_A \otimes V \cap V \otimes I_A = I_B \otimes V \cap V \otimes I_B$, where V is the space of generators of A (and B), I_A and I_B are the spaces of quadratic relations for A and B. But this equality follows immediately from the equality

$$\dim A_3 = \dim B_3 + 2 \dim V$$

which holds since t is strongly free. Now applying Corollary 6.2 we deduce that if $B^!$ is Koszul then $A^!$ is also Koszul. □

7. Koszul algebras in algebraic geometry

Many examples of commutative quadratic algebras appear as homogeneous coordinate rings of projective varieties. Recall that for a closed subscheme $X \subset \mathbb{P}^n$ the homogeneous coordinate ring R_X is defined as the quotient of $\Bbbk[x_0, \ldots, x_n]$ by the homogeneous ideal I_X generated by all homogeneous forms vanishing on X. In this section we are going to discuss several examples of Koszul algebras of this type.

We start with a couple of general observations providing some links between Koszul algebras and projective geometry. The first of them is a geometric criterion for a module of the form $M = \bigoplus_{i \geqslant 0} H^0(X, F(i))$ to be Koszul.

PROPOSITION 7.1 (cf. [**98**], Lemma 2.1). *Let R be a commutative Koszul algebra. Consider the corresponding projective scheme $X = \mathrm{Proj}(R)$. Then for a coherent sheaf F on X such that $H^i(X, F(-i)) = 0$ for all $i \geqslant 1$ the R-module $M = \bigoplus_{i \geqslant 0} H^0(X, F(i))$ is Koszul.*

Proof: Localizing the Koszul complex of R we get an exact complex of sheaves on X of the form

$$\ldots \longrightarrow R^{!*}_2 \otimes \mathcal{O}_X(-2) \longrightarrow R^{!*}_1 \otimes \mathcal{O}_X(-1) \longrightarrow \mathcal{O}_X \longrightarrow 0.$$

Let us tensor it with $F(j)$ and consider the spectral sequence computing the hypercohomology of the obtained complex:

$$E^1_{p,q} = R^{!*}_{-p} \otimes H^q(F(j+p)) \Longrightarrow 0.$$

The assumption on F implies that $E^1_{p,q} = 0$ for $q > 0$ and $j + p + q > 0$. Since the differential d_r has bidegree $(r, -r+1)$, it follows that all the terms $E^2_{p,0}$ with $p > -j$ "survive" in our spectral sequence. Hence, $E^2_{p,0} = 0$ for $p > -j$. It remains to observe that the complex $(E^1_{p,0}, d_1)$ coincides with the component of the internal degree j of the complex

$$\ldots \longrightarrow R_2^{!*} \otimes M(-2) \longrightarrow R_1^{!*} \otimes M(-1) \longrightarrow M$$

obtained by tensoring the Koszul complex of R with M. Therefore, we get an isomorphism $E^2_{p,0} \simeq \mathrm{Tor}^R_{-p}(\Bbbk, M)_j$ and our assertion follows from the above vanishing. □

COROLLARY 7.2 ([**14**]). *Let M be a finitely generated module over a commutative Koszul algebra R. Then for all sufficiently large n the truncated module $M^{[n]} = M_{\geqslant n}(n)$ is Koszul.*

Proof: Let F be a coherent sheaf on $X = \mathrm{Proj}(R)$ obtained from M by localization. Then for sufficiently large n we have $M^{[n]} = \bigoplus_{i\geqslant 0} H^0(X, F(n+i))$. Also for $n \gg 0$ we will have $H^i(X, F(n-i)) = 0$ for all $i \geqslant 1$. It remains to apply the above proposition. □

The next lemma relates the Koszulness property for the coordinate ring of X with that of its hyperplane section. Recall that a closed subscheme $X \subset \mathbb{P}^n$ is called *projectively normal* if for every $i \geqslant 0$ the natural map $H^0(\mathbb{P}^n, \mathcal{O}_{\mathbb{P}^n}(i)) \to H^0(X, \mathcal{O}_X(i))$ is surjective.

LEMMA 7.3. *Let $X \subset \mathbb{P}^n$ be a closed subscheme and let $H \in \mathbb{P}^n$ be a hyperplane. Assume that X is projectively normal and that no irreducible component of X (with reduced scheme structure) is contained in H. Then R_X is Koszul iff $R_{X\cap H}$ is Koszul.*

Proof: Let $f = 0$ be the equation of H. Then for every $i \geqslant 0$ we have an exact sequence

$$0 \to \mathcal{O}_X(i-1) \xrightarrow{f} \mathcal{O}_X(i) \to \mathcal{O}_{X\cap H}(i) \to 0.$$

Projective normality of X implies that $R_{X\cap H}$ coincides with the image of the homomorphism of algebras

$$R_X = \bigoplus_{i\geqslant 0} H^0(X, \mathcal{O}_X(i)) \to \bigoplus_{i\geqslant 0} H^0(X\cap H, \mathcal{O}_{X\cap H}(i)).$$

Hence, $R_{X\cap H} \simeq R_X/(f)$ as graded algebras. Since f has degree one, Koszulness of R_X and $R_X/(f)$ are equivalent (see Example 1 of section 5). □

Remark 1. The assumption of projective normality cannot be omitted in the above lemma. For example, let X be a smooth rational curve of degree 4 in $\mathbb{P}^3$. Then for a generic plane $H \in \mathbb{P}^2$ the algebra $R_{X\cap H}$ is Koszul (e.g., by Lemma 7.4 below, since $X \cap H$ is 4 points in general linear position in $\mathbb{P}^2$). However, R_X is not even quadratic because there is a unique quadric passing through X.

Now we are going to discuss some concrete examples of projective schemes with Koszul coordinate algebras. The simplest class of such examples is provided by complete intersection of quadrics (see Example 2 of section 6). Considering Hilbert series one can easily see that the Koszul dual to these algebras are universal enveloping algebras of quadratic Lie superalgebras L such that $L_3 = 0$ (see Example 3 of section 2).

Next, let us consider the case when $X \subset \mathbb{P}^n$ is a finite set of points. The theorem of Kempf [**73**] asserts that R_X is Koszul provided that $|X| \leqslant 2n$ and points of X are in general linear position. Theorem 7.5 below is a slight generalization of this result. We start with the crucial case when $|X| = 2n$.

LEMMA 7.4. *Let X be a set of $2n$ distinct points in $\mathbb{P}^n$. Assume that $X = X_1 \sqcup X_2$, where $|X_1| = |X_2| = n$ and points in X_1 (resp., X_2) span a hyperplane H_1 (resp., H_2) in $\mathbb{P}^n$ such that $H_1 \cap X = X_1$ (resp., $H_2 \cap X = X_2$). Then the algebra R_X is Koszul and $h_{R_X}(z) = (1+z)(1+(n-1)z)/(1-z)$.*

Proof: Let $A = R_X$. Note that trivializing $\mathcal{O}_{\mathbb{P}^n}(1)$ near X we get natural evaluation maps $A_i \to k^X$. We claim that these maps are isomorphisms for $i \geqslant 2$. Indeed, it suffices to show that for every point $x \in X$ there exists a quadric q such that $q(x) \neq 0$ and $q(X \setminus \{x\}) = 0$. Say, $x \in X_1$. Then we can take $q = l \cdot l_2$, where l is a linear form such that $l(x) \neq 0$ and $l(X_1 \setminus \{x\}) = 0$.

Let $f_1 = 0$ (resp., $f_2 = 0$) be the equation of H_1 (resp., H_2). Let us consider the natural surjective homomorphism $A \longrightarrow B := R_{X_1}$. We claim that its kernel coincides with $f_1 A$. Indeed, since our homomorphism factors through $A/f_1 A$, it suffices to prove that $\dim A_i/f_1 A_{i-1} = n$ for $i \geqslant 1$. This is clear for $i = 1$ since $\dim A_1 = n+1$. Also we have $\{f \in A_1 : f_1 f = 0 \in A_2\} = \Bbbk f_2$, since f_1 does not vanish on X_2. Therefore, $\dim f_1 A_1 = n$ and

$$\dim A_2/f_1 A_1 = 2n - \dim f_1 A_1 = n.$$

Finally, it is easy to see that for $i > 2$ one has $f_1 A_{i-1} = \Bbbk^{X_2} \subset \Bbbk^X = A_i$ which implies our claim. Similar arguments show that the complex

$$\ldots A(-3) \xrightarrow{f_1} A(-2) \xrightarrow{f_2} A(-1) \xrightarrow{f_1} A$$

is a resolution of B as an A-module. Hence, the homomorphism $A \longrightarrow B$ is Koszul. It is easy to see that B is isomorphic to the monomial quadratic algebra with n generators $t_1, \ldots, t_n$ and relations $t_i t_j = 0$ for all $i < j$ (since the points in X_1 are linearly independent). Hence, B is Koszul by Corollary 4.3. It remains to apply Corollary 5.4 to conclude that A is also Koszul. □

THEOREM 7.5 ([**98**], Cor. 0.2). *Let $X \subset \mathbb{P}^n$ be a finite set of points. Assume that $X = X_1 \sqcup X_2$, where X_1 and X_2 are linearly independent and*

$$L_1 \cap X_2 = L_2 \cap X_1 = \emptyset,$$

where $L_i \subset \mathbb{P}^n$ is the linear subspace spanned by S_i ($i = 1, 2$). Then R_X is Koszul.

Proof: We can assume that both X_1 and X_2 are nonempty. Hence, $|X_1| \leqslant n$ and $|X_2| \leqslant n$. Let us enlarge X_1 and X_2 to n-tuples by adding generic points. Then we obtain a set of $2n$ points $\widetilde{X}$ containing X, such that $\widetilde{X}$ satisfies the assumptions of Lemma 7.4. It follows from Lemma 7.4 that the natural map

$$H^0(\mathbb{P}^n, \mathcal{O}_{\mathbb{P}^n}(i)) \to H^0(\widetilde{X}, \mathcal{O}_{\widetilde{X}}(i))$$

is surjective for $i \geqslant 2$. Now let $J \subset R_{\widetilde{X}}$ be the kernel of the natural homomorphism $R_{\widetilde{X}} \longrightarrow R_X$ and let $\mathcal{J} \subset \mathcal{O}_{\widetilde{X}}$ be the ideal sheaf corresponding to $X \subset \widetilde{X}$. The above surjectivity implies that $J_{\geqslant 2} = \bigoplus_{i \geqslant 2} H^0(\widetilde{X}, \mathcal{J}(i))$. Since by Lemma 7.4 the algebra $R_{\widetilde{X}}$ is Koszul, we can apply Proposition 7.1 to the sheaf $\mathcal{J}(2)$ on $\widetilde{X}$

to deduce that $J_{\geqslant 2}(2)$ is a Koszul module. Since $R_{\widetilde{X}}$ is Koszul from the exact sequence

$$0 \longrightarrow J_{\geqslant 2} \longrightarrow J \longrightarrow J_1 \otimes \Bbbk(-1) \longrightarrow 0$$

we derive that $\operatorname{Tor}_i^{R_{\widetilde{X}}}(J, \Bbbk)_j = 0$ for $j > i+2$. Therefore, $\operatorname{Tor}_i^{R_{\widetilde{X}}}(R_X, \Bbbk)_j = 0$ for $j > i+1$. It remains to apply Theorem 5.2 to the homomorphism $R_{\widetilde{X}} \longrightarrow R_X$. □

Remark 2. The above theorem implies the Koszulness property for $2n$ points in general linear position in $\mathbb{P}^n$ (first established by Kempf in [**73**]). Another proof of Kempf's theorem using the notion of Koszul filtration is given in [**40**]. In [**39**], sec.4, it is shown that for points in general linear position this result is optimal: there exist 9 points in $\mathbb{P}^4$ in general linear position, whose coordinate algebra is quadratic but not Koszul. On the other hand, there is a generalization of Theorem 7.5 proved in [**98**] that gives a criterion of Koszulness for a set of more than $2n$ points in $\mathbb{P}^n$ under certain conditions on linear spans of various subsets of this set.

Remark 3. Let us say that a finite set of points $X \subset \mathbb{P}^n$ is *Koszul* if the algebra R_X is Koszul. Theorem 4.2 of [**40**] states that the set of s points in $\mathbb{P}^n$ with generic coordinates is Koszul iff $s \leqslant 1 + n + n^2/4$. The "only if" part can be strengthened as follows: if a Koszul set of s points $X \subset \mathbb{P}^n$ imposes independent conditions on quadrics then $s \leqslant 1 + n + n^2/4$. Indeed, let A be the quotient of R_X by a generic linear form. Then $h_A(z) = 1 + nz + (s - n - 1)z^2$ (here we use the assumption that X imposes independent conditions on quadrics). But A is still a Koszul algebra, so $h_A(z)^{-1}$ should have nonnegative coefficients. The latter condition is equivalent to the required inequality $s - n - 1 \leqslant n^2/4$.

COROLLARY 7.6. *Assume that the characteristic of the ground field is zero. Let $C \subset \mathbb{P}^n$ be a projectively normal irreducible curve of degree $\leqslant 2n-2$, not contained in any hyperplane. Then R_C is Koszul.*

Proof: It is well known that points of a generic hyperplane section of C will be in general linear position (see [**10**], III.1), so we can apply Lemma 7.3 and Theorem 7.5. □

Another remarkable example of a Koszul homogeneous coordinate algebra is the canonical ring of a curve.

THEOREM 7.7 ([**50**]). *Let $C \subset \mathbb{P}^{g-1}$ be a smooth curve of genus g embedded by the canonical linear series $|K|$ (so C is not hyperelliptic). Assume that C is not trigonal and not a plane quintic. Then R_C is Koszul.*

Proof: It is known that under our assumptions on C there exists a divisor $D = p_1 + \ldots + p_{g-1}$ of degree $g-1$ such that $h^0(D) = 2$ and both linear series $|D|$ and $|K-D|$ have no base points (see [**65**]). Note that the linear span L of points $p_1, \ldots, p_{g-1}$ has codimension $h^0(K-D) = 2$. We claim that these points are in general linear position in L. Indeed, it is enough to show that any $g-2$ of them span L, i.e. for every $i = 1, \ldots, g-1$ one has $h^0(K - D + p_i) = 2$. But this is equivalent to the equality $h^0(D - p_i) = 1$ which holds since p_i is not a base point of $|D|$. Since $g \geqslant 5$ we have $g - 1 \leqslant 2(g-3)$, hence, by Theorem 7.5 the algebra R_D is Koszul (see Remark 2 above). Now we are going to check that the assumptions of Theorem 5.6 are satisfied for the homomorphism $R_C \to R_D$. This would imply that this homomorphism is Koszul and hence the algebra R_C is Koszul by Corollary 5.4. Set $V = H^0(D)$, $U = H^0(K-D)$ (here and below the global sections are taken on

C). Since the linear systems $|D|$ and $|K-D|$ are base point free, we have exact sequences

$$0 \longrightarrow \mathcal{O}_C(-D) \longrightarrow V \otimes \mathcal{O}_C \longrightarrow \mathcal{O}_C(D) \longrightarrow 0,$$
$$0 \longrightarrow \mathcal{O}_C(D-K) \longrightarrow U \otimes \mathcal{O}_C \longrightarrow \mathcal{O}_C(K-D) \longrightarrow 0.$$

Tensoring these sequences with powers of the canonical bundle and passing to global sections we obtain exact sequences

$$0 \longrightarrow H^0(nK-D) \xrightarrow{a_n} V \otimes H^0(nK) \xrightarrow{\alpha_n} H^0(nK+D),$$

$$0 \longrightarrow H^0((n-1)K+D) \xrightarrow{b_n} V \otimes H^0(nK) \xrightarrow{\beta_n} H^0((n+1)K-D).$$

Moreover, we have $H^1(nK-D) = H^1((n-1)K+D) = 0$ for $n \geqslant 2$, hence, α_n and β_n are surjective for $n \geqslant 2$. Considering dimensions of the spaces in the above sequences for $n=1$ we see that α_1 and β_1 are also surjective. Now let us consider homomorphisms of R_C-modules

$$f : V \otimes R_C \longrightarrow U \otimes R_C(1),\ g : U \otimes R_C \to V \otimes R_C(1)$$

with grading components given by $b_{n+1} \circ \alpha_n$ and $a_{n+1} \circ \beta_n$. Let also

$$h : U \otimes R_C(-1) \longrightarrow R_C$$

be the homomorphism with grading components given by b_{n-1} followed by the embedding $H^0(nK-D) \longrightarrow H^0(nK)$. The above surjectivity properties of α_n and β_n easily imply that the complex

$$K_\bullet : \ldots \xrightarrow{f} U \otimes R_C(-3) \xrightarrow{g} V \otimes R_C(-2) \xrightarrow{f} U \otimes R_C(-1) \xrightarrow{h} R_C$$

satisfies $H_0(K_\bullet) = R_D$, $H_i(K_\bullet)_j = 0$ for $i \geqslant 1$ and $j > i+1$. □

The above proof is taken from [**96**]. It is different from the original proof in [**50**]. Yet another proof can be found in [**90**]. The fact that under the assumptions of the theorem the algebra R_C is quadratic is a classical theorem due to Petri (see [**10**]). One can ask whether the homogeneous coordinate algebra of a projectively normal smooth connected complex curve is Koszul provided it is quadratic. However, the answer turns out to be negative (see [**116**], Thm.3.1 and [**39**], sec.4).

Other examples of projective varieties with Koszul coordinate rings are:
(1) abelian varieties embedded into projective spaces using L^n where L is an ample line bundle and $n \geqslant 4$ (see [**74**]);
(2) Schubert varieties (see [**29**] and [**67**]);
(3) smooth projective varieties of dimension n embedded by $K \otimes A^d \otimes B$, where K is the canonical class, A is very ample and B is numerically effective, $d \geqslant n+1$ (with one simple exception, see [**89**]);
(4) some toric varieties (see [**34**], [**91**]). Note that there is a conjecture that the coordinate algebra of every projectively normal smooth toric variety is quadratic (see [**115**], Conj. 13.19). Perhaps one should also expect that these algebras are Koszul.

Also, we will show in chapter 3 that the operations of taking the image of a projective variety by a Veronese embedding and taking the Segre product of projective varieties preserve Koszulness. Furthermore, as Backelin proved in [**16**], for any projective variety the Veronese embedding of a sufficiently high degree has Koszul homogeneous coordinate ring (see Corollary 3.4 of chapter 3, for another proof see [**46**]). A similar result with "Koszul" replaced by "quadratic" appears in [**85**] and [**20**].

Last but not least, let us mention the remarkable class of *elliptic Sklyanin algebras* defined by Feigin and Odesskii [**48**]. These are (noncommutative) quadratic algebras associated with elliptic curves (plus some additional data). Their most remarkable feature is that they have the same Hilbert series as the algebras of polynomials. In addition they are Koszul and satisfy other nice homological properties (see [**117**]). We will consider three-dimensional Sklyanin algebras in section 11 of chapter 4.

8. Infinitesimal Hopf algebra associated with a Koszul algebra

In this section we will show that to every Koszul algebra A one can associate a graded *infinitesimal Hopf algebra* containing A as a subalgebra and $A^{!*}$ as a sub-coalgebra. Let us first recall the notion of *infinitesimal bialgebra* introduced by S. Joni and G. Rota in [**68**], sec. XII and studied by M. Aguiar [**3, 4**].

In the following definition and the subsequent discussion by an algebra (resp., coalgebra) we mean an associative algebra (resp., coassociative coalgebra) over $\Bbbk$, not necessarily having a unit (resp., counit).

Definition. An *infinitesimal bialgebra* (abbreviated as ϵ-bialgebra) is a triple (B, μ, Δ), where (B, μ) is an algebra, (B, Δ) is a coalgebra, and Δ is a derivation of B with values in $B \otimes B$, i.e.,

$$\Delta(ab) = (a \otimes 1)\Delta(b) + \Delta(a)(1 \otimes b)$$

for $a, b \in B$.

There is a natural structure of an ϵ-bialgebra on the bar-complex $\bigoplus_{n\geqslant 0} A^{\otimes n}$ of an associative algebra A (not necessarily unital) with the following maps μ and Δ:

$$\mu([a_1|\dots|a_m], [a'_1|\dots|a'_n]) = [a_1 \dots |a_{m-1}|a_m a'_1|a'_2|\dots|a'_n],$$

$$\Delta([a_1|\dots|a_m]) = \sum_{i=1}^{m-1} [a_1|\dots|a_i] \otimes [a_{i+1}|\dots|a_m].$$

The axioms are easily checked. Note that we have $\Delta(A) = \Delta(\Bbbk) = 0$, so there is no counit for Δ (there is an obvious unit for μ).

Now let $A = \bigoplus_{n\geqslant 0} A_n$ be a graded algebra with $A_0 = \Bbbk$. Then we have the above structure of ϵ-bialgebra on the bar-complex $\mathcal{B}ar_\bullet(A) = \bigoplus_{n\geqslant 0} A_+^{\geqslant n}$. Using the grading on A we can define a natural sub-ϵ-bialgebra in $\mathcal{B}ar_\bullet(A)$. Namely, for every (non-empty) collection of positive integers $(n_1, \dots, n_r)$ we define a subspace $V(n_1, \dots, n_r) \subset \mathcal{B}ar_s(A)$ by setting

$$V(n_1, \dots, n_r) = (A_{n_1} \otimes \dots \otimes A_{n_r}) \cap \ker(\partial),$$

where $\partial : \mathcal{B}ar_r(A) \longrightarrow \mathcal{B}ar_{r-1}(A)$ is the bar-differential (see section 1). We define the $\mathbb{Z}$-grading on

$$V = V_A := \bigoplus_{n_1, \dots, n_r} V(n_1, \dots, n_r)$$

by setting $\deg V(n_1, \dots, n_r) = n_1 + \dots + n_r$. Note that every grading component of V is finite-dimensional. Let us denote $V(1^n) = V(1, \dots, 1)$ (1 is repeated n times).

PROPOSITION 8.1. *The subspace $V \subset \mathcal{B}ar_\bullet(A)$ is a graded sub ϵ-bialgebra. Furthermore, $\bigoplus_{n\geq 1} V(n) \subset V$ is a subalgebra isomorphic to A_+ and $\bigoplus_n V(1^n) \subset V$ is a sub-coalgebra isomorphic to $A_+^{!*}$.*

The proof is a straightforward consequence of the definitions.

Recall (see [**3**]) that an ϵ-bialgebra (B, μ, Δ) is called an ϵ-Hopf algebra if it has an antipode, i.e., a $\Bbbk$-linear map $S : B \longrightarrow B$ satisfying the equations

$$\mu \circ (S \otimes \mathrm{id}) \circ \Delta = \mu \circ (\mathrm{id} \otimes S) \circ \Delta = -S - \mathrm{id}\,.$$

In the case of the ϵ-bialgebra V_A we have $\mu \circ \Delta = 0$, hence, we can set $S = -\mathrm{id}$. Thus, V_A is an ϵ-Hopf algebra.

Next, we are going to study V_A in the case when A is Koszul.

LEMMA 8.2. (1) *There is a bijective correspondence between the set of all sequences* $n_1, \ldots, n_r \geqslant 1$, $r \geqslant 1$ *with fixed* $n_1 + \cdots + n_r = n$ *and the set of all subsets* $J \subset [1, n-1]$ *defined by the rule*

$$J_{n_1,\ldots,n_r} = [1, n] \setminus \{n_1, n_1 + n_2, \ldots, n_1 + \cdots + n_r\} \subset [1, n-1].$$

(2) *One has* $J_{n_1,\ldots,n_r} \subset J_{m_1,\ldots,m_s}$ *if and only if there are some* $1 \leqslant t_1 < \cdots < t_{s-1} < r$ *such that* $m_1 = n_1 + \cdots + n_{t_1}, \ldots, m_s = n_{t_{s-1}+1} + \cdots + n_r$.

(3) *For any quadratic algebra* $A = \{V, R\}$, *positive integers* $n_1 + \cdots + n_r = n$, *and the subset* $J = J_{n_1,\ldots,n_r} \subset [1, n-1]$ *there is a natural isomorphism*

$$V^{\otimes n} \Big/ \sum_{j \in J} R_j^{(n)} \simeq A_{n_1} \otimes \cdots \otimes A_{n_r},$$

where $R_j^{(n)} = V^{\otimes j-1} \otimes R \otimes V^{\otimes n-j-1}$.

Proof: The statements (1) and (2) are elementary; (3) follows easily from the fact that

$$A_n = V^{\otimes n} \Big/ \sum_{j=1}^{n-1} R_j^{(n)}.$$

□

PROPOSITION 8.3. *If the algebra* A *is* m*-Koszul then the following sequence is exact for every* $(n_1, \ldots, n_r)$ *such that* $n_1 + \ldots + n_r \leqslant m$:

$$0 \to V(n_1, \ldots, n_r) \longrightarrow C^1(n_1, \ldots, n_r) \longrightarrow \ldots \longrightarrow C^r(n_1, \ldots, n_r) \longrightarrow 0$$

where

$$C^i(n_1, \ldots, n_r) = \bigoplus_{1 \leqslant s_1 < \ldots < s_{r-i} < s_{r+1-i} = r} A_{n_1 + \ldots + n_{s_1}} \otimes A_{n_{s_1+1} + \ldots + n_{s_2}} \otimes \ldots \otimes A_{n_{s_{r-i}+1} + \ldots + n_r},$$

the differential on $C^\bullet(n_1, \ldots, n_r)$ *is induced by the bar-differential and*

$$V(n_1, \ldots, n_r) \longrightarrow A_{n_1} \otimes \ldots \otimes A_{n_r} = C^1(n_1, \ldots, n_r)$$

is the natural embedding.

Proof: Let qA be the quadratic part of the algebra A. Then the homomorphism $qA \longrightarrow A$ is an isomorphism in degree $\leqslant m$. Hence, we can assume that A is a quadratic algebra. Let $A = \{V, R\}$, where $R \subset V \otimes V$ is the space of quadratic relations. Using the notation of Lemma 8.2 we set $J = J_{n_1,\ldots,n_r} \subset [1, n-1]$, where $n = n_1 + \ldots + n_r$. Recall that according to this lemma we have

$$W = V^{\otimes n} / \sum_{j \in J} R_j^{(n)} \simeq A_{n_1} \otimes \ldots \otimes A_{n_r}.$$

Now the images X_i of $R_i^{(n)}$ for $i \notin J$ form a distributive collection of subspaces in W (here we use the assumption that A is m-Koszul). One can easily check that the complex $C^\bullet(n_1,\dots,n_r)$ coincides with the complex considered in Proposition 7.2(c) of chapter 1 for this collection of subspaces in W. This implies that $C^\bullet(n_1,\dots,n_r)$ is exact everywhere except for the first term. It remains to observe that $V(n_1,\dots,n_r)$ coincides with the kernel of the map $C^1(n_1,\dots,n_r) \longrightarrow C^2(n_1,\dots,n_r)$ by the definition. □

COROLLARY 8.4. *If A is a Koszul algebra then for every $n_1,\dots,n_r$ and every $1 \leqslant s < r$ we have the following exact sequence:*

$$0 \longrightarrow V(n_1,\dots,n_r) \longrightarrow V(n_1,\dots,n_s)\otimes V(n_{s+1},\dots,n_r)$$
$$\longrightarrow V(n_1,\dots,n_{s-1},n_s+n_{s+1},n_{s+2}\dots,n_r) \longrightarrow 0,$$

Proof: From the definition of complexes $C^\bullet(n_1,\dots,n_r)$ one can immediately observe that for $1 \leqslant s < r$ there is a natural exact triple of complexes

$$0 \longrightarrow C^\bullet(n_1,\dots,n_s+n_{s+1},\dots,n_r)[-1] \longrightarrow C^\bullet(n_1,\dots,n_r)$$
$$\longrightarrow C^\bullet(n_1,\dots,n_s)\otimes C^\bullet(n_{s+1},\dots,n_r) \longrightarrow 0.$$

Now the required exact sequences are obtained by considering the corresponding long exact sequence of cohomology. □

Remark 1. The exact complex of Proposition 8.3 allows us to express the dimensions of $V(n_1,...,n_k)$ as some universal polynomials in dimensions of the grading components of the algebra A. In section 2 of chapter 7 we will consider corresponding polynomial inequalities on $\dim A_n$. There is a nice interpretation of these formulas in terms of noncommutative generating series (see chapter 7, section 8, Prop. 8.2).

Our interest to bialgebras V_A is due to their nice behavior under Koszul duality proved in the next proposition. Below we use again the notation of Lemma 8.2. We denote by $V_{A^!}^*$ the restricted dual to $V_{A^!}$ using the grading on it.

PROPOSITION 8.5. *There is an isomorphism of graded ϵ-bialgebras $V_A \simeq V_{A^!}^*$ sending $V_A(n_1,\dots,n_r)$ to $V_{A^!}(m_1,\dots,m_s)^*$, where $J_{m_1,\dots,m_s}$ is the complement to $J_{n_1,\dots,n_r} \subset [1,n-1]$ (where $n = n_1+\dots+n_r = m_1+\dots+m_s$).*

Proof: Let us fix n and abbreviate $R_j^{(n)}$ to R_j. For every $J \subset [1,n-1]$ we set $R_J = \sum_{j\in J} R_j$ and $R^J = \bigcap_{j\in J} R_j$. Recall that the quadratic relations for $A^!$ are defined by $R^! = R^\perp \subset (V^*)^{\otimes 2}$. It follows that for every J one has

$$(R_J)^\perp = R^{!J},\ (R^J)^\perp = R^!_J.$$

Let us set $V(J) = V_A(n_1,\dots,n_r)$, where $J = J_{n_1,\dots,n_r}$. Then Lemma 8.2(3) implies that

$$V(J) = \bigcap_{i\notin J} R_{J\cup\{i\}}/R_J. \tag{8.1}$$

Using the distributivity of the lattice generated by (R_j) we compute

$$\bigcap_{i\notin J} R_{J\cup\{i\}} = \bigcap_{i\notin J}(R_J+R_i) = R_J + R^{J^c},$$

where $J^c \subset [1,n-1]$ denotes the complement to J. Therefore,

$$V(J) = (R_J + R^{J^c})/R_J \simeq R^{J^c}/(R_J\cap R^{J^c}).$$

Passing to orthogonal complements we get

$$V(J)^* \simeq (R_J \cap R^{J^c})^\perp/(R^{J^c})^\perp = (R^{!J} + R^!_{J^c})/R^!_{J^c}$$

which gives the identification of $V(J)^*$ with the component of $V_{A^!}$ corresponding to J^c.

It remains check that this isomorphism is compatible with multiplication and comultiplication. For $J = J_{n_1,\dots,n_r}$ let us set $n_J = n_1 + \cdots + n_r$. Consider the following component of the multiplication on V_A:

$$V(I) \otimes V(J) \longrightarrow V(K),$$

where $K = I \cup (n_I + J) \cup \{n_I\}$. It is easy to check that it is induced by the natural isomorphism

$$R^{I^c} \otimes R^{J^c} \simeq R^{K^c}$$

that follows from the equality $K^c = I^c \cup (n_I + J^c)$. Similarly, for $K = I \cup (n_I + J)$ the component of the comultiplication

$$V(K) \longrightarrow V(I) \otimes V(J)$$

is induced by the embedding

$$R^{K^c} \subset R^{K^c \setminus \{n_I\}} \simeq R^{I^c} \otimes R^{J^c}.$$

Using this interpretation one can easily check that the above isomorphism $V_A^* \simeq V_{A^!}$ is compatible with ϵ-bialgebra structures. □

Remark 2. From the above proof one can easily see that as a vector space V_A can be identified with the associated graded space of $\bigoplus_{n\geqslant 1} V^{\otimes n}$ with respect to the decreasing filtration $F_\bullet$ defined as follows:

$$F_m V^{\otimes n} = \sum_{|J|=m} R^J.$$

More precisely, there is a canonical isomorphism

$$F_{r-1}V^{\otimes n}/F_r V^{\otimes n} \simeq \bigoplus_{n_1+\ldots+n_r=n} V_A(n_1,\dots,n_r).$$

Furthermore, $(F_\bullet)$ (resp., $F_{\bullet-1}$) is an algebra (resp., coalgebra) filtration on $\bigoplus_{n\geqslant 1} V^{\otimes n}$ and the above isomorphism is compatible with multiplication (resp., comultiplication).

Example. Let $A = \Bbbk\{x_1,\dots,x_n\}/(x_ix_j,\ (i,j) \in S)$, where $S \subset [1,n]^2$ is a subset. Then we can identify V_A with the augmentation ideal in the free algebra $\Bbbk\{x_1,\dots,x_n\}$, equipped with the following multiplication and comultiplication:

$$\mu(x_{i_1}\dots x_{i_k} \otimes x_{j_1}\dots x_{j_m}) = \begin{cases} x_{i_1}\dots x_{i_k}x_{j_1}\dots x_{j_m}, & (i_k,j_1) \notin S, \\ 0, & (i_k,j_1) \in S; \end{cases}$$

$$\Delta(x_{i_1}\dots x_{i_k}) = \sum_{r:(i_r,i_{r+1})\in S} x_{i_1}\dots x_{i_r} \otimes x_{i_{r+1}}\dots x_{i_k}.$$

It is an interesting problem to find what are possible minimal subsets J such that the vanishing $V(J) = 0$ can occur for some Koszul algebra. There is a related numerical problem for one-dependent stochastic sequences (see section 7 of chapter 7). It suggests that with some simple exceptions the only possibilities should be $[1,n]$ and $\emptyset$.

Finally, we observe that there is a slight generalization of the above construction involving a Koszul module. Namely, if M is a Koszul module over a Koszul algebra A then we can introduce a subspace

$$V_{A,M} = \bigoplus_{n_1,\dots,n_r} V_{A,M}(n_1,\dots,n_r) \subset \mathcal{B}ar_\bullet(A,M)$$

by setting

$$V_{A,M}(n_1,\dots,n_r) = (A_{n_1} \otimes \dots \otimes A_{n_{r-1}} \otimes M_{n_r}) \cap \ker(\partial),$$

where $n_1,\dots,n_{r-1} \geqslant 1$ and $n_r \geqslant 0$. One can check that $V_{A,M}$ is an *infinitesimal Hopf module* over V_A (see section 2 of [**4**]). Furthermore, we have exact sequences

$$0 \longrightarrow V_{A,M}(n_1,\dots,n_r) \longrightarrow V_A(n_1,\dots,n_s) \otimes V_{A,M}(n_{s+1},\dots,n_r)$$
$$\longrightarrow V_{A,M}(n_1,\dots,n_{s-1},n_s+n_{s+1},n_{s+2},\dots,n_r) \longrightarrow 0$$

similar to those of Corollary 8.4.

9. Koszul algebras and monoidal functors

In this section we sketch a more general approach to the notion of a Koszul algebra that includes the generalization to the case of graded algebras A for which A_0 is arbitrary and A_n is not necessarily of finite type over A_0. We also show that a Koszul algebra can be interpreted as a (nonunital) monoidal functor from a certain universal (nonunital) monoidal abelian category to the category of vector spaces. The infinitesimal Hopf algebra considered in the previous section records the values of this monoidal functor on simple objects. Note that the Koszulness property is preserved under exact monoidal functors. One application of this observation is the recent proof in [**76**] of Koszulness of preprojective algebras of quivers (introduced in [**60**]) for quivers not of Dynkin type.

Below we work with nonunital monoidal categories, i.e., categories $\mathcal{C}$ with the functor $\circ : \mathcal{C} \times \mathcal{C} \longrightarrow \mathcal{C}$ equipped with the associativity isomorphisms satisfying the pentagon constraint, but not necessarily possessing a unit object. When considering monoidal *abelian* categories we require the functor $\circ$ to be exact in each argument.

First, let us observe that the definition of a Koszul algebra in terms of distributive lattices makes sense in an arbitrary monoidal abelian category $\mathcal{C}$. Indeed, since the notion of distributivity can be considered for subobjects of an object in any abelian category, we can make the following

Definition. Let $\mathcal{C}$ be a monoidal abelian category. A *Koszul algebra in* $\mathcal{C}$ is a pair (V,R), where V is an object of $\mathcal{C}$, R is a subobject of $V \circ V$, such that for every $n \geqslant 3$ the collection of subobjects

$$R \circ V \circ \dots \circ V, V \circ R \circ V \dots V, \dots, V \circ \dots \circ V \circ R \tag{9.1}$$

in $V^{\circ(n+1)}$ is distributive, i.e., generates a distributive lattice.

Koszul duality in this general context may be viewed as a duality between Koszul algebras in $\mathcal{C}$ and Koszul algebras in $\mathcal{C}^{opp}$. Namely, the dual of a Koszul algebra (V,R) in $\mathcal{C}$ is the Koszul algebra (V,Q) in $\mathcal{C}^{opp}$, where Q is the quotient of $V \circ V$ by R in $\mathcal{C}$ (hence, a subobject of $V \circ V$ in $\mathcal{C}^{opp}$). If the category $\mathcal{C}$ is equipped with an exact monoidal duality functor $\mathcal{C}^{opp} \longrightarrow \mathcal{C}$ then we can apply this functor to (V,Q) to get a Koszul algebra in $\mathcal{C}$. For example, the usual Koszul duality corresponds to applying the standard duality on the category of finite-dimensional vector spaces.

Remark 1. In the case when $\mathcal{C}$ has a unit object $1_{\mathcal{C}}$ one can define a graded algebra $A = \bigoplus_{n \geqslant 0} A_n$ in $\mathcal{C}$ associated with a pair (V, R) by setting $A_0 = 1_{\mathcal{C}}$, $A_1 = V$, $A_2 = V^{\circ 2}/R$, $A_3 = V^{\circ 3}/(R \circ V + V \circ R)$, etc. It is easy to see that Koszulness in this case is still equivalent to the condition that the object $1_{\mathcal{C}}$ considered as an A-module admits a linear free resolution.

Example 1. Let A_0 be a semisimple ring and let $A_0 - \mathrm{Mod} - A_0$ denote the category of A_0-bimodules. It has a natural monoidal structure given by the tensor product over A_0. Then a Koszul algebra in $A_0 - \mathrm{Mod} - A_0$ can be viewed as a graded algebra $A = \bigoplus_{n \geqslant 0} A_n$, such that A_0 viewed as a left A-module has a linear free resolution. Most of the theory of Koszul algebras that we presented in the case when A_0 is a field can be generalized to this setup (see [**24**]). Let $A_0 - \mathrm{Mod}^{lf} - A_0 \subset A_0 - \mathrm{Mod} - A_0$ (resp., $A_0 - \mathrm{Mod}^{rf} - A_0 \subset A_0 - \mathrm{Mod} - A_0$) be the full subcategory of A_0-bimodules that are finitely generated as left (resp., right) A_0-modules. Then we have a natural duality

$$(A_0 - \mathrm{Mod}^{lf} - A_0)^{\oplus} \widetilde{\longrightarrow} A_0 - \mathrm{Mod}^{rf} - A_0 : V \longmapsto V^*,$$

where $V^* = \mathrm{Hom}_{A_0-\mathrm{Mod}}(M, A_0)$ with the bimodule structure given by $af(v) = f(va)$, $fa(v) = f(v)a$ for $f \in V^*$, $a \in A_0$, $v \in V$. Moreover, it is easy to check that this duality respects tensor products, hence, it induces the duality between Koszul algebras in $A_0 - \mathrm{Mod}^{lf} - A_0$ and Koszul algebras in $A_0 - \mathrm{Mod}^{rf} - A_0$ (called *left finite* and *right finite* Koszul algebras, respectively). The dual algebra to a left finite Koszul algebra A can be identified with $\mathrm{Ext}^*_A(A_0, A_0)^{opp}$ (see [**24**], Thm. 2.10.1).

Example 2. An important particular case of the previous example is when $A_0 = \mathbb{k}^n$, the direct sum of n copies of a ground field. Then A_n is a bimodule over $\mathbb{k}^n$, i.e., a vector space equipped with a decomposition $A_n = \bigoplus_{ij} (A_n)_{ij}$. For such an algebra let $A^{\mathbb{k}}$ denote the graded algebra with $A^{\mathbb{k}}_0 = \mathbb{k}$ and $A^{\mathbb{k}}_i = A_i$ for $i > 0$. An explicit lattice consideration shows that *the algebras A and $A^{\mathbb{k}}$ are Koszul simultaneously*. Note that in the case when A_n are finite-dimensional one can define a matrix-valued Hilbert series of such an algebra as

$$h_A(t) = \sum_{n \geqslant 0} (\dim (A_n)_{ij}) t^n.$$

The matrix-valued analogue of the relation (2.2) still holds in this case. Algebras of this kind appear as endomorphisms of strongly exceptional collections in triangulated categories. Recall that such collections $(E_1, \ldots, E_n)$ are characterized by the condition that $\mathrm{Ext}^*(E_i, E_j) = 0$ for $i > j$, $\mathrm{Ext}^n(E_i, E_j) = 0$ for $i \leqslant j$ and $n > 0$, and $\mathrm{Hom}(E_i, E_i) = \mathbb{k}$ for all i. The corresponding endomorphism algebra $A = \mathrm{End}(\bigoplus_i E_i) = \bigoplus_{i \leqslant j} \mathrm{Hom}(E_i, E_j)$ has a natural grading given by $A_n = \bigoplus_{j-i=n} \mathrm{Hom}(E_i, E_j)$ (note that A is an example of a *mixed* algebra in the terminology of [**23**]). Let S_i be the simple A-module corresponding to the i-th summand in the decomposition $A_0 = \bigoplus_{i=1}^n \mathrm{Hom}(E_i, E_i)$. Then Koszulness of A is equivalent to the condition $\mathrm{Ext}^n_A(S_i, S_j) = 0$ for $n \neq i - j$. This leads to a natural interpretation of Koszulness of A in terms of mutations (see [**30**], Cor. 7.2).

Next, we are going to explain a way of thinking about Koszul algebras in $\mathbb{k}$-linear monoidal categories in terms of monoidal functors from a certain universal category. Let Cube_n denote the category of n-dimensional commutative cubes of finite-dimensional vector spaces over $\mathbb{k}$. An object of Cube_n is a collection of finite-dimensional vector spaces $(V_I, I \subset [1, n])$ and morphisms $\alpha_{I,J} : V_I \longrightarrow V_J$ for

$I \subset J$, such that for $I \subset J \subset K$ one has $\alpha_{I,K} = \alpha_{J,K} \circ \alpha_{I,J}$. For $n = 0$ we define Cube_0 to be the category of finite-dimensional $\Bbbk$-vector spaces.

There is a natural tensor product operation

$$\otimes : \mathrm{Cube}_m \times \mathrm{Cube}_n \to \mathrm{Cube}_{m+n}$$

given by

$$(U_\bullet) \otimes (V_\bullet) = (W_\bullet),$$

where $W_{I \cup (J+m)} = U_I \otimes V_J$ and $I \subset [1,m]$, $J \subset [1,n]$. Now let $P(1)$ be the object of Cube_1 corresponding to the diagram $\Bbbk \xrightarrow{\mathrm{id}} \Bbbk$. Then we can define a new monoidal structure on $\mathrm{Cube}_\bullet := \bigoplus_{n \geqslant 0} \mathrm{Cube}_n$ by setting

$$X * Y = X \otimes P(1) \otimes Y,$$

so that $\mathrm{Cube}_m * \mathrm{Cube}_n \subset \mathrm{Cube}_{m+n+1}$. Note that this new monoidal structure has no unit object.

For every subset $I \subset [1,n]$ we denote by $S^I = S^I(n)$ the simple object of Cube_n with $S^I_I = \Bbbk$ and $S^I_J = 0$ for $J \neq I$. Let $P^I(n) = P^I$ be the projective cover of S^I, so that $P^I_J = \Bbbk$ for $I \subset J$ and $P^I_J = 0$ for $I \not\subset J$. For $i = 1, \dots, n$ we denote $P^i = P^{\{i\}}$. It is clear that $P^1, \dots, P^n$ is a distributive collection of subobjects in $P^\emptyset$. We claim that this collection is universal in the following sense.

LEMMA 9.1. *A collection of subobjects $R_1, \dots, R_n \subset X$ of an object in a $\Bbbk$-linear abelian category $\mathcal{C}$ is distributive iff there exists an exact functor $F : \mathrm{Cube}_n \longrightarrow \mathcal{C}$ such that $F(P^\emptyset) = X$ and $F(P^i \longrightarrow P^\emptyset) = R_i \hookrightarrow X$.*

Proof: The "if" part follows immediately from the fact that $(P^1, \dots, P^n)$ is distributive in $P^\emptyset$. Conversely, for an arbitrary collection of subobjects $(R_1, \dots, R_n)$ in X we define the exact functor from Cube_n to the category of complexes over $\mathcal{C}$ by sending an object $(V_J, J \subset [1,n])$ to the following complex:

$$C(V_\bullet)^i = \oplus_{J \cup K = [1,n], |J \cap K| = i} V_J \otimes X / \sum_{j \in K} R_j,$$

where the differential components are of the form

$$V_J \otimes X / \sum_{j \in K} R_j \xrightarrow{\epsilon} V_{J'} \otimes X / \sum_{j \in K'} R_j,$$

for $J \subset J'$, $K \subset K'$, and $|J' \setminus J| + |K' \setminus K| = 1$, where the above map differs from the natural one by the sign $\epsilon = (-1)^{|J| \cdot |K' \setminus K|}$. Proposition 7.2 (c) (or rather, its analogue for abelian categories) implies that for every simple object $S \in \mathrm{Cube}_n$ one has $H^i C(S) = 0$ for $i > 0$. Hence, this is true for every object of Cube_n and the functor $V_\bullet \longmapsto H^0 C(V_\bullet)$ is exact. One can also easily check that $H^0 C(P^i) \simeq R_i$. □

Remark 2. In the case when $\mathcal{C} = \mathcal{V}ect_\Bbbk$, the category of finite-dimensional $\Bbbk$-vector spaces, the exact functor $F : \mathrm{Cube}_n \longrightarrow \mathcal{V}ect_\Bbbk$ corresponding to a distributive collection $(R_1, \dots, R_n)$ of subspaces in X is representable. Namely, we have $F(V_\bullet) = \mathrm{Hom}_{\mathrm{Cube}_n}(P, V_\bullet)$, where the projective object $P \in \mathrm{Cube}_n$ is defined as follows. For every $I \subset [1,n]$ we set $P_I = (X / \sum_{i \notin I} R_i)^*$ with natural maps $P_I \longrightarrow P_J$ for $I \subset J$. Note that we can define this object $P_I \in \mathrm{Cube}_n$ for an arbitrary collection $(R_1, \dots, R_n)$. The above lemma implies that P_I is projective iff the collection is distributive.

THEOREM 9.2. *Let $\mathcal{C}$ be a $\Bbbk$-linear monoidal abelian category. Then the category of Koszul algebras in $\mathcal{C}$ is equivalent to the category of exact monoidal functors* $(\mathrm{Cube}_\bullet, *) \to \mathcal{C}$.

Proof: Let $S(0) = \Bbbk$ be the ground field viewed as an object of Cube_0. Then $S(0) * S(0) = P^\emptyset(1) \in \mathrm{Cube}_1$ and we have an exact sequence

$$0 \longrightarrow P^1(1) \longrightarrow P^\emptyset(1) \longrightarrow S^\emptyset(1) \longrightarrow 0.$$

Now for an exact monoidal functor $F : (\bigoplus_{n\geqslant 0} \mathcal{C}_n, *) \to \mathcal{C}$ we set

$$V = F(S(0)),\ R = F(P^1(1)) \subset F(P^\emptyset(1)) \simeq V \circ V.$$

We claim that the quadratic algebra (V, R) is Koszul. Indeed, note that $S(0)^{*(n+1)} = P^\emptyset(n) \in \mathrm{Cube}_n$ and

$$S(0)^{*(i-1)} * P^1(1) * S(0)^{*(n-i)} \simeq P^i(n) \subset P^\emptyset(n).$$

This immediately implies that the collection of subspaces (9.1) is distributive.

Conversely, if (V, R) is a Koszul algebra in $\mathcal{C}$ then for every n by Lemma 9.1 we have an exact functor $F_n : \mathrm{Cube}_n \to \mathcal{C}$ sending $P^\emptyset(n)$ to $V^{\circ(n+1)}$ and $P^i(n)$ to $R_i^{(n+1)} := V^{\circ(i-1)} \circ R \circ V^{\circ(n-i)}$. We claim that the induced functor $(F_n) : \mathrm{Cube}_\bullet \to \mathcal{C}$ has a monoidal structure. Indeed, if $U_\bullet \in \mathrm{Cube}_m$ and $W_\bullet \in \mathrm{Cube}_n$ then we have a natural morphism of complexes

$$C(U_\bullet) \circ C(W_\bullet) \to C(U_\bullet * W_\bullet)$$

with components

$$[U_J \otimes V^{\circ(m+1)} / \sum_{k\in K} R_k^{(m+1)}] \circ [W_{J'} \otimes V^{\circ(n+1)} / \sum_{k'\in K'} R_{k'}^{(n+1)}] \longrightarrow$$
$$U_J \otimes W_{J'} \otimes V^{\circ(m+n+2)} / \sum_{j\in K\cup(K'+m+1)} R_j^{(m+n+2)}.$$

One can easily check that this morphism is a quasi-isomorphism, hence it induces an isomorphism

$$F_m(U_\bullet) \circ F_n(W_\bullet) \xrightarrow{\sim} F_{m+n+1}(U_\bullet * W_\bullet).$$

□

Remark 4. The infinitesimal bialgebra V_A associated with a Koszul algebra $A = (V, R)$ (see section 8) can be defined also in the more general context considered above. Moreover, it has a natural interpretation in terms of the corresponding monoidal functor $F : \mathrm{Cube}_\bullet \to \mathcal{C}$. Namely, we claim that it is the direct sum of the values of F on all simple objects of $\mathrm{Cube}_\bullet$. Indeed, the construction of Lemma 9.1 shows that for $I \subset [1, n]$ one has

$$F(S^I) = \ker(V^{\circ(n+1)} / \sum_{j\notin I} R_j^{(n+1)} \longrightarrow \bigoplus_{i\in I} V^{\circ(n+1)} / [R_i^{(n+1)} + \sum_{j\notin I} R_j^{(n+1)}]$$

(where we use the notation from the proof of Theorem 9.2). These are exactly the components of V_A (see (8.1)).

Remark 4. According to the theorem any exact monoidal functor $G : \mathrm{Cube}_\bullet \longrightarrow \mathrm{Cube}_\bullet$ generates an operation on Koszul algebras which acts on the corresponding monoidal functors by $F \longmapsto F \circ G$. Also, G itself can be viewed as a Koszul algebra in $\mathrm{Cube}_\bullet$. Similarly, a Koszul algebra in $\mathrm{Cube}_\bullet \otimes_\Bbbk \mathrm{Cube}_\bullet$ gives rise to a binary operation on Koszul algebras. Here is an example of a Koszul algebra in $\mathrm{Cube}_\bullet$

associated with a positive integer d: $V = P^{\emptyset}(d)$, $R = \ker(P^{\emptyset}(d) \longrightarrow S^{\emptyset}(d))$. The corresponding operation is the $(d+1)$-th Veronese power (see chapter 3).

Example 3. Malkin, Ostrik and Vybornov in [**76**] use a certain simple Koszul algebra in the category of representations of the quantum group $SL_q(2)$ (where q is not a root of unity) to derive Koszulness of the preprojective algebra of a quiver not of Dynkin type (without loops). Here is a sketch of their proof (note that a different proof of a particular case is given in [**81**]). Fix $q \in \Bbbk^*$ and let $(V_{ij})_{1\leqslant i,j\leqslant n}$ be a collection of finite-dimensional vector spaces equipped with nondegenerate bilinear forms $E_{ij} : V_{ij} \otimes V_{ji} \longrightarrow \Bbbk$ such that for every i one has

$$\sum_{j=1}^{n} \mathrm{Tr}(E_{ij}(E_{ji})^{-1}) = -q - q^{-1}.$$

Since E_{ij} identifies V_{ji} with V_{ij}^* we have a canonical tensor $e_{ij} \in V_{ij} \otimes V_{ji}$. Consider the quadratic algebra Π with $\Pi_0 = \Bbbk^n$, $\Pi_1 = \bigoplus_{i,j} V_{ij}$ and with n defining relations

$$\sum_{j=1}^{n} e_{ij}e_{ji} = 0, \ \ i = 1, \dots, n.$$

The main assertion is that the algebra Π is Koszul provided q is not a root of unity (all preprojective algebras of quivers that are not of Dynkin type appear in this family of algebras). The main idea of the proof is that by the work of Etingof and Ostrik [**47**] the above data gives a tensor functor from the tensor category $\mathcal{C}_q$ of representations of $SL_q(2)$ to the category of $\Bbbk^n$-bimodules. One easily checks that Π is the image under this functor of a certain quadratic algebra $\widetilde{\Pi}$ in $\mathcal{C}_q$. When q is not a root of unity the algebra $\widetilde{\Pi}$ turns out to be the analogue of the algebra of polynomials in two variables: $\widetilde{\Pi} = \bigoplus_{n\geqslant 0} L_n$ where L_n is the irreducible $(n+1)$-dimensional representation of $SL_q(2)$. Its Koszulness is checked in the same way as for the usual algebra of polynomials: one uses exactness of the Koszul complex

$$0 \longrightarrow \widetilde{\Pi}(-2) \longrightarrow L_1 \otimes \widetilde{\Pi}(-1) \longrightarrow \widetilde{\Pi} \longrightarrow 1_{\mathcal{C}_q} \longrightarrow 0.$$

Since exact monoidal functors preserve Koszulness, this implies that the algebra Π is also Koszul.

10. Relative Koszulness of modules

In this section we explain the notion of relative Koszulness introduced by R. Bezrukavnikov (see [**29**]).

PROPOSITION 10.1. *Let A be a graded algebra, R (resp., L) a nonnegatively graded right (resp., left) A-module. Then nonzero spaces $\mathrm{Tor}^A_{ij}(R, L)$ are concentrated in the region $i \leqslant j$ and for the diagonal part one has*

$$\mathrm{Tor}^A_{ii}(R, L) \simeq (R^! \otimes_{A^!} L^!)^*_i$$

where $L^! = (\mathrm{q}_A L)^!_{(\mathrm{q}A)}$ and $R^! = (\mathrm{q}_{A^{\mathrm{op}}} R)^!_{(\mathrm{q}A^{\mathrm{op}})}$.

Proof: The spaces $\mathrm{Tor}^A_{ij}(R, L)$ are computed as homology of the bar-complex $\mathcal{B}ar_\bullet(R, A, L)$, where

$$\mathcal{B}ar_{ij}(R, A, L) = \sum_{p+k_1+\cdots+k_i+q=j,\ k_s\geqslant 1,\ p,q\geqslant 0} R_p \otimes A_{k_1} \otimes \cdots \otimes A_{k_i} \otimes L_q.$$

So, the complex itself is concentrated in the region required and the diagonal Tor-spaces are just the kernels of the morphisms $R_0 \otimes A_1^{\otimes i} \otimes L_0 \longrightarrow R \otimes A^{\otimes i-1} \otimes L$. □

THEOREM 10.2. *Let $A = \{V, I\}$ be a Koszul algebra, $R = \langle H', K' \rangle_{A^{\mathrm{op}}}$ a Koszul right A-module, and $L = \langle H'', K'' \rangle_A$ a Koszul left A-module. Then for every $n \geqslant 0$ the following conditions are equivalent:*

(a) $\mathrm{Tor}^A_{i,n}(R, L) = 0$ *for all* $0 < i < n$;

(b) *the comultiplication maps* $\mathrm{Tor}^A_i(R, L) \longrightarrow \mathrm{Tor}^A_{i'}(R, \Bbbk) \otimes \mathrm{Tor}^A_{i''}(\Bbbk, L)$ *are injective in the internal degree n for all $i' + i'' = i$, where $i', i'' \geqslant 0$;*

(c) *the degree-n component of the Koszul complex $K^A_\bullet(R, L^!_A)$*

(10.1)
$$0 \longrightarrow R_0 \otimes L_n^{!*} \longrightarrow \cdots \longrightarrow R_{n-1} \otimes L_1^{!*} \longrightarrow R_n \otimes L_0^{!*} \longrightarrow 0$$

is exact at all terms except for the first and the last;

(d) *the collection of subspaces*

$$X_i = \begin{cases} K' \otimes V^{\otimes n-1} \otimes H'', & i = 0 \\ H' \otimes V^{\otimes i-1} \otimes I \otimes V^{\otimes n-i-1} \otimes H'', & i = 1, \ldots, n-1 \\ H' \otimes V^{\otimes n-1} \otimes K'', & i = n \end{cases}$$

in $H' \otimes V^{\otimes n} \otimes H''$ is distributive.

Proof: (a) $\Longrightarrow$ (b). This follows from Proposition 10.1.
(b) $\Longrightarrow$ (a). Since R and L are Koszul modules, the spaces $\mathrm{Tor}^A_{i',j'}(R, \Bbbk)$ and $\mathrm{Tor}^A_{i'',j''}(\Bbbk, L)$ are concentrated on the diagonals.
(a) $\Longleftrightarrow$ (c). The complex $K^A_\bullet(A, L^!)$ is a free graded resolution of a Koszul A-module L. Therefore, the homology spaces of the complex $K^A_\bullet(R, L^!) = R \otimes_A K^A_\bullet(A, L^!)$ are isomorphic to $\mathrm{Tor}^A_{ij}(R, L)$.
(c) $\Longleftrightarrow$ (d). Set $W = H' \otimes V^{\otimes n} \otimes H''$ Since the modules R and L are Koszul, any proper subcollection of the collection $X_0, \ldots, X_n$ in W is distributive. Now we observe that the Koszul complex (10.1) coincides with the complex $K_\bullet(W; X_0, \ldots, X_n)$ with the first and the last terms deleted (see Proposition 7.2 of chapter 1).
(a) $\Longleftrightarrow$ (d). The degree-n component of the bar-complex $\mathcal{B}ar_\bullet(R, A, L)$ coincides with the complex $B_\bullet(W; X_0, \ldots, X_n)$ with the rightmost term deleted (see Proposition 7.2 of chapter 1). □

Definition. If the equivalent conditions of the above theorem are satisfied for all $n \geqslant 0$ then we say that the pair (R, L) is *relatively Koszul.*

A nice example of a relatively Koszul pair of modules is given in [**29**], Rem. 2.5. In this example A, R and L are homogeneous coordinate rings of X, Y and Z, where X is the projectivization of the orbit of the highest weight vector, $Y \subset X$ is a Schubert variety and $Z \subset X$ is the opposite Schubert variety.

CHAPTER 3

Operations on graded algebras and modules

The class of Koszul algebras is closed under a large set of operations on graded algebras [**20, 77, 78**]. In section 1 we will consider the simplest of these operations such as the direct sum, the free product and a family of tensor products. Then in section 2 we will discuss Segre products and Veronese powers. Almost all the results of sections 1 and 2 are due to J. Backelin and R. Fröberg [**20**]. In section 3 we study homological properties of Segre products and Veronese powers following the ideas of J. Backelin [**16**]. In section 4 we discuss the internal cohomomorphism operation (introduced by Manin) closely related to the Segre product. Generalizing computations of [**101**] we show that this operation can be used to describe cohomology of the Segre product of two algebras one of which is Koszul. Finally, in section 5, following Piontkovskii [**92**], we apply our study of operations to produce a pair of quadratic algebras A and A' such that A is Koszul, A' is not Koszul, but $h_A(z) = h_{A'}(z)$ and $h_{A^!}(z) = h_{A'^!}(z)$.

1. Direct sums, free products and tensor products

Let A and B be graded algebras (as before, we assume that $A_i = B_i = 0$ for $i < 0$ and $A_0 = B_0 = \Bbbk$). Then there are essentially three different constructions of graded algebras $A \star B$ with $(A \star B)_1 = A_1 \oplus B_1$.

Definition 1. The *direct sum* (categorical product) $A \sqcap B$ is the graded algebra with $(A \sqcap B)_0 = \Bbbk$ and $(A \sqcap B)_i = A_i \oplus B_i$ for $i > 0$, where the products $A_+ \cdot B_+$ and $B_+ \cdot A_+$ are set to be zero. The *free product* (categorical coproduct) $A \sqcup B$ is the associative algebra generated freely by A and B. Explicitly, we have

$$(1.1) \qquad A \sqcup B = \bigoplus_{i \geqslant 0;\varepsilon_1,\,\varepsilon_2 \in \{0,1\}} A_+^{\otimes \varepsilon_1} \otimes (B_+ \otimes A_+)^{\otimes i} \otimes B_+^{\otimes \varepsilon_2}.$$

Finally, there is a family of q-*tensor products* $A \otimes^q B$ with $q \in \mathbb{P}^1_\Bbbk$. By definition, $A \otimes^q B$ is the quotient algebra of $A \sqcup B$ by the relations $ba - q^{\tilde{b}\tilde{a}} ab = 0$ for $a \in A_{\tilde{a}}$ and $b \in B_{\tilde{b}}$ (for $q = \infty$ the relation becomes $ab = 0$). We will use a special notation for the *one-sided products* $A \vartriangle B := A \otimes^0 B$ and $A \triangledown B := A \otimes^\infty B$.

For $q \neq \infty$ the multiplication map $A \otimes B \longrightarrow A \otimes^q B$ is an isomorphism of graded vector spaces. In terms of this identification, the multiplication in $A \otimes^q B$ is given by the formula $(a_1 \otimes^q b_1)(a_2 \otimes^q b_2) = q^{\widetilde{b_1}\widetilde{a_2}}(a_1 a_2 \otimes^q b_1 b_2)$. Note that the complete family of algebras $A \otimes^q B$, $q \in \mathbb{P}^1_\Bbbk$ cannot be viewed as a family of multiplications on the same graded vector space: the grading components $(A \otimes^q B)_n$ form a nontrivial vector bundle $\bigoplus_{j+k=n} A_j \otimes B_k \otimes \mathcal{O}(jk)$ over $\mathbb{P}^1_\Bbbk$.

All of the above operations have a computable effect on Hilbert series. Namely, one can immediately check that

$$h_{A \sqcap B}(z) = h_A(z) + h_B(z) - 1,$$

$$h_{A\otimes^q B}(z) = h_A(z)h_B(z).$$

Also, from (1.1) one can easily derive that

$$h_{A\sqcup B}(z) = (h_A(z)^{-1} + h_B(z)^{-1} - 1)^{-1}. \tag{1.2}$$

Definition 2(M). Let M and N be nonnegatively graded modules over graded algebras A and B equipped with isomorphisms $M_0 \simeq N_0 \simeq H$, where H is a fixed vector space. For each of the operations $\star = \sqcap$, $\sqcup$, $\vartriangle$, $\triangledown$, we are going to construct a nonnegatively graded $A\star B$-module $M\star_H N$ with $(M\star_H N)_0 = H$. Let us denote $M_+ = \bigoplus_{i=1}^{\infty} M_i$ (resp., $N_+ = \bigoplus_{i=1}^{\infty} N_i$).
(i) For $\star = \sqcap$ we set

$$M \sqcap_H N = H \oplus M_+ \oplus N_+ = M \oplus N_+ = M_+ \oplus N.$$

The $A\sqcap B$-module structure is defined by letting A act by zero on N_+ and B act by zero on M_+. Note that there is an embedding of $A\sqcap B$-modules $M\sqcap_H N \longrightarrow M\oplus N$, where M (resp., N) is an $A\sqcap B$-module with the trivial action of B (resp., A).
(ii) For $\star = \sqcup$ we set

$$M \sqcup_H N = (B \sqcup M_+) \oplus H \oplus (A \sqcup N_+),$$

where for an A-module P (resp., B-module Q) we denote $B \sqcup P = (A \sqcup B) \otimes_A P$ (resp., $A \sqcup Q = (A \sqcup B) \otimes_B Q$). It is easy to see that $B \sqcup P = P \oplus (A \sqcup (B_+ \otimes P))$ as an A-module. Hence, we get an identification

$$M \sqcup_H N = M \oplus (A \sqcup (B_+ \otimes M_+ \oplus N_+))$$

that allows us to view $M \sqcup_H N$ as an A-module. Similarly, we get a B-module structure from the identification

$$M \sqcup_H N = (B \sqcup (M_+ \oplus A_+ \otimes N_+)) \oplus N.$$

In fact, these actions come from an $A \sqcup B$-module structure. There is a natural surjection of $A\sqcup B$-modules $A\sqcup N \oplus B\sqcup M \longrightarrow M\sqcup_H N$. Its kernel is the free $A\sqcup B$-submodule generated by the image of the embedding $H \to M_0\oplus N_0 : h \longmapsto (h, -h)$.
(iii) For $\star = \vartriangle$ we set

$$M \vartriangle_H N = M \oplus A \otimes N_+ = M_+ \oplus A_+ \otimes N_+ \oplus N,$$

where A acts on $A \otimes N_+$ as on the free module generated by N_+ and B acts on $M_+ \oplus A_+ \otimes N_+$ by zero. As an A-module (resp., B-module) $M \vartriangle_H N$ can be identified with a quotient (resp., submodule) of the $A \vartriangle B$-module $M \oplus A \otimes N$. Similarly, for $\star = \triangledown$ we set

$$M \triangledown_H N = M \oplus B_+ \otimes M_+ \oplus N_+ = B \otimes M_+ \oplus N,$$

so that $M \triangledown_H N \simeq N \vartriangle_H M$ as a module over $A \triangledown B \simeq B \vartriangle A$.

Definition 3(M). For a graded A-module M, a graded B-module N and any $q \in \Bbbk^*$ we define an $A \otimes^q B$-module structure $M \otimes^q N$ on the graded vector space $M \otimes N$ by the rule $(a \otimes^q b)(m \otimes^q n) = q^{\widetilde{b}\widetilde{m}}(am \otimes^q bn)$. Note that from our point of view the analogous construction for $q = 0$ is badly behaved. For example, it is easy to see that the $A \otimes^0 B$-module $M \otimes^0 N$ is *never* quadratic, unless the A-module M is free or the B-module N is trivial.

PROPOSITION 1.1. *For any graded algebras A and B and any $q \in \mathbb{P}^1_{\Bbbk}$, we have natural isomorphisms of bigraded algebras*

$$\operatorname{Ext}^*_{A\sqcap B}(\Bbbk,\Bbbk) \simeq \operatorname{Ext}^*_A(\Bbbk,\Bbbk) \sqcup \operatorname{Ext}^*_B(\Bbbk,\Bbbk),$$
$$\operatorname{Ext}^*_{A\sqcup B}(\Bbbk,\Bbbk) \simeq \operatorname{Ext}^*_A(\Bbbk,\Bbbk) \sqcap \operatorname{Ext}^*_B(\Bbbk,\Bbbk),$$
$$\operatorname{Ext}^*_{A\otimes^q B}(\Bbbk,\Bbbk) \simeq \operatorname{Ext}^*_A(\Bbbk,\Bbbk)\, \overline{\otimes}^{q^{-1}}\, \operatorname{Ext}^*_B(\Bbbk,\Bbbk).$$

Now let M (resp., N) be a nonnegatively graded A-module (resp., B-module) equipped with an identification $M_0 \simeq H$ (resp., $N_0 \simeq H$) for a fixed vector space H. Then for any $q \in \Bbbk^*$ there are natural isomorphisms of bigraded modules

$$\operatorname{Ext}^*_{A\sqcap B}(M \sqcap_H N,\Bbbk) \simeq \operatorname{Ext}^*_A(M,\Bbbk) \sqcup_{H^*} \operatorname{Ext}^*_B(N,\Bbbk),$$
$$\operatorname{Ext}^*_{A\sqcup B}(M \sqcup_H N,\Bbbk) \simeq \operatorname{Ext}^*_A(M,\Bbbk) \sqcap_{H^*} \operatorname{Ext}^*_B(N,\Bbbk),$$
$$\operatorname{Ext}^*_{A\vartriangle B}(M \vartriangle_H N,\Bbbk) \simeq \operatorname{Ext}^*_A(M,\Bbbk) \triangledown_{H^*} \operatorname{Ext}^*_B(N,\Bbbk),$$
$$\operatorname{Ext}^*_{A\otimes^q B}(M \otimes^q N,\Bbbk) \simeq \operatorname{Ext}^*_A(M,\Bbbk)\, \overline{\otimes}^{q^{-1}}\, \operatorname{Ext}^*_B(N,\Bbbk),$$

where $\operatorname{Ext}^*_A(M,\Bbbk)$ *and* $\operatorname{Ext}^*_B(N,\Bbbk)$ *are considered as modules over* $\operatorname{Ext}^*_A(\Bbbk,\Bbbk)$ *and* $\operatorname{Ext}^*_B(\Bbbk,\Bbbk)$, *respectively. The symbol $\overline{\otimes}^z$ denotes the multiplication rule*

$$(\xi_1 \,\overline{\otimes}^z\, \eta_1)(\xi_2 \,\overline{\otimes}^z\, \eta_2) = (-1)^{\bar\eta_1\bar\xi_2} z^{\widetilde{\eta_1}\widetilde{\xi_2}} (\xi_1\xi_2 \,\overline{\otimes}^z\, \eta_1\eta_2),$$

where $\bar\xi \in \mathbb{Z}_{\geqslant 0}$ is the homological grading of ξ and $\widetilde{\xi} \in \mathbb{Z}_{\geqslant 0}$ is the internal grading.

Sketch of proof: First of all, it is not difficult to extend our operations to differential algebras and modules and check that they commute with passing to the cohomology. Then one should construct natural morphisms of differential algebras

$$\mathcal{C}ob^\bullet(A \star B) \longrightarrow \mathcal{C}ob^\bullet(A) \,\bar\star\, \mathcal{C}ob^\bullet(B)$$

and morphisms of differential modules

$$\mathcal{C}ob^\bullet(A \star B,\, M \star_H N) \longrightarrow \mathcal{C}ob^\bullet(A,M) \,\bar\star_{H^*}\, \mathcal{C}ob^\bullet(B,N),$$
$$\mathcal{C}ob^\bullet(A \otimes^q B,\, M \otimes^q N) \longrightarrow \mathcal{C}ob^\bullet(A,M)\, \overline{\otimes}^{q^{-1}}\, \mathcal{C}ob^\bullet(B,N),$$

where $\mathcal{C}ob^\bullet(A,M)$ and $\mathcal{C}ob^\bullet(B,N)$ are considered as differential modules over $\mathcal{C}ob^\bullet(A)$ and $\mathcal{C}ob^\bullet(B)$ and the correspondence $\star \longmapsto \bar\star$ is defined by the statements of the proposition. The last step is to check that the morphisms of complexes constructed induce isomorphisms of the cohomology.

Let us consider in detail the most difficult case of the algebra $A \otimes^q B$ and the module $M \otimes^q N$. The so-called "shuffle product" morphism

$$\mathrm{sh} : \mathcal{B}ar_{i'}(A,M) \otimes \mathcal{B}ar_{i''}(B,N) \longrightarrow \mathcal{B}ar_{i'+i''}(A \otimes^q B,\, M \otimes^q N)$$

is given by the rule

$$\mathrm{sh} : (a_1 \otimes \cdots \otimes a_{i'} \otimes m) \otimes (b_1 \otimes \cdots \otimes b_{i''} \otimes n)$$
$$\longmapsto q^{-(\widetilde{b_1}+\cdots+\widetilde{b_{i''}})\widetilde{m}} a_1 \otimes \cdots \otimes a_{i'} \otimes b_1 \otimes \cdots \otimes b_{i''} \otimes m \otimes n$$
$$- q^{-(\widetilde{b_1}+\cdots+\widetilde{b_{i''}})\widetilde{m} - \widetilde{a_{i'}}\widetilde{b_1}} a_1 \otimes \cdots \otimes a_{i'-1} \otimes b_1 \otimes a_{i'} \otimes b_2 \otimes \cdots \otimes b_{i''} \otimes m \otimes n + \cdots,$$

where the summation is over all the $\binom{i'+i''}{i'}$ permutations of the tensor components $a_1, \ldots, a_{i'}$ and $b_1, \ldots, b_{i''}$ preserving the order within each group (i.e., over shuffles). The components m and n always stay at the very end. The rule for the coefficients is the following: any permutation of components a_s and b_t generates the factor $-q^{-\widetilde{a_s b_t}}$ in the coefficient. The permutation of m through $b_1 \otimes \cdots \otimes b_{i''}$ generates the factor that appears with the first summand in the above formula. The desired morphism for the cobar-complexes is the dual map to sh. It is straightforward to check that it commutes with the differential and the multiplication. The following trick helps to prove that it is a quasi-isomorphism. Using the same formula as above one can define a morphism

$$\widetilde{\mathcal{B}ar}_\bullet(A, M) \otimes \widetilde{\mathcal{B}ar}_\bullet(B, N) \longrightarrow \widetilde{\mathcal{B}ar}_\bullet(A \otimes^q B,\, M \otimes^q N) \tag{1.3}$$

of the bar-resolutions. The tensor product of complexes on the left has a natural $A \otimes^q B$-module structure. Moreover, both sides are free resolutions of the $A \otimes^q B$-module $M \otimes^q N$. Hence, (1.3) is a homotopy equivalence. It remains to observe that our morphism of cobar-complexes is obtained by applying the functor $\mathrm{Hom}_{A\otimes^q B}(-, \Bbbk)$ to the morphism (1.3).

For other operations $\star$, the construction of the morphism on the level of cobar-complexes is clear. For $A \sqcap B$ and $M \sqcap_H N$ we even get an isomorphism of complexes. In the remaining cases an explicit homotopy can be constructed. For $\star = \sqcup$ one can also use the same argument as above, since the operation $M \sqcup_H N$ takes free graded modules to free graded modules. □

COROLLARY 1.2. *We keep the notation of Proposition* (1.1). *Each of the algebras $A \sqcap B$, $A \sqcup B$, $A \otimes^q B$, $q \in \mathbb{P}^1_\Bbbk$ is quadratic (resp., Koszul) iff both algebras A and B are quadratic (resp., Koszul). The quadratic dual algebras are given by the formulas $(A \sqcap B)^! = A^! \sqcup B^!$ and $(A \otimes^q B)^! = A^! \otimes^{-q^{-1}} B^!$. If A and B are quadratic (resp., Koszul) then each of the modules $M \sqcap_H N$, $M \sqcup_H N$, $M \vartriangle_H N$ is quadratic (resp., Koszul) iff both modules M and N are quadratic or (resp., Koszul). If $M \neq 0$ and $N \neq 0$ then the analogous statement is true for the $A \otimes^q B$-module $M \otimes^q N$, where $q \in \Bbbk^*$.* □

For example, for each of the operations $\star = \sqcap, \sqcup, \vartriangle$ we have

$$\langle H, K' \rangle_A \star_H \langle H, K'' \rangle_B \simeq \langle H, K' \oplus K'' \rangle_{A \star B},$$

where $K' \oplus K'' \subset (A_1 \oplus B_1) \otimes H$. Note that for all the above operations except for $\otimes^q$, where $q \in \Bbbk^*$, it is easy to verify the statements of the above corollary explicitly: first, one needs to check that these operations preserve the classes of quadratic algebras and modules, and then use Backelin's criterion of Koszulness in terms of distributive lattices (Theorems 4.1 and 4.2 of chapter 2). Indeed, the space $(A \star B)_1^{\otimes n} = (A_1 \oplus B_1)^{\otimes n}$ is the direct sum of several tensor product spaces, including $A_1^{\otimes n}$ and $B_1^{\otimes n}$. It is easy to see that each of the subspaces $X_i \subset (A \star B_1)^{\otimes n}$ from Theorem 4.1 of chapter 2 is a direct sum of the corresponding subspaces in the direct summands of $(A \oplus B)_1^{\otimes n}$. The subspaces we get in $A_1^{\otimes n}$ and $B_1^{\otimes n}$ coincide with the subspaces X_i for the algebras A and B. The collections of subspaces in other summands are distributive provided that the collections of subspaces in $A_1^{\otimes j}$ and $B_1^{\otimes j}$ corresponding to the algebras A and B are distributive for $j < n$ (see the remark on distributivity of direct sum collections in the beginning of section 7 of chapter 1).

Koszulness of the tensor product of Koszul algebras $A \otimes^q B$ can be derived from Koszulness of $A \otimes^0 B$ using the standard spectral sequence associated with a filtration (see Example 3 in section 7 of chapter 4). We do not know any simple lattice-theoretic explanation for Koszulness of the tensor product of modules $M \otimes^q N$.

2. Segre products and Veronese powers. I

Definition 1. The *Segre product* of two graded algebras A and B is the graded algebra $A \circ B = \bigoplus_{n=0}^{\infty} A_n \otimes B_n$. For a pair of graded modules M and N over A and B, respectively, their *Segre product* is the graded $A \circ B$-module $M \circ N = \bigoplus_{n \in \mathbb{Z}} M_n \otimes N_n$. By definition, $A \circ B$ (resp., $M \circ N$) is a subalgebra of the tensor product $A \otimes B$ (resp., submodule of $M \otimes N$). However, these embeddings *are not* morphisms of *graded* algebras or modules, since they double the grading.

One can define similarly q-Segre products $A \circ^q B \subset A \otimes^q B$ for $q \in \Bbbk^*$; however, for simplicity we restrict ourselves to the case $q = 1$.

PROPOSITION 2.1. *(i) If both algebras A and B are one-generated, quadratic, or Koszul, then the algebra $A \circ B$ is of the same type. More generally, if the algebra A is one-generated then the generator spaces of the algebra $A \circ B$ are given by the formula $\mathrm{Tor}^{A\circ B}_{1,n}(\Bbbk, \Bbbk) = A_n \otimes \mathrm{Tor}^B_{1,n}(\Bbbk, \Bbbk)$. If both algebras are one-generated then there are natural exact triples for the relation spaces*

$$0 \longrightarrow \mathrm{Tor}^A_{2,n}(\Bbbk, \Bbbk) \otimes \mathrm{Tor}^B_{2,n}(\Bbbk, \Bbbk) \longrightarrow \mathrm{Tor}^{A\circ B}_{2,n}(\Bbbk, \Bbbk) \\ \longrightarrow \mathrm{Tor}^A_{2,n}(\Bbbk, \Bbbk) \otimes B_n \oplus A_n \otimes \mathrm{Tor}^B_{2,n}(\Bbbk, \Bbbk) \longrightarrow 0.$$

(ii)(M) If both A-module M and B-module N are generated in degree zero (resp., quadratic, resp., Koszul) then the same is true for the $A \circ B$-module $M \circ N$. More precisely, if the algebra A is one-generated and the A-module M is generated in degree zero then the generator spaces of the module $M \circ N$ are given by the formula $\mathrm{Tor}^{A\circ B}_{0,n}(M \circ N, \Bbbk) = M_n \otimes \mathrm{Tor}^B_{0,n}(N, \Bbbk)$. If both algebras are one-generated and both modules are generated in degree zero then there are natural exact triples

$$0 \longrightarrow \mathrm{Tor}^A_{1,n}(M, \Bbbk) \otimes \mathrm{Tor}^B_{1,n}(N, \Bbbk) \longrightarrow \mathrm{Tor}^{A\circ B}_{1,n}(M \circ N, \Bbbk) \\ \longrightarrow \mathrm{Tor}^A_{1,n}(M, \Bbbk) \otimes N_n \oplus M_n \otimes \mathrm{Tor}^B_{1,n}(N, \Bbbk) \longrightarrow 0.$$

Proof: (i) For any graded algebra C one has $\mathrm{Tor}^C_{1,n}(\Bbbk, \Bbbk) = (C_+/C_+C_+)_n$. Hence, in the case when A is one-generated we have

$$\mathrm{Tor}^{A\circ B}_{1,n}(\Bbbk, \Bbbk) = A_n \otimes B_n \Big/ \Big(\sum\nolimits_{j+k=n}^{j,k\geqslant 1} A_j A_k \otimes B_j B_k\Big) \\ = A_n \otimes B_n \Big/ \Big(\sum\nolimits_{j+k=n}^{j,k\geqslant 1} A_n \otimes B_j B_k\Big) = A_n \otimes \mathrm{Tor}^B_{1,n}(\Bbbk, \Bbbk).$$

For the rest of the proof we assume that both A and B are one-generated. Assume that A and B have no defining relations of degree k. Let us check that the same is true for the algebra $A \circ B$. For a graded algebra C let us denote

$$J^C_i := \ker(C_1^{\otimes i} \longrightarrow C_i).$$

Note that C has no defining relations of degree k iff the natural map

$$C_1 \otimes J^C_{k-1} \oplus J^C_{k-1} \otimes C_1 \longrightarrow J^C_k \tag{2.1}$$

is surjective. Now for every $i \geqslant 1$ we have an exact sequence

$$0 \longrightarrow J^A_i \otimes J^B_i \longrightarrow J^{A\circ B}_i \longrightarrow J^A_i \otimes B_i \oplus A_i \otimes J^B_i \longrightarrow 0.$$

Using these sequences it is easy to show that surjectivity of (2.1) for both algebras $C = A$ and $C = B$ implies its surjectivity for $C = A \circ B$.

For a graded algebra C let $C^{(n)}$ denote the graded algebra with the same generators and relations of degree $< n$ as in C and with no generators or defining relations of degree $\geqslant n$. The above argument applied to $A^{\langle n\rangle}$ and $B^{\langle n\rangle}$, where $n \geqslant 2$, shows that $A^{\langle n\rangle} \circ B^{\langle n\rangle}$ has no defining relations of degree $\geqslant n$, and hence

$$(A \circ B)^{\langle n\rangle} \simeq A^{\langle n\rangle} \circ B^{\langle n\rangle}$$

for every $n \geqslant 2$. Note that for a one-generated algebra C there is an exact sequence

$$0 \longrightarrow \mathrm{Tor}^C_{2,n}(\Bbbk, \Bbbk) \longrightarrow C_n^{\langle n\rangle} \longrightarrow C_n \longrightarrow 0$$

(see the end of section 5 of chapter 1). Applying this to $C = A \circ B$ we see that $\mathrm{Tor}^C_{2,n}(\Bbbk, \Bbbk)$ is isomorphic to the kernel of the natural map $A_n^{\langle n\rangle} \otimes B^{\langle n\rangle} \to A_n \otimes B_n$. Using the above exact sequence for $C = A$ and $C = B$ we obtain the required exact sequence for the relation spaces.

Finally, to prove Koszulness, we use Backelin's criterion (see ch. 2, Theorems 4.1 and 4.2). Distributivity of the relevant collections can be easily checked using Proposition 7.1 of chapter 1.

The proof of (ii) is completely analogous. □

Note that in contrast with operations considered in section 1, Koszulness of the Segre product $A \circ B$ does not imply Koszulness of A and B. Corollary 4.9 below describes completely the situation assuming that one of the factors is Koszul.

Definition 2. For any integer $d \geqslant 1$, the *Veronese subalgebra* of degree d of a graded algebra A is the graded algebra $A^{(d)}$ with the components $A_n^{(d)} = A_{dn}$. The *Veronese submodule* of a graded A-module M is the graded $A^{(d)}$-module $M^{(d)} = \bigoplus_{n\in\mathbb{Z}} M_{dn}$. By definition, $A^{(d)}$ is a subalgebra of A and $M^{(d)}$ is an $A^{(d)}$-submodule of M but these embeddings are not compatible with the grading.

Proposition 2.2. *(i) Veronese subalgebras of a one-generated (resp., quadratic, resp., Koszul) algebra A are still one-generated (resp., quadratic, resp., Koszul). Moreover, if A is one-generated and has no defining relations of degree $> (k-1)d+1$ then $A^{(d)}$ has no defining relations of degree $> k$. In particular, if A is one-generated and has no defining relations of degree $> d+1$ then $A^{(d)}$ is quadratic.*
(ii)(M) Assume that A is one-generated. If an A-module M is generated in degree zero (resp., A and M are quadratic, resp., A and M are Koszul) then all Veronese submodules $M^{(d)}$ are also generated in degree zero (resp., quadratic, resp., Koszul). If M is generated in degrees $\leqslant kd$ then $M^{(d)}$ is generated in degrees $\leqslant k$. Finally, assume that M has no defining relations of degree $> ld$, and either M is generated in degree zero, or M is generated in degrees $\leqslant kd$ and A has no defining relations of degree $> (l-k)d+1$. Then $M^{(d)}$ has no defining relations of degree $> l$.

Proof: Assume that A is a one-generated algebra. Then all the maps $A_i \otimes A_j \longrightarrow A_{i+j}$ are surjective; hence the algebra $A^{(d)}$ is also one-generated. Assume in addition that A has no defining relations of degree $> (k-1)d+1$. Let us show $A^{(d)}$ has no defining relations of degree $> k$. Using the notation from the proof of Proposition 2.1 we have to show surjectivity of the morphisms

$$A_1^{(d)} \otimes J_{n-1}^{A^{(d)}} \oplus J_{n-1}^{A^{(d)}} \otimes A_1^{(d)} \longrightarrow J_n^{A^{(d)}}$$

for all $n > k$. The natural projections $A_1^{\otimes dn} \longrightarrow A_d^{\otimes n} = A_1^{(d)\otimes n}$ map J^A_{nd} surjectively onto $J_n^{A^{(d)}}$. Therefore, it suffices to check that the morphisms

$$A_1^{\otimes d} \otimes J^A_{(n-1)d} \oplus J^A_{(n-1)d} \otimes A_1^{\otimes d} \longrightarrow J^A_{nd} \tag{2.2}$$

are surjective for $n > k$. Our assumption about A implies that the morphisms

$$A_1 \otimes J^A_{m-1} \oplus J^A_{m-1} \otimes A_1 \longrightarrow J^A_m$$

are surjective for $m > (k-1)d+1$. This easily implies surjectivity of the morphisms

$$\bigoplus\nolimits_{t=0}^{s} A_1^{\otimes t} \otimes J_{m-s} \otimes A_1^{\otimes s-t} \longrightarrow J_m$$

for $m - s \geqslant (k-1)d+1$. It remains to observe that for $s = 2d-1$ and $m = nd$ the latter morphism factors through (2.2).

Note that if A is one-generated then an A-module M is generated in degree $\leqslant n$ iff the maps $A_1 \otimes M_{m-1} \longrightarrow M_m$ are surjective for $m > n$. In this case the maps $A_s \otimes M_{m-s} \longrightarrow M_m$ are surjective for $m - s \geqslant n$. To show that the module $M^{(d)}$ is generated in degrees $\leqslant k$, take $s = d$ and $m \geqslant (k+1)d$ and use the assumption that M is generated in degrees $\leqslant kd$.

Furthermore, for an A-module M generated in degree zero consider the A-module $M' = \ker(A \otimes M_0 \longrightarrow M)$. Then the relation spaces of the module M can be identified with the generator spaces of the module M'. Since $M'^{(d)} = M^{(d)\prime}$, we get the desired estimate for the relations of the Veronese submodule of the module generated in degree zero.

Now assume that we have a surjective morphism $A \otimes X \longrightarrow M$ with the kernel M', where the graded vector space X is concentrated in the degrees $\leqslant kd$ and the A-module M' is generated in degrees $\leqslant ld$. Consider the composition of surjective morphisms

$$\bigoplus\nolimits_j A^{(d)} \otimes A_{-j \bmod d} \otimes X_j \xrightarrow{\ f\ } (A \otimes X)^{(d)} \xrightarrow{\ g\ } M^{(d)},$$

where $0 \leqslant -j \bmod d \leqslant d-1$. We have to prove that its kernel $M^{(d)\prime}$ is generated in degrees $\leqslant l$. Clearly, it suffices to prove that both $A^{(d)}$-modules $\ker(f)$ and $\ker(g)$ are generated in degrees $\leqslant l$.

We have $\ker(g) \simeq M'^{(d)}$; hence it is generated in degrees $\leqslant l$ (since M' is generated in degrees $\leqslant ld$). For the module $L = \ker(f)$ we have

$$L = \bigoplus_j L_{(-j \bmod d)} \otimes X_j,$$

where $L_{(r)} = \ker(A^{(d)} \otimes A_r \longrightarrow A)$. It remains to show that the $A^{(d)}$-modules $L_{(r)}$ are generated in degrees $\leqslant l-k$ provided A has no defining relations of degree $> (l-k)d+1$. Let us check that the maps $A_d \otimes L_{(r)n-1} \longrightarrow L_{(r)n}$ are surjective for $n > l-k$. The restriction of the multiplication map $A_1^{\otimes nd+r} \longrightarrow A_{nd} \otimes A_r$ provides a surjective map $J^A_{nd+r} \longrightarrow L_{(r)n}$. For $n > l-k$ the multiplication map

$$\bigoplus_{s+t=d+r-1} A_1^{\otimes s} \otimes J^A_{(n-1)d+1} \otimes A_1^{\otimes t} \longrightarrow J^A_{nd+r}$$

is also surjective. Using the embedding $L_{(r)n} \longrightarrow A_{nd} \otimes A_r$ it is easy to see that the composition $A_1^{\otimes s} \otimes J^A_{(n-1)d+1} \otimes A_1^{\otimes t} \longrightarrow L_{(r)n}$ is zero for $t \geqslant r$, and factors through $A_d \otimes L_{(d)n-1}$ for $s \geqslant d$. At least one of these conditions holds for any $s + t = d + r - 1$, so the assertion follows.

To prove Koszulness we apply Backelin's criterion. For a quadratic algebra A and a quadratic A-module M, the collection of subspaces $X_1^{(d)}, \dots, X_n^{(d)}$ of the vector space $W^{(d)} = A_d^{\otimes n} \otimes M_0$ corresponding to the algebra $A^{(d)}$ and module $M^{(d)}$

is connected with the collection of subspaces $X_1, \ldots, X_{nd} \subset W = A_1^{\otimes nd} \otimes M_0$ corresponding to the algebra A and the module M in the following way:

$$W^{(d)} = W/\sum\nolimits_{k\not\equiv 0\,(d)} X_k; \qquad X_i^{(d)} = X_{di} \bmod \sum\nolimits_{k\not\equiv 0\,(d)} X_k.$$

Clearly, such a procedure preserves distributivity. □

Remark 1. The complexity of the operations $A \circ B$ and $A^{(d)}$ in comparison with the operations considered in section 1 is manifested by the fact that it is in general impossible to recover the spaces $\mathrm{Ext}^{ij}_{A\circ B}(\Bbbk,\Bbbk)$ or $\mathrm{Ext}^{ij}_{A^{(d)}}(\Bbbk,\Bbbk)$ from the Ext-spaces of the algebras A and B. For example, the space $\mathrm{Ext}^{2,j}_{A^{(2)}}(\Bbbk,\Bbbk)$ for a one-generated algebra A can be described as follows. It has a natural 3-step filtration $0 \subset F_1 \subset F_2 \subset F_3 = \mathrm{Ext}^{2,j}_{A^{(2)}}(\Bbbk,\Bbbk)$ such that $F_1 \simeq \mathrm{Ext}^{2,2j}_A(\Bbbk,\Bbbk)$, F_2/F_1 is isomorphic to the kernel of the multiplication map

$$\mathrm{Ext}^{1,1}_A(\Bbbk,\Bbbk) \otimes \mathrm{Ext}^{2,2j-1}_A(\Bbbk,\Bbbk) \oplus \mathrm{Ext}^{2,2j-1}_A(\Bbbk,\Bbbk) \otimes \mathrm{Ext}^{1,1}_A(\Bbbk,\Bbbk) \longrightarrow \mathrm{Ext}^{3,2j}_A(\Bbbk,\Bbbk),$$

and F_3/F_2 is isomorphic to the kernel of the Massey product morphism from the kernel of the multiplication map

$$\begin{aligned}&\mathrm{Ext}^{1,1}_A(\Bbbk,\Bbbk) \otimes \mathrm{Ext}^{2,2j-2}_A(\Bbbk,\Bbbk) \otimes \mathrm{Ext}^{1,1}_A(\Bbbk,\Bbbk)\\ &\qquad \longrightarrow \mathrm{Ext}^{1,1}_A(\Bbbk,\Bbbk) \otimes \mathrm{Ext}^{3,2j-1}_A(\Bbbk,\Bbbk) \oplus \mathrm{Ext}^{3,2j-1}_A(\Bbbk,\Bbbk) \otimes \mathrm{Ext}^{1,1}_A(\Bbbk,\Bbbk)\end{aligned}$$

to $\mathrm{Ext}^{3,2j}_A(\Bbbk,\Bbbk)$ (see Appendix).

Now let us describe the space

$$\mathrm{Tor}^{A\circ B}_{1,3}(\Bbbk,\Bbbk) = (A_3 \otimes B_3)/(A_1A_2 \otimes B_1B_2 + A_2A_1 + B_2B_1)$$

(cf. Remark 1 in section 2 of chapter 2). It has a natural subspace coming from $(A_1A_2 + A_2A_1) \otimes (B_1B_2 + B_2B_1) \subset A_3 \otimes B_3$. The quotient by this subspace can be identified with the kernel of the map

$$A_3 \otimes B_3 \longrightarrow \mathrm{Tor}^A_{1,3}(\Bbbk,\Bbbk) \otimes \mathrm{Tor}^B_{1,3}(\Bbbk,\Bbbk),$$

so its dimension is determined by the dimensions of the homology spaces of A and B. However, the subspace itself is isomorphic to

$$(A_1A_2/A_2A_1) \otimes (B_2B_1/B_1B_2) \oplus (A_2A_1/A_1A_2) \otimes (B_1B_2/B_2B_1),$$

where we denote $U/V = (U+V)/V$. It is easy to see that the space A_1A_2/A_2A_1 is isomorphic to the cokernel of the comultiplication map

$$\mathrm{Tor}^A_{2,3}(\Bbbk,\Bbbk) \longrightarrow \mathrm{Tor}^A_{1,1}(\Bbbk,\Bbbk) \otimes \mathrm{Tor}^A_{1,2}(\Bbbk,\Bbbk).$$

On the other hand, as we will show in the next section, the estimates for the generators and relations of Veronese subalgebras and submodules obtained in Proposition 2.2 can be generalized to higher homology spaces $\mathrm{Tor}^{A^{(d)}}_{ij}(\Bbbk,\Bbbk)$ and $\mathrm{Tor}^{A^{(d)}}_{ij}(M^{(d)},\Bbbk)$. The situation here is similar (although much more complicated) to the case of truncated modules $M^{[r]}$ considered in section 1 of chapter 2. Roughly speaking, we will show that the algebras $A^{(d)}$ and modules $M^{(d)}$ for a one-generated algebra A and an A-module M generated in degree zero become closer and closer to being Koszul as d grows to infinity (one can consider Proposition 1.1 in section 1 of chapter 2 as a toy version of this result). Note that the condition that A is one-generated cannot be omitted here: for the free associative algebra A with one generator of degree 1 and one generator of degree 2 all the algebras $A^{(d)}$, $d \geqslant 2$ are infinitely generated. Similarly, we will estabish in the next section estimates for the vanishing region of higher Tor-spaces of the Segre product algebra $A \circ B$ and module $M \circ N$.

Remark 2. For a *commutative* or *skew-commutative* one-generated algebra A one can get slightly stronger estimates for the degrees of relations of the Veronese subalgebras and submodules. Namely, the Veronese subalgebra $A^{(d)}$ has no defining relations of degree $> k$, $k \geqslant 2$, provided the algebra A has no defining relations of degree $> kd$. The module $M^{(d)}$ has no defining relations of degree $> l$ provided M has no generators of degree $> (l-1)d$ and no defining relations of degree $> ld$ and $a(M) + b(A) \leqslant ld$, where $a(M)$ is the maximal degree of a generator of M not divisible by d and $b(A)$ is the maximal degree of a defining relation of A. The proof is analogous to that of Proposition 2.2, with the following changes. To prove the first assertion one should consider the algebra A as a quotient algebra of the symmetric algebra $\mathbb{S}(A_1)$ (or the corresponding exterior algebra). Since the symmetric algebra is quadratic, it suffices to estimate the degrees of generators of the corresponding ideal in $\mathbb{S}(A_1)$. For the second statement one has to show that the $A^{(d)}$-module $\bigoplus_n A_{nd+r}$ has no relations of degree $> t$, $t \geqslant 1$ (that is, the module $L_{(r)}$ has no generators of degree $> t$), if the algebra A has no relations of degree $> td + r$. To do this consider the algebra $A^{(d)}$ as a quotient algebra of $\mathbb{S}(A_1)^{(d)}$ and the module $\bigoplus_n A_{nd+r}$ as a quotient module of $\bigoplus_n \mathbb{S}^{nd+r}(A_1)$ over $\mathbb{S}(A_1)^{(d)}$. Then observe that the latter module is quadratic and estimate the degrees of generators of the kernel.

3. Segre products and Veronese powers. II

In this section we obtain general estimates for the vanishing region of the homology spaces of Veronese subalgebras and submodules and Segre products of algebras and modules. We use the following numerical measure introduced by J. Backelin.

Definition 1. The rate of growth of the homology spaces of a graded algebra A (or simply the rate of A) is defined as the finite or infinite supremum

$$\operatorname{rate}(A) = \sup\left\{ \frac{j-1}{i-1} \mid \operatorname{Tor}^A_{i,j}(\Bbbk,\Bbbk) \neq 0 \right\}.$$

If A is not one-generated then by definition $\operatorname{rate}(A) = +\infty$.

THEOREM 3.1. *For any graded algebra A and positive integer d one has*

$$\operatorname{rate}(A^{(d)}) \leqslant \left\lceil \frac{\operatorname{rate}(A)}{d} \right\rceil.$$

Proof: Obviously, we can assume that A is one-generated. Consider the spectral sequence

$$E^2_{pq} = \operatorname{Tor}^A_p(\operatorname{Tor}^{A^{(d)}}_q(\Bbbk, A), \Bbbk) \implies \operatorname{Tor}^{A^{(d)}}_{p+q}(\Bbbk,\Bbbk); \qquad d^r_{pq}\colon E^r_{p,q} \longrightarrow E^r_{p-r,q+r-1}$$

associated with the homomorphism of augmented algebras $f\colon A^{(d)} \longrightarrow A$. Note that the homomorphism f multiplies the grading by d. Therefore, the internal grading of terms E^∞_{pq} should correspond to the internal grading of the Tor over $A^{(d)}$ multiplied by d. In particular, it follows that the internal grading of E^∞_{pq} is concentrated in degrees divisible by d. On the other hand, the $A^{(d)}$-module A splits into a direct sum of modules numbered by residues modulo d, and the submodule corresponding to the residue 0 is free. Therefore, the internal grading of $\operatorname{Tor}^{A^{(d)}}_q(\Bbbk, A)$ lives entirely outside of values divisible by d for all $q > 0$.

Denote by $b(q)$ the maximal internal grading of $\mathrm{Tor}_q^{A^{(d)}}(\Bbbk, A)$ and by $c(q)$ the maximal internal grading of $E^2_{0,q} = \mathrm{Tor}_q^{A^{(d)}}(\Bbbk, A) \otimes_A \Bbbk$. Since A is one-generated, it follows easily that $\lceil c(q)/d \rceil = \lceil b(q)/d \rceil$ for any q. Finally, denote by $\mathrm{hg}_A(p)$ the maximal internal grading of $\mathrm{Tor}_p^A(\Bbbk, \Bbbk)$, and analogously for $A^{(d)}$. Notice that the maximal internal grading of E^2_{pq} does not exceed $\mathrm{hg}_A(p) + b(q)$.

It follows from the divisibility observations above that the terms $E^2_{0,q}$ are killed in the spectral sequence for all $q > 0$. So we have

$$c(q) \leqslant \max_{1 \leqslant s \leqslant q} \mathrm{hg}_A(s+1) + b(q-s).$$

Dividing by d and applying the upper integral parts, we obtain

$$\begin{aligned} \left\lceil \frac{b(q)}{d} \right\rceil = \left\lceil \frac{c(q)}{d} \right\rceil &\leqslant \max_{1 \leqslant s \leqslant q} \left\lceil \frac{\mathrm{hg}_A(s+1) + b(q-s)}{d} \right\rceil \\ &\leqslant \max_{1 \leqslant s \leqslant q} \left(\left\lceil \frac{\mathrm{hg}_A(s+1) - 1}{d} \right\rceil + \left\lceil \frac{b(q-s)+1}{d} \right\rceil \right) \\ &= \max_{1 \leqslant s \leqslant q} \left(\left\lceil \frac{\mathrm{hg}_A(s+1) - 1}{d} \right\rceil + \left\lceil \frac{b(q-s)}{d} \right\rceil \right) \end{aligned}$$

since $b(q-s)$ is never divisible by d. Analogously,

$$\begin{aligned} \mathrm{hg}_{A^{(d)}}(n) &\leqslant \max_{1 \leqslant p \leqslant n} \left\lfloor \frac{\mathrm{hg}_A(p) + b(n-p)}{d} \right\rfloor \\ &\leqslant \max_{1 \leqslant p \leqslant n} \left(\left\lfloor \frac{\mathrm{hg}_A(p) - 1}{d} \right\rfloor + \left\lceil \frac{b(n-p)+1}{d} \right\rceil \right) \\ &= \max_{1 \leqslant p \leqslant n} \left(\left\lfloor \frac{\mathrm{hg}_A(p) - 1}{d} \right\rfloor + \left\lceil \frac{b(n-p)}{d} \right\rceil \right) \end{aligned}$$

since $\lfloor x+y \rfloor \leqslant \lfloor x \rfloor + \lceil y \rceil$ for all real x, y.

Now let us prove by induction in q that $\lceil b(q)/d \rceil \leqslant q \lceil \mathrm{rate}(A)/d \rceil + 1$. By definition, $\mathrm{hg}_A(s+1) - 1 \leqslant s \cdot \mathrm{rate}(A)$. Using the induction assumption for $q-s$ we get

$$\begin{aligned} \left\lceil \frac{b(q)}{d} \right\rceil &\leqslant \max_{1 \leqslant s \leqslant q} \left(\left\lceil \frac{s \cdot \mathrm{rate}(A)}{d} \right\rceil + (q-s) \left\lceil \frac{\mathrm{rate}(A)}{d} \right\rceil + 1 \right) \\ &\leqslant \max_{1 \leqslant s \leqslant q} \left(s \left\lceil \frac{\mathrm{rate}(A)}{d} \right\rceil + (q-s) \left\lceil \frac{\mathrm{rate}(A)}{d} \right\rceil + 1 \right) = q \left\lceil \frac{\mathrm{rate}(A)}{d} \right\rceil + 1. \end{aligned}$$

Finally, we can obtain the desired estimate for $\mathrm{hg}_{A^{(d)}}(n)$:

$$\begin{aligned} &\mathrm{hg}_{A^{(d)}}(n) \leqslant \max_{1 \leqslant p \leqslant n} \left(\left\lfloor \frac{\mathrm{hg}_A(p) - 1}{d} \right\rfloor + \left\lceil \frac{b(n-p)}{d} \right\rceil \right) \\ &\leqslant \max_{1 \leqslant p \leqslant n} \left(\left\lfloor \frac{(p-1)\,\mathrm{rate}(A)}{d} \right\rfloor + (n-p) \left\lceil \frac{\mathrm{rate}(A)}{d} \right\rceil + 1 \right) \leqslant (n-1) \left\lceil \frac{\mathrm{rate}(A)}{d} \right\rceil + 1. \end{aligned}$$

□

Definintion 2(M). Let A be a graded algebra and M a nonnegatively graded left A-module. We denote by $\mathrm{hg}_{A,M}(n)$ the maximal internal grading of $\mathrm{Tor}_n^A(\Bbbk, M)$ and define $\mathrm{rate}(A, M)$ as $\sup_p(\mathrm{hg}_{A,M}(p)/p)$. If M is not generated in degree zero then by definition $\mathrm{rate}(A, M) = +\infty$.

PROPOSITION 3.2.(M). *Assume that* $\operatorname{rate}(A) \leqslant c_1$ *and* $\operatorname{hg}_{A,M}(p) \leqslant c_1 p + c_2$ *for some constants* c_1, c_2 *and all* p. *Then* $\operatorname{hg}_{A^{(d)},M^{(d)}}(n) \leqslant \lceil c_1/d \rceil n + \lceil c_2/d \rceil$. *In particular, when* $c_2 = 0$ *we obtain the estimate*

$$\operatorname{rate}(A^{(d)}, M^{(d)}) \leqslant \left\lceil \frac{\max(\operatorname{rate}(A), \operatorname{rate}(A, M))}{d} \right\rceil.$$

Proof: Consider the spectral sequence

$$E^2_{pq} = \operatorname{Tor}^A_p(\operatorname{Tor}^{A^{(d)}}_q(\Bbbk, A), M) \implies \operatorname{Tor}^{A^{(d)}}_{p+q}(\Bbbk, M).$$

Notice that $\operatorname{Tor}^{A^{(d)}}_n(\Bbbk, M^{(d)})$ is a direct summand of $\operatorname{Tor}^{A^{(d)}}_n(\Bbbk, M)$; hence

$$d \operatorname{hg}_{A^{(d)},M^{(d)}}(n) \leqslant \max\nolimits_{0 \leqslant p \leqslant n}(\operatorname{hg}_{A,M}(p) + b(n-p)),$$

where we keep the notation $b(q)$ for the maximal internal grading of $\operatorname{Tor}^{A^{(d)}}_q(\Bbbk, A)$ from the previous proof. Calculating as above, we obtain

$$\operatorname{hg}_{A^{(d)},M^{(d)}}(n) \leqslant \max_{0 \leqslant p \leqslant n} \left(\left\lfloor \frac{\operatorname{hg}_{A,M}(p) - 1}{d} \right\rfloor + \left\lceil \frac{b(n-p)}{d} \right\rceil \right)$$

$$\leqslant \max_{0 \leqslant p \leqslant n} \left(\left\lfloor \frac{c_1 p + c_2 - 1}{d} \right\rfloor + (n-p)\left\lceil \frac{c_1}{d} \right\rceil + 1 \right) = \max_{0 \leqslant p \leqslant n} \left(\left\lceil \frac{c_1 p + c_2}{d} \right\rceil + (n-p)\left\lceil \frac{c_1}{d} \right\rceil \right)$$

$$\leqslant \max_{0 \leqslant p \leqslant n} \left(p\left\lceil \frac{c_1}{d} \right\rceil + \left\lceil \frac{c_2}{d} \right\rceil + (n-p)\left\lceil \frac{c_1}{d} \right\rceil \right) = n\left\lceil \frac{c_1}{d} \right\rceil + \left\lceil \frac{c_2}{d} \right\rceil.$$

□

The condition $\operatorname{rate}(A) \leqslant d$ is sufficient for $A^{(d)}$ to be Koszul but by no means is it necessary. Using the above proposition we get another sufficient condition.

COROLLARY 3.3. *Let* $B \to A$ *be a homomorphism of graded algebras. Assume that* B *is Koszul and* $\operatorname{hg}_{B,A}(p) \leqslant (p+1)d$ *for all* p. *Then* $A^{(d)}$ *is Koszul.*

Proof: Indeed, from Proposition 3.2 we obtain that $\operatorname{hg}_{B^{(d)},A^{(d)}}(p) \leqslant p+1$ for all p. Since $B^{(d)}$ is Koszul, this implies Koszulness of $A^{(d)}$ by Theorem 5.2 of chapter 2. □

For example, if A is commutative and one-generated then we can take $B = \mathbb{S}(A_1)$. Since $\operatorname{rate}(B, A)$ is finite, we derive the following result.

COROLLARY 3.4. **[16]** *Let* A *be a commutative one-generated graded algebra. Then* $A^{(d)}$ *is Koszul for* $d \gg 0$.

Further discussion of homological properties of Veronese subrings can be found in [**16, 17, 20, 46**].

Now we turn to homology estimates for Segre products.

THEOREM 3.5. *For any graded algebras* A *and* B, *one has*

$$\operatorname{rate}(A \circ B) \leqslant \max\{\operatorname{rate} A, \operatorname{rate} B\}.$$

Proof: We can assume that both A and B are one-generated. Consider the spectral sequence

$$E^2_{pq} = \operatorname{Tor}^{A \otimes B}_p(\operatorname{Tor}^{A \circ B}_q(\Bbbk, A \otimes B), \Bbbk) \implies \operatorname{Tor}^{A \circ B}_{p+q}(\Bbbk, \Bbbk) \quad (3.1)$$

corresponding to the homomorphism of algebras $f \colon A \circ B \longrightarrow A \otimes B$. The tensor product $A \otimes B$ is equipped with a natural *internal bigrading* (α, β), and the Segre product $A \circ B$ is exactly the part of $A \otimes B$ where $\alpha = \beta$. So f is a homomorphism of bigraded algebras, where on the first algebra both gradings coincide. Therefore, E^r_{pq}

is a bigraded spectral sequence and on the limit term E^∞_{pq} two gradings coincide, $E^\infty_{pq,\alpha\beta} = 0$ for $\alpha \neq \beta$, since the limit term corresponds to the homology of $A \circ B$. On the other hand, we have the decomposition

$$A \otimes B = \bigoplus_{i\in\mathbb{Z}} (\bigoplus_{\alpha-\beta=i} A_\alpha \otimes B_\beta)$$

of $A\otimes B$ viewed as an $A\circ B$-module, where the summand corresponding to $\alpha-\beta=0$ is a free $A\circ B$-module. It follows that $\mathrm{Tor}^{A\circ B}_q(\Bbbk, A\otimes B)_{\alpha,\beta}$ vanishes for $\alpha=\beta$ except for the case $q=0=\alpha=\beta$. Hence, for any $q>0$ the $A\otimes B$-module $\mathrm{Tor}^{A\circ B}_q(\Bbbk, A\otimes B)$ splits into the direct sum of $\mathrm{Tor}^{A\circ B}_q(\Bbbk, A\otimes B)_{\alpha<\beta}$ and $\mathrm{Tor}^{A\circ B}_q(\Bbbk, A\otimes B)_{\alpha>\beta}$.

We will need two more spectral sequences:

$$'E^2_{pq} = \mathrm{Tor}^B_p(\mathrm{Tor}^{A\circ B}_q(\Bbbk, A\otimes B), \Bbbk) \implies \mathrm{Tor}^{A\circ B}_{p+q}(\Bbbk, A) = \mathrm{Tor}^{A\circ B}_{p+q}(\Bbbk,\Bbbk)\otimes_\Bbbk A \text{ and}$$

$$''E^2_{pq} = \mathrm{Tor}^A_p(\mathrm{Tor}^{A\circ B}_q(\Bbbk, A\otimes B), \Bbbk) \implies \mathrm{Tor}^{A\circ B}_{p+q}(\Bbbk, B) = \mathrm{Tor}^{A\circ B}_{p+q}(\Bbbk,\Bbbk)\otimes_\Bbbk B.$$

The limit term $'E^\infty_{pq}$ is concentrated in the internal bidegrees $\alpha \geqslant \beta$. Hence, the direct summand $\mathrm{Tor}^B_p(\mathrm{Tor}^{A\circ B}_q(\Bbbk, A\otimes B)_{\alpha<\beta}, \Bbbk)$ of $'E^2_{pq}$ is killed somewhere in $'E^r_{pq}$ for all $p\geqslant 0$, $q \geqslant 1$.

Now let us define

$$\begin{aligned} a(q) &= \max\{\alpha \mid \exists\beta>\alpha : \mathrm{Tor}^{A\circ B}_q(\Bbbk, A\otimes B)_{\alpha\beta}\neq 0\} \\ &= \max\{\alpha \mid \exists\beta>\alpha : \mathrm{Tor}^B_0(\mathrm{Tor}^{A\circ B}_q(\Bbbk, A\otimes B)_{\alpha\beta}, \Bbbk)\neq 0\} \end{aligned}$$

and analogously

$$\begin{aligned} b(q) &= \max\{\beta \mid \exists\alpha>\beta : \mathrm{Tor}^{A\circ B}_q(\Bbbk, A\otimes B)_{\alpha\beta}\neq 0\} \\ &= \max\{\beta \mid \exists\alpha>\beta : \mathrm{Tor}^A_0(\mathrm{Tor}^{A\circ B}_q(\Bbbk, A\otimes B)_{\alpha\beta}, \Bbbk)\neq 0\}. \end{aligned}$$

Assume that for given $a<b$ the component $\mathrm{Tor}^B_0(\mathrm{Tor}^{A\circ B}_q(\Bbbk, A\otimes B)_{ab}, \Bbbk)$ is nonzero. Since it gets killed in the spectral sequence $'E^r_{p,q}$, there should exist $s=r-1$ such that $'E^2_{s+1,q-s,a,b} = \mathrm{Tor}^B_{s+1}(\mathrm{Tor}^{A\circ B}_{q-s}(\Bbbk, A\otimes B), \Bbbk)_{ab}\neq 0$. The latter component splits into a direct sum of two parts according to the splitting of $\mathrm{Tor}^{A\circ B}_{q-s}(\Bbbk, A\otimes B)$ into pieces with internal bigrading $\alpha<\beta$ and $\alpha>\beta$. Hence, one of these parts has to be nonzero. If $\mathrm{Tor}^B_{s+1}(\mathrm{Tor}^{A\circ B}_{q-s}(\Bbbk, A\otimes B)_{\alpha<\beta}, \Bbbk)_{ab}\neq 0$ then we conclude that $a \leqslant a(q-s)$, since applying Tor over B does not change the α-grading. If $\mathrm{Tor}^B_{s+1}(\mathrm{Tor}^{A\circ B}_{q-s}(\Bbbk, A\otimes B)_{\alpha>\beta}, \Bbbk)_{ab}\neq 0$ then $b<\mathrm{hg}_B(q-s)+b(q-s)$ (where we keep the notation $\mathrm{hg}_A(q)$ from the proof of Theorem 1). In any case we obtain

$$a(q) \leqslant \max_{1\leqslant s\leqslant q} \max\{\, a(q-s),\, b(q-s)+\mathrm{hg}_B(s+1)-1\,\}.$$

As $\mathrm{hg}_A(s+1)-1 \leqslant s\,\mathrm{rate}(A)$ and the same for B, it follows immediately by induction simultaneous for $a(q)$ and $b(q)$ that

$$a(q),\, b(q) \leqslant q\max\{\mathrm{rate}(A), \mathrm{rate}(B)\}.$$

Now let us prove the desired estimate for $\mathrm{hg}_{A\circ B}(n)$. Here we will use the spectral sequence E^r_{pq} (see (3.1)). Let c be a number such that $E^\infty_{pq,cc}\neq 0$. Then the component $E^2_{pq,cc} = \mathrm{Tor}^{A\otimes B}_p(\mathrm{Tor}^{A\circ B}_q(\Bbbk, A\otimes B), \Bbbk)_{cc}$ should also be nonzero. As above, for $q\neq 0$ this component splits into two direct summands corresponding to the two direct summands of $\mathrm{Tor}^{A\circ B}_q(\Bbbk, A\otimes B)$. In the case $q=0$ one needs an estimate for the internal bigrading of $\mathrm{Tor}^{A\otimes B}_{n+1}(\Bbbk,\Bbbk)$, which can be easily obtained.

In the case $q \neq 0$ assume that $\mathrm{Tor}_p^{A\otimes B}(\mathrm{Tor}_q^{A\circ B}(\Bbbk, A\otimes B)_{\alpha<\beta}, \Bbbk)_{cc} \neq 0$. It follows that the tensor product of $\mathrm{Tor}_p^{A\otimes B}(k,k)$ and $\mathrm{Tor}_q^{A\circ B}(\Bbbk, A\otimes B)_{\alpha<\beta}$ will also have nonzero component in bigrading (c,c). Hence $c \leqslant \max_{1\leqslant t\leqslant p}(\mathrm{hg}_A(t)+a(q))$ and generally

$$\mathrm{hg}_{A\circ B}(n) \leqslant \max_{1\leqslant t\leqslant p\leqslant n} \max\{\mathrm{hg}_A(t)+a(n-p), \mathrm{hg}_B(t)+b(n-p)\}.$$

Recall that $\mathrm{hg}_A(s)-1 \leqslant (s-1)\,\mathrm{rate}(A)$ and we finally obtain the desired estimate

$$\mathrm{hg}_{A\circ B}(n)-1 \leqslant (n-1)\max\{\mathrm{rate}(A), \mathrm{rate}(B)\}.$$

□

PROPOSITION 3.6.(M). *Let A and B be graded algebras, M (resp., N) be a nonnegatively graded A-module (resp., B-module). Assume that*

$$\max(\mathrm{rate}(A), \mathrm{rate}(B)) \leqslant c_1 \text{ and}$$
$$\max(\mathrm{hg}_{A,M}(p), \mathrm{hg}_{B,N}(p)) \leqslant c_1 p + c_2$$

for some constants c_1 and c_2. Then $\mathrm{hg}_{A\circ B,\, M\circ N}(n) \leqslant c_1 n + c_2$. In particular, in the case $c_2 = 0$ we have

$$\mathrm{rate}(A\circ B,\, M\circ N) \leqslant \max\{\mathrm{rate}(A), \mathrm{rate}(A,M), \mathrm{rate}(B), \mathrm{rate}(B,N)\}.$$

Proof: Let us keep the notation $a(q)$ and $b(q)$ from the previous proof. Clearly, we can assume that A and B are one-generated. Consider the spectral sequence

$$E^2_{pq} = \mathrm{Tor}_p^{A\otimes B}(\mathrm{Tor}_q^{A\circ B}(\Bbbk, A\otimes B), M\otimes N) \implies \mathrm{Tor}_{p+q}^{A\circ B}(M\otimes N).$$

Clearly, this spectral sequence carries the bigrading (α,β) used in the proof of Theorem 3.5. As before we use the splitting of the $A\circ B$-module $M\otimes N$ into the direct sum of submodules obtained by fixing the value of $\alpha-\beta$. Note that the summand corresponding to $\alpha=\beta$ is $M\circ N$. Therefore, $\mathrm{hg}_{A\circ B,\, M\circ N}(n)$ is the supremum over $p+q=n$ of γ such that $E^\infty_{pq,\gamma\gamma} \neq 0$. Thus, we have to show that for such p,q,γ one has $\gamma \leqslant c_1 n + c_2$. We are going to deduce this from the nonvanishing of $E^2_{pq,\gamma\gamma}$. As before, we decompose the term E^2_{pq} into the direct sum of two parts corresponding to the direct decomposition of the $A\otimes B$-module $\mathrm{Tor}_q^{A\circ B}(\Bbbk, A\otimes B)$ into two summands, one with $\alpha<\beta$, the other with $\beta<\alpha$. At least one of these summands of E^2_{pq} should have a nonzero component of internal bidegree (γ,γ), say

$$\mathrm{Tor}_p^{A\otimes B}(\mathrm{Tor}_q^{A\circ B}(\Bbbk, A\otimes B)_{\alpha<\beta}, M\otimes N)_{\gamma,\gamma} \neq 0.$$

As in the proof of Theorem 3.5 this implies that the tensor product of $\mathrm{Tor}_q^{A\circ B}(\Bbbk, A\otimes B)_{\alpha<\beta}$ with $\mathrm{Tor}_p^{A\otimes B}(\Bbbk, M\otimes N)$ has a nonzero component of bidegree (γ,γ). Using the formula $\mathrm{Tor}_*^{A\otimes B}(\Bbbk, M\otimes N) = \mathrm{Tor}_*^{A}(\Bbbk, M)\otimes \mathrm{Tor}_*^{B}(\Bbbk, N)$ we conclude that

$$\gamma \leqslant \max_{0\leqslant t\leqslant p} \mathrm{hg}(A,M)(t) + a(q) \leqslant c_1 t + c_2 + c_1 q \leqslant c_1 n + c_2.$$

□

4. Internal cohomomorphism

After defining the operation $\operatorname{cohom}(A, B)$ (due to Yu. I. Manin [**77, 79**]) and establishing its relation to the Segre product, we proceed to study the cohomology of $\operatorname{cohom}(A, B)$. We show that in the case when B is Koszul, passing to cohomology switches the operations $\operatorname{cohom}(A, B)$ and $A \circ B$ (see Theorem 4.6).

Definition 1. For any integer $n \geqslant 0$ we define the category Θ_n as follows. An object of Θ_n is a sequence of integers $\theta = (\theta_1, \dots, \theta_{l(\theta)})$ with $l(\theta) \geqslant 0$, $\theta_i \geqslant 1$, and $\sum_{i=1}^{l(\theta)} \theta_i = n$. Morphisms have the form

$$\theta = (\theta_1, \dots, \theta_{l(\theta)}) \longrightarrow \theta' = (\theta_1 + \dots + \theta_{i_1}, \dots, \theta_{i_{l(\theta')-1}+1} + \dots + \theta_{l(\theta)}).$$

There is at most one morphism between any two objects. For any $n, m \geqslant 0$ there is a canonical functor $\rho_{n,m} \colon \Theta_n \times \Theta_m \longrightarrow \Theta_{n+m}$ defined by

$$(\theta'_1, \dots, \theta'_{l(\theta')}) \times (\theta''_1, \dots, \theta''_{l(\theta'')}) \longmapsto (\theta'_1, \dots, \theta'_{l(\theta')}, \theta''_1, \dots, \theta''_{l(\theta'')}).$$

Also, for any $n \in \mathbb{Z}$ we define the category $\overline{\Theta}_n$ with objects $\theta = (\theta_1, \dots, \theta_{l(\theta)})$, where $l(\theta) \geqslant 1$, $\theta_i \geqslant 1$ for $i < l(\theta)$, $\theta_{l(\theta)} \in \mathbb{Z}$, and $\sum_{i=1}^{l(\theta)} \theta_i = n$. Morphisms are defined in the same way as above. Note that for any $n \geqslant 0$ and $m \in \mathbb{Z}$ we have a natural functor $\bar\rho_{n,m} \colon \Theta_n \times \overline{\Theta}_m \longrightarrow \overline{\Theta}_{n+m}$.

For any small category[1] Σ let us consider the category of covariant functors from Σ to the category $\mathcal{V}ect$ of finite-dimensional vector spaces over $\Bbbk$. For a pair of such functors $\Phi, \Psi \colon \Sigma \longrightarrow \mathcal{V}ect$ the set $\operatorname{Hom}^{\Sigma}(\Phi, \Psi)$ of morphisms of functors $\Phi \to \Psi$ has a natural structure of a vector space over $\Bbbk$ (infinite-dimensional, in general). We have a natural isomorphism

$$\operatorname{Hom}^{\Sigma}(\Phi, \Psi) \simeq \operatorname{Cohom}^{\Sigma}(\Phi, \Psi)^{\vee},$$

where

$$\operatorname{Cohom}^{\Sigma}(\Phi, \Psi) = \operatorname{coker}(\bigoplus\nolimits_{\sigma_1 \longrightarrow \sigma_2} \Phi(\sigma_1) \otimes \Psi(\sigma_2)^* \to \bigoplus\nolimits_{\sigma \in \Sigma} \Phi(\sigma) \otimes \Psi(\sigma)^*).$$

A covariant functor $\pi \colon \Sigma' \longrightarrow \Sigma''$ induces natural linear maps

$$\pi_* \colon \operatorname{Cohom}^{\Sigma'}(\pi \circ \Phi, \pi \circ \Psi) \longrightarrow \operatorname{Cohom}^{\Sigma''}(\Phi, \Psi)$$

for any pair of functors $\Phi, \Psi \colon \Sigma'' \longrightarrow \mathcal{V}ect$.

For a pair of functors as above $\Phi' \colon \Sigma' \longrightarrow \mathcal{V}ect$ and $\Phi'' \colon \Sigma'' \longrightarrow \mathcal{V}ect$, we can define the functor $\Phi' \otimes \Phi'' \colon \Sigma' \otimes \Sigma'' \longrightarrow \mathcal{V}ect$ sending $\sigma' \times \sigma''$ to $\Phi'(\sigma') \otimes \Phi''(\sigma'')$. It is easy to check that for a quadruple of functors $\Phi', \Psi' \colon \Sigma' \longrightarrow \mathcal{V}ect$ and $\Phi'', \Psi'' \colon \Sigma'' \longrightarrow \mathcal{V}ect$ there is a natural isomorphism

$$\operatorname{Cohom}^{\Sigma'}(\Phi', \Psi') \otimes \operatorname{Cohom}^{\Sigma''}(\Phi'', \Psi'') \simeq \operatorname{Cohom}^{\Sigma' \times \Sigma''}(\Phi' \otimes \Phi'', \Psi' \otimes \Psi'').$$

Definition 2. For a graded algebra A we define a sequence of functors

$$\Phi_A^{(n)} \colon \Theta_n \longrightarrow \mathcal{V}ect$$

by setting $\Phi_A^{(n)}(\theta_1, \dots, \theta_{l(\theta)}) = A_{\theta_1} \otimes \dots \otimes A_{\theta_{l(\theta)}}$ where the action of $\Phi^{(n)}$ on morphisms is defined using the multiplication in A. Similarly, for a graded A-module M we have a sequence of functors

$$\Phi_{A,M}^{(n)} \colon \overline{\Theta}_n \longrightarrow \mathcal{V}ect$$

[1] i.e., a category in which objects form a set.

such that $\Phi^{(n)}_{A,M}(\theta_1,\dots,\theta_{l(\theta)}) = A_{\theta_1}\otimes\cdots\otimes A_{\theta_{l(\theta)-1}}\otimes M_{\theta_{l(\theta)}}$.

Clearly, there are natural functor isomorphisms

$$\Phi^{(n+m)}_A\circ\rho_{n,m} \simeq \Phi^{(n)}_A\otimes\Phi^{(m)}_A$$
$$\Phi^{(n+m)}_{A,M}\circ\bar\rho_{n,m} \simeq \Phi^{(n)}_A\otimes\Phi^{(m)}_{A,M}.$$

Now we are ready to define the cohomomorphism operations on graded algebras and modules.

Definition 3. For a pair of graded algebras A and B the *internal cohomomorphism* (or *universal coacting*) algebra $\operatorname{cohom}(A,B)$ is defined by

$$\operatorname{cohom}(A,B)_n = \operatorname{Cohom}^{\Theta_n}(\Phi^{(n)}_A,\Phi^{(n)}_B),\qquad n\geqslant 0.$$

The multiplication is given by the composition

$$\begin{aligned}\operatorname{Cohom}^{\Theta_n}&(\Phi^{(n)}_A,\Phi^{(n)}_B)\otimes\operatorname{Cohom}^{\Theta_m}(\Phi^{(m)}_A,\Phi^{(m)}_B)\\ &\simeq \operatorname{Cohom}^{\Theta_n\times\Theta_m}(\Phi^{(n)}_A\otimes\Phi^{(m)}_A,\ \Phi^{(n)}_B\otimes\Phi^{(m)}_B)\\ &\simeq \operatorname{Cohom}^{\Theta_n\times\Theta_m}(\Phi^{(n+m)}_A\circ\rho_{n,m},\ \Phi^{(n+m)}_B\circ\rho_{n,m})\\ &\xrightarrow{(\rho_{n,m})_*}\operatorname{Cohom}^{\Theta_{n+m}}(\Phi^{(n+m)}_A,\Phi^{(n+m)}_B).\end{aligned}$$

Similarly, for a graded A-module M and a B-module N, one can define the *internal cohomomorphism module* $\operatorname{cohom}^{A,B}(M,N)$ over $\operatorname{cohom}(A,B)$ such that

$$\operatorname{cohom}_{A,B}(M,N)_n = \operatorname{Cohom}^{\overline\Theta_n}(\Phi^{(n)}_{A,M},\Phi^{(n)}_{B,M}),\qquad n\in\mathbb{Z}.$$

Note that $\operatorname{cohom}(A,B)_0=\Bbbk$ and $\operatorname{cohom}(A,B)_1 = A_1\otimes B_1^*$. If $M_i=N_i=0$ for $i<0$ then $\operatorname{cohom}_{A,B}(M,N)_0 = M_0\otimes N_0^*$. Since the category Θ_n has a finite number of objects, the vector spaces $\operatorname{cohom}(A,B)_n$ are always finite-dimensional. The same is true for the vector spaces $\operatorname{cohom}(M,N)_n$, because the functors $\Phi^{(n)}_{A,M}$ have nonzero values on a finite number of objects of $\overline\Theta_n$.

PROPOSITION 4.1. *(i) For any triple of graded algebras A, B, C one has a natural bijection between the sets of graded algebra homomorphisms*

(4.1) $$\operatorname{Hom}(A,\ B\circ C)\simeq\operatorname{Hom}(\operatorname{cohom}(A,B),\ C).$$

In particular, for a pair of graded algebras A and B there is a canonical morphism

(4.2) $$A\longrightarrow B\circ\operatorname{cohom}(A,B)$$

corresponding to the identity endomorphism of $\operatorname{cohom}(A,B)$.
(ii)(M) For any triple of graded modules M, N, and Q over the algebras A, B, and $\operatorname{cohom}(A,B)$, *respectively, one has a natural isomorphism*

$$\operatorname{Hom}^0_A(M,\ N\circ Q)\simeq\operatorname{Hom}^0_{\operatorname{cohom}(A,B)}(\operatorname{cohom}_{A,B}(M,N),\ Q),$$

where $N\circ Q$ is equipped with an A-module structure via the homomorphism (4.2). *In particular, for any graded modules M and N over A and B there is a natural morphism of graded A-modules*

$$M\longrightarrow N\circ\operatorname{cohom}_{A,B}(M,N)$$

corresponding to the identity endomorphism of $\operatorname{cohom}_{A,B}(M,N)$.

Proof: A morphism of graded algebras $f\colon A \longrightarrow B \circ C$ is given by a sequence of linear maps $A_n \longrightarrow B_n \otimes C_n$, or equivalently, linear maps

$$f_n\colon C_n^* \longrightarrow \mathrm{Hom}(A_n, B_n).$$

On the other hand, a morphism $g\colon \mathrm{cohom}(A,B) \longrightarrow C$ is presented by a collection of maps $\mathrm{Cohom}^{\Theta_m}(\Phi_A^{(m)}, \Phi_B^{(m)}) \longrightarrow C_m$, which is the same as

$$g_m\colon C_m^* \longrightarrow \mathrm{Hom}^{\Theta_m}(\Phi_A^{(m)}, \Phi_B^{(m)}).$$

Now given a sequence (g_m) we define the corresponding maps f_n as compositions of g_n with the natural maps

$$\mathrm{Hom}^{\Theta_n}(\Phi_A^{(n)}, \Phi_B^{(n)}) \longrightarrow \mathrm{Hom}(A_n, B_n)$$

sending a morphism of functors to its value on the object $\theta_0 = (n) \in \Theta_n$, $l(\theta_0) = 1$. Conversely, for a sequence (f_n) one can recover (g_m) by looking at compositions

$$C_m^* \longrightarrow C_{\theta_1}^* \otimes \cdots \otimes C_{\theta_{l(\theta)}}^* \xrightarrow{f_{\theta_1}\otimes\ldots\otimes f_{\theta_{l(\theta)}}} \mathrm{Hom}(A_{\theta_1} \otimes \cdots \otimes A_{\theta_{l(\theta)}}, B_{\theta_1} \otimes \cdots \otimes B_{\theta_{l(\theta)}}).$$

It is straightforward to check that the above correspondence leads to the isomorphism (4.1). The proof for the case of modules is completely analogous. □

Remark. Dualizing the n-th component of the homomorphism (4.2) we get a linear map

$$A \circ B^* \to \mathrm{cohom}(A,B) \tag{4.3}$$

of graded vector spaces, where $A \circ B^* = \bigoplus_n A_n \otimes B_n^*$. Below we will show that $\mathrm{cohom}(A,B)$ is generated as an algebra by the image of this morphism.

COROLLARY 4.2. *(i) For any triple of graded algebras A, B, and C there are natural (iso)morphisms of graded algebras*

$$\mathrm{cohom}(\mathrm{cohom}(A,B),\, C) \simeq \mathrm{cohom}(A,\, B \circ C) \tag{4.4}$$

$$\mathrm{cohom}(A,C) \longrightarrow \mathrm{cohom}(B,C) \circ \mathrm{cohom}(A,B) \tag{4.5}$$

$$\mathrm{cohom}(A,\, \mathrm{cohom}(B,C)) \longrightarrow \mathrm{cohom}(A,B) \circ C. \tag{4.6}$$

(ii)(M) For any triple of graded modules M, N, and P over A, B, and C there are natural (iso)morphisms of graded modules

$$\mathrm{cohom}_{\mathrm{cohom}(A,B),C}(\mathrm{cohom}_{A,B}(M,N),\, P) \simeq \mathrm{cohom}_{A,\,B\circ C}(M,\, N \circ P)$$
$$\mathrm{cohom}_{A,C}(M,P) \longrightarrow \mathrm{cohom}_{B,C}(N,P) \circ \mathrm{cohom}_{A,B}(M,N)$$
$$\mathrm{cohom}_{A,\,\mathrm{cohom}(B,C)}(M,\, \mathrm{cohom}(N,P)) \longrightarrow \mathrm{cohom}_{A,B}(M,N) \circ P$$

compatible with the above homomorphisms of algebras.

Proof: Let X by either the left-hand side or the right-hand side of (4.4). Then from the universal property given in Proposition 4.1 we get a natural isomorphism $\mathrm{Hom}(X,D) = \mathrm{Hom}(A,\, B \circ C \circ D)$. This immediately implies (4.4). Both morphisms (4.5) and (4.6) correspond via the universal property to the composition

$$A \longrightarrow B \circ \mathrm{cohom}(A,B) \longrightarrow C \circ \mathrm{cohom}(B,C) \circ \mathrm{cohom}(A,B).$$

The case of modules is similar. □

In Proposition 4.3 below we will give an alternative description of the internal cohomomorphism algebras and modules in terms of generators and relations (see

[**77, 79**]). In the case of quadratic algebras and modules this description becomes especially simple and is related to the following construction.

Definition 4. The operation *black circle product* $A \bullet B := (A^! \circ B^!)^!$ (see [**77**]), quadratic dual to the Segre product of quadratic algebras, can also be defined in terms of generators and quadratic relations as follows:

$$\{V', I'\} \bullet \{V'', I''\} = \{V' \otimes V'', \, I' \otimes I''\}.$$

Note that this definition makes sense for quadratic algebras only. There is an analogous operation for quadratic modules: $M \bullet N = (M^!_A \circ N^!_B)^!_{A^! \circ B^!}$, or explicitly,

$$\langle H', K' \rangle_A \bullet \langle H'', K'' \rangle_B = \langle H' \otimes H'', \, K' \otimes K'' \rangle_{A \bullet B}.$$

For a pair of graded vector spaces Z and T let us denote $Z \circ T = \bigoplus_n Z_n \otimes T_n$.

PROPOSITION 4.3. *(i) For a quadratic algebra A and an arbitrary graded algebra B there is a natural isomorphism*

$$\operatorname{cohom}(A, B) \simeq A \bullet (\mathrm{q}B)^!.$$

More generally, let A be a graded algebra generated by a graded vector subspace $X \subset A$ with the ideal of relations generated by a graded vector subspace $Y \subset \mathbb{T}(X)$. Then for any graded algebra B one has

$$\operatorname{cohom}(A, B) \simeq \operatorname{coker}(Y \circ B^* \to \mathbb{T}(X \circ B^*)),$$

where the map $Y \circ B^* \to \mathbb{T}(X \circ B^*)$ has components

$$Y_n \circ B_n^* \longrightarrow \bigoplus_{n_1 + \cdots + n_s = n} X_{n_1} \otimes B^*_{n_1} \otimes \cdots \otimes X_{n_s} \otimes B^*_{n_s}$$

given by the tensor products of the inclusion maps $Y_n \longrightarrow \mathbb{T}^n(X)$ and the comultiplication maps $B_n^* \longrightarrow \mathbb{T}^n(B)^*$.
(ii)(M) For a quadratic module M over a quadratic algebra A and a graded module N over a graded algebra B there is a natural isomorphism of $\operatorname{cohom}(A, B)$*-modules*

$$\operatorname{cohom}_{A,B}(M, N) \simeq M \bullet (\mathrm{q}N)^!_{\mathrm{q}B}.$$

More generally, let M be a graded A-module generated by a graded subspace $X \subset M$ with the submodule of relations generated by a graded subspace $Y \subset A \otimes X$. Then for any graded B-module N one has

$$\operatorname{cohom}_{A,B}(M, N) \simeq \operatorname{coker}(Y \circ N^* \to \operatorname{cohom}(A, B) \otimes (X \circ N^*)),$$

where the map $Y \circ N^* \to \operatorname{cohom}(A, B) \otimes (X \circ N^*)$ is given by the composition

$$Y \circ N^* \xrightarrow{\ f\ } (A \circ B^*) \otimes (X \circ N^*) \xrightarrow{\ g\ } \operatorname{cohom}(A, B) \otimes (X \circ N^*),$$

where f is induced by the embedding $Y \longrightarrow A \otimes X$ and the coaction $B^* \longrightarrow B^* \otimes N^*$, while g comes from (4.3).

Proof: It suffices to check that the quotient algebra (resp., module) satisfies the universal property from Proposition 4.1. There is a bijective correspondence between homomorphisms of graded algebras $\mathbb{T}(X \circ B^*) \longrightarrow C$ and maps of graded vector spaces $X \longrightarrow B \circ C$. It is easy to see that a homomorphism $\mathbb{T}(X \circ B^*) \longrightarrow C$ kills the image of $Y \circ B^*$ iff the corresponding map of vector spaces can be extended to an algebra homomorphism $A \longrightarrow B \circ C$. The argument for modules is analogous. □

COROLLARY 4.4. *(i) For any graded algebras A and B there are natural maps of the* Ext*-spaces*

$$\mathrm{Ext}^{ij}_{\mathrm{cohom}\,(A,B)}(\Bbbk,\Bbbk) \longrightarrow \mathrm{Ext}^{ij}_A(\Bbbk,\Bbbk)\otimes B_j \tag{4.7}$$

compatible with products. The map (4.7) *is an isomorphism if $i=1$ and a monomorphism if $i=2$. If an algebra B is one-generated then* (4.7) *is also an isomorphism for $i=2$.*
(ii)(M) For any pair consisting of a graded A-module M and a graded B-module N there are natural maps

$$\mathrm{Ext}^{ij}_{\mathrm{cohom}\,(A,B)}(\mathrm{cohom}_{A,B}(M,N),\,\Bbbk) \longrightarrow \mathrm{Ext}^{ij}_A(M,\Bbbk)\otimes N_j \tag{4.8}$$

compatible with products by elements of the spaces appearing in (4.7)*. These maps are isomorphisms for $i=0$ and monomorphisms for $i=1$.*

Proof: For any graded algebras B and C and graded modules N and P over B and C, respectively, there are natural morphisms of bar-complexes

$$\mathcal{B}ar_\bullet(B\circ C) \longrightarrow B\circ \mathcal{B}ar_\bullet(C)$$
$$\mathcal{B}ar_\bullet(B\circ C,\,N\circ P) \longrightarrow N\circ \mathcal{B}ar_\bullet(C,P),$$

where we use the internal grading on the bar-complexes when forming Segre products, so that $B\circ\mathcal{B}ar_\bullet(C)=\bigoplus_{i,j}B_j\otimes\mathcal{B}ar_{ij}(C)$. The above morphisms are defined using the natural multiplication maps $B_+^{\otimes i}\longrightarrow B$ and $B_+^{\otimes i}\otimes N\longrightarrow N$. The desired maps between Ext-spaces are induced by the morphisms of cobar-complexes dual to the compositions

$$\mathcal{B}ar_\bullet(A) \longrightarrow \mathcal{B}ar_\bullet(B\circ \mathrm{cohom}(A,B)) \longrightarrow B\circ\mathcal{B}ar_\bullet(\mathrm{cohom}(A,B)),$$

$$\begin{aligned}\mathcal{B}ar_\bullet(A,M) &\longrightarrow \mathcal{B}ar_\bullet(B\circ\mathrm{cohom}(A,B),\,N\circ\mathrm{cohom}_{A,B}(M,N))\\ &\longrightarrow N\circ\mathcal{B}ar_\bullet(\mathrm{cohom}(A,B),\,\mathrm{cohom}_{A,B}(M,N)).\end{aligned}$$

It is easy to check compatibility of the obtained maps with products.

To check the statements on the behavior of our maps for $i=1$ and $i=2$ we will use Proposition 4.3. Let us choose a minimal generator space $X\subset A$ and let $Y\subset\mathbb{T}(X)$ be a minimal relation space, so that $X\simeq\mathrm{Tor}_1^A(\Bbbk,\Bbbk)$ and $Y\simeq\mathrm{Tor}_2^A(\Bbbk,\Bbbk)$ (see Proposition 5.2 of chapter 1). Then the composition

$$Y\circ B^* \longrightarrow \mathbb{T}(X\circ B^*)_+/\mathbb{T}(X\circ B^*)_+^2 = X\circ B^*$$

is zero, hence $X\circ B^*\simeq\mathrm{Tor}_1^{\mathrm{cohom}\,(A,B)}(\Bbbk,\Bbbk)$. It is also easy to see that the subspace $\mathrm{im}\,Y\circ B^*\subset\mathbb{T}(X\circ B^*)$ is a minimal generating subspace for the two-sided ideal it generates, hence $\mathrm{im}\,Y\circ B^*\simeq\mathrm{Tor}_2^{\mathrm{cohom}\,(A,B)}(\Bbbk,\Bbbk)$. Note that if B is one-generated then the map $Y\circ B^*\longrightarrow\mathbb{T}(X\circ B^*)$, so in this case $Y\circ B^*\simeq\mathrm{Tor}_2^{\mathrm{cohom}\,(A,B)}(\Bbbk,\Bbbk)$. One can check that these identifications are compatible with the maps of Ext spaces defined above. The proof of the assertion for modules is similar. □

Next, we are going to compute the cohomology of the algebras $A\circ B$ and $\mathrm{cohom}(A,B)$ (resp., modules $M\circ N$ and $\mathrm{cohom}_{A,B}(M,N)$) assuming that *the algebra B (resp., the module N) is Koszul.*

We start with an auxiliary lemma. Let $\mathcal{V}ect^\bullet$ denote the category of bounded complexes of finite-dimensional vector spaces. For any functor $\Phi^\bullet\colon\Sigma\longrightarrow\mathcal{V}ect^\bullet$ let $H(\Phi^\bullet)$ denote the corresponding cohomology functor (taking values in the category of graded vector spaces). The construction of the spaces $\mathrm{Cohom}^\Sigma(\Phi,\Psi)$ can be extended to the case of functors to $\mathcal{V}ect^\bullet$ in the obvious way. Below for a set S we denote by $\Bbbk\langle S\rangle$ the vector space over $\Bbbk$ with the basis S.

LEMMA 4.5. *Let Σ be a finite partially ordered set considered as a category with morphisms $\sigma_1 \longrightarrow \sigma_2$ corresponding to ordered pairs $\sigma_1 \geqslant \sigma_2$. Then for a functor $\Psi\colon \Sigma \longrightarrow \mathcal{V}ect$ the following conditions are equivalent:*

(a) *For any functor $\Phi^\bullet\colon \Sigma \longrightarrow \mathcal{V}ect^\bullet$ there is a natural isomorphism of graded vector spaces*

$$H(\mathrm{Cohom}^{\Sigma}(\Phi^\bullet, \Psi)) \simeq \mathrm{Cohom}^{\Sigma}(H(\Phi^\bullet), \Psi).$$

(b) *The following three conditions hold:*
 (i) *all the maps $\Psi(\sigma_1 \longrightarrow \sigma_2)\colon \Psi(\sigma_1) \longrightarrow \Psi(\sigma_2)$ are surjective;*
 (ii) *for $\sigma_0 \geqslant \sigma_1$, $\sigma_0 \geqslant \sigma_2$ one has*

$$\ker\Psi(\sigma_0 \longrightarrow \sigma_1) + \ker\Psi(\sigma_0 \longrightarrow \sigma_2) \;=\; \bigcap\nolimits_{\sigma_1,\,\sigma_2 \geqslant \sigma_3} \ker\Psi(\sigma_0 \longrightarrow \sigma_3);$$

 (iii) *for any object $\sigma_0 \in \Sigma$ the collection of subspaces*

$$(\ker\Psi(\sigma_0 \longrightarrow \sigma_1) \subset \Psi(\sigma_0), \qquad \sigma_0 \geqslant \sigma_1 \in \Sigma)$$

 in $\Psi(\sigma_0)$ is distributive.

(c) *The functor Ψ is isomorphic to a direct sum of functors of the form $\Psi^{\sigma_0}(\sigma) = \Bbbk\langle \mathrm{Mor}_\Sigma(\sigma, \sigma_0)\rangle^*$ (we call these functors "corepresentable").*

Proof: It is more convenient to consider the dual situation using the functor $\Psi^* : \Sigma^{\mathrm{op}} \longrightarrow \mathcal{V}ect : \sigma \longmapsto \Psi(\sigma)^*$. We have to check the equivalence of the following conditons for Ψ^*:

(a) For any functor $\Phi^\bullet\colon \Sigma^{\mathrm{op}} \longrightarrow \mathcal{V}ect^\bullet$ there is a natural isomorphism of graded vector spaces

$$H(\mathrm{Hom}^{\Sigma^{\mathrm{op}}}(\Psi^*, \Phi^\bullet)) \simeq \mathrm{Hom}^{\Sigma^{\mathrm{op}}}(\Psi^*, H(\Phi^\bullet)).$$

(b) The following three conditions hold:
 (i) all the maps $\Psi^*(\sigma_1 \longrightarrow \sigma_2)\colon \Psi^*(\sigma_1) \longrightarrow \Psi^*(\sigma_2)$ are injective;
 (ii) for $\sigma_0 \geqslant \sigma_1$, $\sigma_0 \geqslant \sigma_2$ one has

$$\mathrm{im}\,\Psi^*(\sigma_1) \cap \mathrm{im}\,\Psi^*(\sigma_2) \;=\; \sum\nolimits_{\sigma_3 \leqslant \sigma_1,\,\sigma_2} \mathrm{im}\,\Psi^*(\sigma_3) \;\subset\; \Psi^*(\sigma_0);$$

 (iii) for any object σ_0 the collection of subspaces

$$(\mathrm{im}\,\Psi^*(\sigma_1 \longrightarrow \sigma_0) \subset \Psi^*(\sigma_0), \qquad \sigma_0 \geqslant \sigma_1)$$

 in $\Psi^*(\sigma_0)$ is distributive .

(c) The functor Ψ^* is isomorphic to a direct sum of representable functors, i.e., functors of the form $\Psi^*_{\sigma_0}(\sigma) = \Bbbk\langle \mathrm{Mor}_{\Sigma^{\mathrm{op}}}(\sigma_0, \sigma)\rangle$.

(c) $\Longrightarrow$ (a). This follows immediately from the natural isomorphism

$$\mathrm{Hom}^{\Sigma^{\mathrm{op}}}(\Psi^*_{\sigma_0}, \Phi^\bullet) \simeq \Phi^\bullet(\sigma_0).$$

(a) $\Longrightarrow$ (b). The condition (a) means that the functor Ψ^* is a projective object in the abelian category $\mathrm{Funct}(\Sigma^{\mathrm{op}}, \mathcal{V}ect)$ of functors $\Sigma^{\mathrm{op}} \longrightarrow \mathcal{V}ect$. It is easy to see that $\mathrm{Funct}(\Sigma^{\mathrm{op}}, \mathcal{V}ect)$ is equivalent to the category of modules over a certain ring and that the free module with one generator over this ring corresponds to the functor $\bigoplus_{\sigma_0 \in \Sigma} \Psi^*_{\sigma_0}$. Since a projective module is a direct summand of a free one, the validity of condition (b) for Ψ^* follows immediately from its validity for the functors $\Psi^*_{\sigma_0}$.

(b) $\Longrightarrow$ (c). This is a generalization of Proposition 7.1 of chapter 1, (a) $\Longrightarrow$ (b), and the proof is analogous. For simplicity of notation we identify $\Psi^*(s_1)$ with its image under the embedding $\Psi^*(\sigma_1) \longrightarrow \Psi^*(\sigma_2)$ for $\sigma_1 \leqslant \sigma_2$. For every $\sigma_0 \in \Sigma$ let us choose a direct complement $V_{\sigma_0} \subset \Psi^*(\sigma_0)$ to the subspace $\sum_{\sigma_1<\sigma_0} \Psi^*(\sigma_1) \subset \Psi^*(\sigma_0)$. We have a natural morphism of functors

$$\bigoplus_{\sigma_0} V_{\sigma_0} \otimes \Psi^*_{\sigma_0} \longrightarrow \Psi^*.$$

To prove that it is an isomorphism we have to check that the map $\bigoplus_{\sigma<\sigma_0} V_\sigma \longrightarrow \Psi^*(\sigma_0)$ is an isomorphism for any $\sigma_0 \in \Sigma$. It is easy to prove by increasing induction in σ_0 that this map is surjective. It remains to prove its injectivity. Assume that we have $\sum_{i=1}^k v_i = 0$ in $\Psi^*(\sigma_0)$ for a collection $(\sigma_1, \dots, \sigma_k)$, where $\sigma_i < \sigma_0$ for all i, and some nonzero vectors $v_i \in V_{\sigma_i}$. Without loss of generality we can assume that there is no $i \geqslant 2$ with $\sigma_1 \leqslant \sigma_i$. Then the inclusion $v_1 \in \Psi^*(\sigma_1) \cap \sum_{i\geqslant 2} \Psi^*(\sigma_i) \subset \Psi^*(\sigma_0)$ together with conditions (ii) and (iii) imply that $v_1 \in \sum_{\sigma<\sigma_1} \Psi^*(\sigma) \subset \Psi^*(\sigma_0)$, which is a contradiction. □

THEOREM 4.6. *(i) For a graded algebra A and a Koszul algebra B one has natural isomorphisms of bigraded algebras*

$$\mathrm{Ext}_{A\circ B}(\Bbbk,\Bbbk) \simeq \mathrm{cohom}(\mathrm{Ext}_A(\Bbbk,\Bbbk),\, B),$$
$$\mathrm{Ext}_{\mathrm{cohom}(A,B)}(\Bbbk,\Bbbk) \simeq \mathrm{Ext}_A(\Bbbk,\Bbbk)\circ B,$$

where the operations in the right-hand side use internal grading on $\mathrm{Ext}_A(\Bbbk,\Bbbk)$.
(ii)(M) Let A and B be as in (i). For a graded A-module M and a Koszul B-module N one has natural isomorphisms of bigraded modules

$$\mathrm{Ext}_{A\circ B}(M\circ N,\, \Bbbk) \simeq \mathrm{cohom}_{\mathrm{Ext}_A(\Bbbk,\Bbbk),\,B}(\mathrm{Ext}_A(M,\Bbbk),\, N),$$
$$\mathrm{Ext}_{\mathrm{cohom}(A,B)}(\mathrm{cohom}_{A,B}(M,N),\, \Bbbk) \simeq \mathrm{Ext}_A(M,\Bbbk)\circ N.$$

Proof: First, let us prove the formulas for $\mathrm{Ext}_{A\circ B}(\Bbbk,\Bbbk)$ and $\mathrm{Ext}_{A\circ B}(M\circ N,\, \Bbbk)$. We claim that for any graded algebras A and B and modules M and N there are natural isomorphisms of differential algebras and modules

$$\mathcal{C}ob^\bullet(A\circ B) \simeq \mathrm{cohom}(\mathcal{C}ob^\bullet(A),\, B),$$
$$\mathcal{C}ob^\bullet(A\circ B,\, M\circ N) \simeq \mathrm{cohom}_{\mathcal{C}ob^\bullet(A),\,B}(\mathcal{C}ob^\bullet(A,M),\, N),$$

where we extend the operation cohom to DG algebras and DG-modules in the obvious way. Indeed, let us ignore the differentials for a moment and consider $\mathcal{C}ob(A)$ and $\mathcal{C}ob(A,M)$ as a graded algebra and a graded module over it using the internal grading. Then we can write $\mathcal{C}ob(A) = \mathbb{T}(A_+^*)$ and $\mathcal{C}ob(A,M) = \mathbb{T}(A_+^*)\otimes M^*$. Hence, from Proposition 4.3 we get

$$\mathrm{cohom}(\mathbb{T}(A_+^*),\, B) \simeq \mathbb{T}(A_+^*\circ B^*) \simeq \mathbb{T}((A\circ B)_+^*) \text{ and}$$

$$\begin{aligned}\mathrm{cohom}_{\mathbb{T}(A_+^*),\,B}(\mathbb{T}(A_+^*)\otimes M^*,\, N) &\simeq \mathrm{cohom}(\mathbb{T}(A_+^*),\, B)\otimes(M^*\circ N^*)\\ &\simeq \mathbb{T}((A\circ B)_+^*)\otimes(M\otimes N)^*.\end{aligned}$$

It is not difficult to check that these isomorphisms commute with the differentials.

Now if the algebra B and the B-module N are Koszul then the operations $\mathrm{cohom}(C^\bullet, B)$ and $\mathrm{cohom}_{C^\bullet,B}(P^\bullet, N)$ commute with passing to cohomology of a DG-algebra $C^\bullet$ and a DG-module $P^\bullet$. Indeed, this follows from Lemma 4.5, (b) $\Longrightarrow$ (a), applied to the functors $\Phi_B^{(n)}\colon \Theta_n \longrightarrow \mathcal{V}ect$ and $\Phi_{B,N}^{(n)}\colon \bar\Theta_n \longrightarrow \mathcal{V}ect$.

One just has to observe that condition (b.i) means that the algebra B (resp., B-module N) is one-generated (resp., generated in degree zero), (b.ii) means that B and N are quadratic, and (b.iii) is equivalent to Backelin's Koszulness condition for B and N.

It remains to prove that morphisms (4.7) and (4.8) constructed in Corollary 4.4 are isomorphisms provided B and N are Koszul. Since $\mathrm{cohom}_{A,B}(\Bbbk, B) = \Bbbk$, it suffices to do this for (4.8). Consider the bar-resolution $\widetilde{\mathcal{B}ar}_\bullet(A, M)$ of the A-module M. Similarly to the proof of Corollary 4.4 we have a morphism of complexes of A-modules

$$\widetilde{\mathcal{B}ar}_\bullet(A, M) \longrightarrow \widetilde{\mathcal{B}ar}_\bullet(B \circ \mathrm{cohom}(A, B),\ N \circ \mathrm{cohom}_{A,B}(M, N))$$
$$\longrightarrow N \circ \widetilde{\mathcal{B}ar}_\bullet(\mathrm{cohom}(A, B),\ \mathrm{cohom}_{A,B}(M, N)).$$

By Proposition 4.1 it induces a morphism of complexes of $\mathrm{cohom}(A, B)$-modules

$$\mathrm{cohom}_{A,B}(\widetilde{\mathcal{B}ar}_\bullet(A, M),\ N) \longrightarrow \widetilde{\mathcal{B}ar}_\bullet(\mathrm{cohom}(A, B),\ \mathrm{cohom}_{A,B}(M, N)). \tag{4.9}$$

As we have seen above for Koszul B and N the operation $P^\bullet \longmapsto \mathrm{cohom}_{A,B}(P^\bullet, N)$ commutes with passing to cohomology. Hence, the complex

$$\mathrm{cohom}_{A,B}(\widetilde{\mathcal{B}ar}_\bullet(A, M),\ N)$$

is a $\mathrm{cohom}(A, B)$-module resolution of $\mathrm{cohom}_{A,B}(M, N)$. Furthermore, since $\widetilde{\mathcal{B}ar}(A, M) \simeq A \otimes \mathcal{B}ar(A, M)$ as a bigraded A-module, by Proposition 4.3 we get an isomorphism of bigraded $\mathrm{cohom}(A, B)$-modules

$$\mathrm{cohom}_{A,B}(\widetilde{\mathcal{B}ar}(A, M),\ N) \simeq \mathrm{cohom}(A, B) \otimes (\mathcal{B}ar(A, M) \circ N^*).$$

Therefore, (4.9) is a morphism of free resolutions of the module $\mathrm{cohom}_{A,B}(M, N)$, so it should be a homotopy equivalence. Applying the functor $\mathrm{Hom}_{\mathrm{cohom}(A,B)}(-, \Bbbk)$ to (4.9) we get the morphism of cobar-complexes from the proof of Corollary 4.4. Hence, the induced map on cohomology is an isomorphism. □

COROLLARY 4.7. *(i) For any graded algebra A and a Koszul algebra B the Hilbert series of the algebra* $\mathrm{cohom}(A, B)$ *is given by the formula*

$$h_{\mathrm{cohom}(A,B)}(z) = (h_A^{-1} \circ h_B)(z)^{-1},$$

where we denote $(f \circ g)(z) = \sum_i f_i g_i z^i$ *for* $f(z) = \sum_i f_i z^i$ *and* $g(z) = \sum_i g_i z^i$. *If in addition A is quadratic then*

$$h_{A \bullet B}(z) = (h_A^{-1} \circ h_{B^!})(z)^{-1} = (h_A^{-1} \circ h_B^{-1})(-z)^{-1}. \tag{4.10}$$

(ii)(M) Let A and B be as in (i). For any graded A-module M and a Koszul B-module N one has

$$h_{\mathrm{cohom}_{A,B}(M,N)}(z) = (h_A^{-1} \circ h_B)(z)^{-1} \cdot (h_A^{-1} h_M \circ h_N)(z).$$

If in addition A and M are quadratic then

$$h_{M \bullet N}(z) = (h_A^{-1} \circ h_B^{-1})(-z)^{-1} \cdot (h_A^{-1} h_M \circ h_B^{-1} h_N)(-z).$$

Proof: Recall that we denote $P_C(u, z) = \sum_{i,j} u^i z^j \dim \mathrm{Ext}_C^{ij}(\Bbbk, \Bbbk)$. By Theorem 4.6 we have $P_{\mathrm{cohom}(A,B)}(u, z) = P_A(u, z) \circ h_B(z)$. Hence, using Proposition 2.1 of chapter 2 we get

$$h_{\mathrm{cohom}(A,B)}(z) = P_{\mathrm{cohom}(A,B)}(-1, z)^{-1} = (P_A(-1, q) \circ h_B)(z)^{-1} = (h_A^{-1} \circ h_B)(z)^{-1},$$

The formula for $h_{\mathrm{cohom}_{A,B}(M,N)}$ is derived similarly. To compute the Hilbert series of $A \bullet B$ and $M \bullet N$ we apply Proposition 4.3 and use the relation $(f \circ g)(-z) = f(z) \circ g(-z)$. □

Note that for a *fixed* Koszul algebra B and a Koszul B-module N not only the dimensions of the grading components of $\mathrm{cohom}(A,B)$ and $\mathrm{cohom}_{A,B}(M,N)$ are determined by the dimensions of the components of A and M, but the vector spaces $\mathrm{cohom}(A,B)_n$ and $\mathrm{cohom}_{A,B}(M,N)_n$ themselves can be recovered functorially from the vector spaces A_i and M_i. Indeed, this follows from the fact that for any functor $\Psi\colon \Sigma \longrightarrow \mathcal{V}ect$ satisfying the equivalent conditions of Lemma 4.5 and for any functor $\Phi : \Sigma \longrightarrow \mathcal{V}ect$ the vector space $\mathrm{Cohom}^{\Sigma}(\Phi,\Psi)$ is determined by the values of Φ on objects (since this is true for a corepresentable functor Ψ).

COROLLARY 4.8. *(i) Let us fix an integer $n \geqslant 0$ and a graded algebra B with $B_i = 0$ for all $i > n$. Then the algebra $\mathrm{cohom}(A,B)$ is determined by the quotient algebra $A_{\leqslant n} = A/A_{>n} = \bigoplus_{i=0}^n A_i$ of A. Furthermore, if the algebra B is Koszul then the cohomology of the algebras $A \circ B$ and $\mathrm{cohom}(A,B)$ can be recovered from the spaces $\mathrm{Ext}_A^{ij}(\Bbbk,\Bbbk)$ with $j \leqslant n$.*

(ii)(M) Let us fix an integer $n \geqslant 0$, a graded algebra B and a graded B-module N with $N_i = 0$ for all $i > n$. Then the module $\mathrm{cohom}_{A,B}(M,N)$ is determined by the quotient A-module $M_{\leqslant n} = M/M_{>n} = \bigoplus_{i \leqslant n} M_i$. If B and N are Koszul then the cohomology of the $A \circ B$-module $M \circ N$ can be recovered from the cohomology of the algebra A and the spaces $\mathrm{Ext}_A^{ij}(M,\Bbbk)$ with $j \leqslant n$. The cohomology of the $\mathrm{cohom}(A,B)$-module $\mathrm{cohom}_{A,B}(M,N)$ can be recovered from the spaces $\mathrm{Ext}_A^{ij}(M,\Bbbk)$ with $j \leqslant n$.

Proof: It is clear that the algebra $A \circ B$ and the module $M \circ N$ are determined by $A_{\leqslant n}$ and $M_{\leqslant n}$ under our assumptions. Proposition 4.1 immediately implies the same assertion for $\mathrm{cohom}(A,B)$ and $\mathrm{cohom}_{A,B}(M,N)$. The statements about cohomology follow from Theorem 4.6. □

COROLLARY 4.9. *(i) Let B be a Koszul algebra with $B_n \neq 0$ and $B_{n+1} = 0$ for some $n \geqslant 0$, and let A be a graded algebra. Then the algebra $A \circ B$ (resp., $\mathrm{cohom}(A,B)$) is Koszul iff A is n-Koszul.*

(ii)(M) Let B be a Koszul algebra, N a Koszul B-module with $N_n \neq 0$ and $N_{n+1} = 0$. Assume that a graded algebra A is n-Koszul and the algebras $A \circ B$ and $\mathrm{cohom}(A,B)$ are Koszul. Then a graded A-module N is n-Koszul iff the $A \circ B$-module $M \circ N$ (resp., $\mathrm{cohom}(A,B)$-module $\mathrm{cohom}_{A,B}(M,N)$) is Koszul.

Proof: (i) By Theorem 4.6 we have $\mathrm{Ext}_{\mathrm{cohom}(A,B)}(\Bbbk,\Bbbk) \simeq \mathrm{Ext}_A(\Bbbk,\Bbbk) \circ B$. Since $B_i \neq 0$ for $0 \leqslant i \leqslant n$ and $B_i = 0$ for $i > n$, this implies that $\mathrm{cohom}(A,B)$ is Koszul iff A is n-Koszul. To prove the similar result for $A \circ B$ we use the fact that the algebra $A \circ B$ is Koszul iff the algebra $\mathrm{Ext}_{A \circ B}(\Bbbk,\Bbbk)$ is one-generated with respect to the internal grading. By Theorem 4.6 we have $\mathrm{Ext}_{A\circ B}(\Bbbk,\Bbbk) \simeq \mathrm{cohom}(\mathrm{Ext}_A(\Bbbk,\Bbbk),B)$. To find generators of the latter algebra we use Theorem 4.6 again:

$$\mathrm{Ext}^{1j}_{\mathrm{cohom}(\mathrm{Ext}_A(\Bbbk,\Bbbk),B)} \simeq \mathrm{Ext}^{1j}_{\mathrm{Ext}_A(\Bbbk,\Bbbk)}(\Bbbk,\Bbbk) \otimes B_j.$$

It follows that $\mathrm{cohom}(\mathrm{Ext}_A(\Bbbk,k),B)$ is one-generated iff the algebra

$$\mathrm{Ext}_A(\Bbbk,\Bbbk)/(\bigoplus_{i\geqslant 1, j>n} \mathrm{Ext}_A^{ij}(\Bbbk,\Bbbk))$$

is one-generated with respect to the internal grading, which is equivalent to n-Koszulness of A. The proof of (ii) is analogous. □

For arbitrary quadratic algebras A and B the Hilbert series of $A \bullet B$ in general cannot be recovered from the Hilbert series of A and B (and therefore, the same is true about $\mathrm{cohom}(A, B)$). More precisely, we have the following result that was stated in [**101**].

PROPOSITION 4.10. *If quadratic algebras A and B are not 4-Koszul then $\dim(A \bullet B)_4$ is bigger than the dimension given by formula* (4.10).

Proof: First, let us explain another point of view on the formula expressing $h_{A \bullet B}$ in terms of h_A and h_B in the case when one of the algebras is Koszul. For an n-tuple of subspaces $(V_1, \ldots, V_n)$ in a vector space V consider the function on subsets $I \subset [1, n]$ given by

$$\mathcal{N}(V_1, \ldots, V_n)(I) = \dim V / \sum_{i \in I} V_i.$$

If $(W_1, \ldots, W_n)$ is another n-tuple of subspaces in a vector space W then one can consider the n-tuple of tensor products $(V_1 \otimes W_1, \ldots, V_n \otimes W_n)$ in $V \otimes W$. It is easy to see that if $(V_1, \ldots, V_n)$ is distributive then there exist universal formulas expressing values of $\mathcal{N}(V_1 \otimes W_1, \ldots, V_n \otimes W_n)$ as polynomials in values of $\mathcal{N}(V_1, \ldots, V_n)$ and $\mathcal{N}(W_1, \ldots, W_n)$. Indeed, this follows from the fact that every distributive n-tuple $(V_1, \ldots, V_n)$ is a direct sum of one-dimensional n-tuples and the multiplicities of each isomorphism type of one-dimensional n-tuples are universal linear functions of values of $\mathcal{N}(V_1, \ldots, V_n)$. Specializing to the case of n-tuples of the form $(R_1^{(n)}, \ldots, R_{n-1}^{(n)})$ in $V^{\otimes n}$, where $R \subset V^{\otimes 2}$ we obtain the formula for $\dim(A \bullet B)_n$ in the Koszul case. Thus, our statement is a consequence of the following claim: for every pair of non-distributive triples of subspaces (V_1, V_2, V_3) and (W_1, W_2, W_3) in V and W the dimension of

$$V \otimes W / (V_1 \otimes W_1 + V_2 \otimes W_2 + V_3 \otimes W_3)$$

is bigger than the value of the universal expression for this dimension in the distributive case, evaluated at the values of $\mathcal{N}(V_1, V_2, V_3)$ and $\mathcal{N}(W_1, W_2, W_3)$. Since every indecomposable triple of subspaces is either one-dimensional, or the triple of distinct lines in the plane (the only non-distributive indecomposable triple), it suffices to prove the above claim for one pair of non-distributive triples of subspaces. To this end let us take (V_1, V_2, V_3) to be a triple of distinct planes passing through a line in a three-dimensional space V. Then it is easy to check that

$$\dim V^{\otimes 2} / (V_1^{\otimes 2} + V_2^{\otimes 2} + V_3^{\otimes 2}) = 1$$

(since the same is true for a triple of distinct lines in a plane). On the other hand, consider a generic triple of planes (V_1^0, V_2^0, V_3^0) in V. Then it is distributive and

$$V^{\otimes 2} / ((V_1^0)^{\otimes 2} + (V_2^0)^{\otimes 2} + (V_3^0)^{\otimes 3}) = 0.$$

Since $\mathcal{N}(V_1, V_2, V_3) = \mathcal{N}(V_1^0, V_2^0, V_3^0)$ this proves our claim. □

5. Koszulness cannot be checked using Hilbert series

As we have seen in section 2 of chapter 2 for every Koszul algebra A one has $h_A(z) h_{A^!}(-z) = 1$. Examples of quadratic non-Koszul algebras satisfying this relation were constructed in [**101**] and [**106**]. Below we will present a theorem due to Piontkovskii [**92**] showing that in fact it is impossible to tell looking only at Hilbert series of A and $A^!$ whether A is Koszul or not.

Using identity (2.1) from chapter 2 one can easily see that for an $(n-1)$-Koszul quadratic algebra A one has

$$h_A(z)h_{A^!}(-z) = 1 + O(z^n).$$

The following lemma gives a way to construct $(n-1)$-Koszul but not n-Koszul algebras for which $O(z^n)$ can be replaced with $o(z^n)$ in the above identity.

LEMMA 5.1. *Let $n > 4$ be an odd integer, and let D be a quadratic algebra which is $(n-1)$-Koszul but not n-Koszul. Then the quadratic algebra $B = D \otimes D^!$ is $(n-1)$-Koszul but not n-Koszul and satisfies*

$$h_B(z)h_{B^!}(-z) = 1 + o(z^n). \tag{5.1}$$

Proof: Proposition 1.1 easily implies that B is $(n-1)$-Koszul but not n-Koszul. Consider the formal series $f(z) = h_B(z)h_{B^!}(-z)$. Since $B^! = D^! \otimes^{-1} D$, we have

$$f(z) = h_D(z)h_{D^!}(z)h_{D^!}(-z)h_D(-z),$$

hence $f(-z) = f(z)$. On the other hand, B is $(n-1)$-Koszul, so $f(z) = 1 + O(z^n)$. Since n is odd, this is possible only when $f(z) = 1 + o(z^n)$. □

The following lemma is the technical heart of the construction.

LEMMA 5.2. *Let $n \geqslant 0$ be an integer, and let B be a quadratic algebra such that* (5.1) *holds. Then there exist quadratic monomial aglebras M and N such that for $N' = M \sqcup B$ one has*

$$h_{N^!}(z) - h_{N'^!}(z) = o(z^n),$$
$$h_{N'}(z)h_{N'^!}(-z) = 1 + o(z^n),$$
$$h_N(z) - h_{N'}(z) = o(z^n).$$

Proof: If $n \leqslant 2$ then the assertion is clear (take $M = \Bbbk$ and N any quadratic monomial algebra that has the same number of generators and relations as B), so we can assume that $n \geqslant 3$. Note also that since N is Koszul, we have $h_N(z) = h_{N^!}(-z)^{-1}$. Hence, the last equality follows from the first two.

The main building blocks in the construction of M and N are the following two series of quadratic algebras:

$$E^s = \Bbbk\{x_1, \dots, x_s\}/(x_1x_2, \dots, x_{s-2}x_{s-1}, x_{s-1}x_s),$$
$$G^s = \Bbbk\{x_1, \dots, x_s\}/(x_1x_2, \dots, x_{s-2}x_{s-1}, x_{s-2}x_s),$$

where $s \geqslant 3$. One can easily compute the Hilbert series of the dual algebras (cf. section 6 of chapter 4):

$$h_{E^{s!}}(z) = 1 + sz + (s-1)z^2 + \dots + 2z^{s-1} + z^s,$$
$$h_{G^{s!}}(z) = 1 + sz + (s-1)z^2 + \dots + 2z^{s-1}.$$

Therefore,

$$h_{E^{s!}}(z) - h_{G^{s!}}(z) = z^s. \tag{5.2}$$

Let us pick a monomial quadratic algebra Q that has the same number of generators and relations as B. Then

$$h_{B^!}(z) = h_{Q^!}(z) + z^3P(z) + o(z^n)$$

where P is a polynomial of degree $n-3$ with integer coefficients. Let us write

$$z^3P(z)=a_1z^{i_1}+\ldots+a_pz^{i_p}-b_1z^{j_1}-\ldots-b_rz^{j_r},$$

where a_i and b_j are positive integers. Let us define our monomial quadratic algebras M and N as the following free products:

$$M=(G^{i_1})^{\sqcup a_1}\sqcup\ldots\sqcup(G^{i_p})^{\sqcup a_p}\sqcup(E^{j_1})^{\sqcup b_1}\sqcup\ldots\sqcup(E^{j_r})^{\sqcup b_r},$$

$$N=Q\sqcup(E^{i_1})^{\sqcup a_1}\sqcup\ldots\sqcup(E^{i_p})^{\sqcup a_p}\sqcup(G^{j_1})^{\sqcup b_1}\sqcup\ldots\sqcup(G^{j_r})^{\sqcup b_r}.$$

We claim that with this choice of M and N the required equalities with Hilbert series are satisfied. Indeed, since $N'^! = M^!\sqcap B^!$, we have

$$h_{N'^!}(z)=h_{M^!}(z)+h_{B^!}(z)-1. \tag{5.3}$$

Similarly, we can express Hilbert series $h_{M^!}$ and $h_{N^!}$ in terms of $h_{E^{s!}}$, $h_{G^{s!}}$ and $h_{Q^!}$. This leads to

$$\begin{aligned}h_{N'^!}(z)-h_{N^!}(z)&=h_{B^!}(z)+\sum_{m=1}^{p}a_mh_{G^{i_m!}}(z)+\sum_{l=1}^{r}b_lh_{E^{j_l!}}(z)\\&-h_{Q^!}(z)-\sum_{m=1}^{p}a_mh_{E^{i_m!}}(z)-\sum_{l=1}^{r}b_lh_{G^{j_l!}}(z)\\&=z^3P(z)-\sum_{m=1}^{p}a_mz^{i_m}+\sum_{l=1}^{r}b_lz^{j_l}+o(z^n)=o(z^n),\end{aligned}$$

where we used (5.2). This proves the first equality. To prove the second equality we note that by (1.2) one has

$$h_{N'}(z)=(h_M(z)^{-1}+h_B(z)^{-1}-1)^{-1}.$$

Combining this with (5.3) we get

$$h_{N'}(z)h_{N'^!}(-z)=(h_M(z)^{-1}+h_B(z)^{-1}-1)^{-1}(h_M(z)^{-1}+h_{B^!}(-z)-1)=1+o(z^n),$$

since $h_{B^!}(-z)=h_B(z)^{-1}+o(z^n)$ by our assumption. □

THEOREM 5.3. *There exist quadratic algebras A and A' such that $h_A(z)=h_{A'}(z)$, $h_{A^!}(z)=h_{A'^!}(z)$, and A is Koszul but A' is not.*

Proof: The construction has as an input an odd integer $n>4$ and a quadratic algebra D such that D is $(n-1)$-Koszul but not n-Koszul. For example, one can take $n=5$ and the algebra D with three generators x,y,z and two relations: $xy=0$, $yx+z^2=0$. It is not difficult to check that this algebra is 4-Koszul but not 5-Koszul (see [**15**]).

Starting from D as above we set $B=D\otimes D^!$. By Lemma 5.1 the algebra B satisfies the assumptions of Lemma 5.2. Therefore, we can find monomial quadratic algebras M and N satisfying the conclusion of this lemma. Now we set $A=N\circ C$ and $A'=N'\circ C$, where $N'=M\sqcup B$ and C is any Koszul algebra with $C_{n+1}=0$ and $C_n\neq 0$ (say, the exterior algebra with n generators). Note that A is Koszul, being the Segre product of Koszul algebras (see Corollary 1.2). On the other hand, $N'=M\sqcup B$ is not n-Koszul since B is not n-Koszul. By Corollary 4.9 this implies that A' is not Koszul.

Since $C_{n+1} = 0$, the relation $h_N(z) - h_{N'}(z) = o(z^n)$ implies that $h_A(z) = h_{A'}(z)$. On the other hand, since $C^!$ is Koszul we can apply Corollary 4.7 to compute the Hilbert series of $A^! = N^! \bullet C^!$:

$$h_{A^!}(z) = h_{N^! \bullet C^!}(z) = (h_{N^!}^{-1} \circ h_C)^{-1}(z).$$

Similarly, $h_{A'^!}(z) = (h_{N'^!}^{-1} \circ h_C)^{-1}(z)$. Using the relation $h_{N^!}(z) - h_{N'^!}(z) = o(z^n)$ we deduce that $h_{A^!}(z) = h_{A'^!}(z)$. □

CHAPTER 4

Poincaré–Birkhoff–Witt Bases

In this chapter we study quadratic algebras for which a (homogeneous) analogue of the PBW-theorem holds (called PBW-algebras). In another terminology these are algebras admitting generators for which the noncommutative Gröbner basis of relations consists of elements of degree 2 (see [**26, 36**]). These algebras form an important class of Koszul algebras preserved under various operations on quadratic algebras. We consider some generalizations of this class in section 7. These include the case of commutative PBW-algebras (also known as G-quadratic, see [**38**]) considered in more detail in section 8. We also define a notion of $\mathbb{Z}$-PBW basis by replacing graded algebras with a wider class of $\mathbb{Z}$-algebras (see section 9 and 10). Finally, in section 11 we show that a three-dimensional Sklyanin algebra is Koszul but does not admit a $\mathbb{Z}$-PBW-basis.

1. PBW-bases

Let $A = \{V, I\} = \mathbb{T}(V)/J$ be a quadratic algebra with a fixed basis $\{x_1, \ldots, x_m\}$ of the space of generators V. For a multiindex $\alpha = (i_1, \ldots, i_n)$, where $i_k \in [1, m]$, we denote by x^α the monomial $x_{i_1} x_{i_2} \ldots x_{i_n} \in \mathbb{T}(V)$. For $\alpha = \emptyset$ we set $x^\emptyset = 1$. Let us consider the lexicographical order on the set of multiindices of length n: $(i_1, \ldots, i_n) < (j_1, \ldots, j_n)$ iff there exists k such that $i_1 = j_1, \ldots, i_{k-1} = j_{k-1}$ and $i_k < j_k$. The following lemma is the starting point for the construction of a PBW-basis.

LEMMA 1.1. *Let W be a vector space with a basis w_α numbered by a totally ordered set of indices $\mathcal{A}$, and let $K \subset W$ be a subspace. Consider the subset $S_K \subset \mathcal{A}$ consisting of all $\alpha \in \mathcal{A}$ such that w_α cannot be presented as a linear combination of w_β with $\beta < \alpha$ modulo K. Then the images of the elements w_α with $\alpha \in S_K$ form a basis of W/K. The subset $S = S_K$ is also uniquely characterized by the property that there is a basis of K of the form*

$$u_\beta = w_\beta - \sum_{\alpha<\beta} c_{\beta\alpha} w_\alpha, \qquad \beta \in \bar{S} = \mathcal{A} \setminus S.$$

There exists a unique basis of K of this form with the additional property that $c_{\beta\alpha} = 0$ for $\alpha \notin S$. □

Note that the correspondence $K \longleftrightarrow (S, c_{\beta\alpha})$ is nothing else but the Schubert stratification of the Grassmannian $\mathbb{G}(W)$.

Now let us equip the space $V \otimes V$ with the basis consisting of the monomials $x_{i_1} x_{i_2}$. Applying the above lemma to the subspace of quadratic relations $I \subset V \otimes V$ we obtain the set of pairs of indices $S = S_I \subset [1, m]^2$. Hence, the relations in A

can be written in the following form:

$$x_{i_1}x_{i_2} = \sum_{\substack{(j_1,j_2)<(i_1,i_2)\\(j_1,j_2)\in S}} c_{i_1,i_2}^{j_1,j_2} x_{j_1}x_{j_2}, \qquad (i_1,i_2)\in \overline{S}.$$

Definition. Consider the following sets of multiindices:

$$S^{(n)} = \{\,(i_1,\dots,i_n) \mid (i_1,i_2)\in S,\ (i_2,i_3)\in S,\ \dots,\ (i_{n-1},i_n)\in S\,\}, \qquad n\geqslant 2.$$

We also set $S^{(1)} = [1,m]$ and $S^{(0)} = \{\emptyset\}$. Elements $x_1, \dots, x_m \in V$ are called *PBW-generators* of A if the monomials (x^α) with $\alpha \in \bigcup_{n\geqslant 0} S^{(n)}$ form a basis of A (called a *PBW-basis* of A). A *PBW-algebra* is a quadratic algebra admitting a PBW-basis.

Note that in any case the monomials $(x^\alpha, \alpha\in \bigcup_{n\geqslant 0} S^{(n)})$ linearly span the entire A. Indeed, every monomial $x_{i_1}\dots x_{i_n}$ with $(i_k,i_{k+1})\notin S$ can be expressed as a linear combination of smaller monomials modulo J_n. Let us denote by $\pi\colon V\otimes V \longrightarrow V\otimes V$ the projection on the first summand in the direct decomposition $V\otimes V = \langle\, x^\alpha \mid \alpha\in S\,\rangle \oplus I$. Explicitly,

$$\pi(x_{i_1}x_{i_2}) = \begin{cases} x_{i_1}x_{i_2}, & (i_1,i_2)\in S\\ \sum_{\substack{(j_1,j_2)<(i_1,i_2)\\(j_1,j_2)\in S}} c_{i_1,i_2}^{j_1,j_2}x_{j_1}x_{j_2}, & (i_1,i_2)\in\overline{S}.\end{cases}$$

An element $x\in \mathbb{T}^n(V)$ is congruent modulo J to the element

$$\cdots\pi^{12}\pi^{23}\cdots\pi^{n-1,n}\pi^{12}\pi^{23}\cdots\pi^{n-1,n}x,$$

where $\pi^{k,k+1} = \mathrm{id}^{\otimes k-1}\otimes\pi\otimes\mathrm{id}^{\otimes n-k-1} : \mathbb{T}^n(V)\longrightarrow\mathbb{T}^n(V)$. This infinite product is well defined since π decreases the order. It maps $\mathbb{T}^n(V)$ to $\langle x^\alpha,\ \alpha\in S^{(n)}\rangle$. Similarly, one can consider the infinite composition

$$\cdots\pi^{i_3,i_3+1}\pi^{i_2,i_2+1}\pi^{i_1,i_1+1}$$

associated with any sequence i_1, i_2, i_3, $\dots$ containing every index $1, \dots, n-1$ infinitely many times. *The PBW condition is satisfied iff all such products of projections are equal to each other.*

Remark. The PBW property depends on the choice of generators and on their ordering. For example, (x,y) are PBW-generators of the algebra $\Bbbk\{x,y\}/(x^2-xy)$, while (y,x) are not (see [**119**]). However, it is not difficult to see that the subset S and the PBW property are determined by the filtration $\langle x_1\rangle\subset\cdots\subset\langle x_1,\dots,x_{m-1}\rangle \subset V$ spanned by $x_1,\dots,x_m$ (see section 7 below). Also, the *existence* of a PBW-basis can depend on the ground field. For example, consider the quadratic algebra $\mathbb{T}(V)/(f)$ with one relation $f\in V^{\otimes 2}$, where $\dim V>1$ and $f\notin U^{\otimes 2}$ for any proper subspace $U\subset V$. This algebra admits a set of PBW-generators iff the corresponding quadratic form $\xi\longmapsto\langle\xi\otimes\xi, f\rangle$ on V^* represents zero.

2. PBW-theorem

The following result is a particular case of the diamond lemma for noncommutative Gröbner bases [**26, 36**]. We keep the notation of the previous section.

THEOREM 2.1. *If the cubic monomials $(x_{i_1}x_{i_2}x_{i_3}, (i_1,i_2,i_3)\in S^{(3)})$ are linearly independent in A_3 then the same is true in any degree n. Therefore, in this case the elements $(x_1,\dots,x_m)$ are PBW-generators of A.*

Equivalently, elements $(x_1,\dots,x_m)$ are PBW-generators of a quadratic algebra A iff the following equation holds:

$$\cdots\pi^{12}\pi^{23}\pi^{12}=\cdots\pi^{23}\pi^{12}\pi^{23}. \tag{2.1}$$

Sketch of proof: Let us denote

$$\mathbb{T}^n(V)_{\leqslant\alpha}=\langle x^\beta \mid \beta\leqslant\alpha\rangle,$$

$$J^n_{\leqslant\alpha}=\sum_k \mathbb{T}^n(V)_{\leqslant\alpha}\cap V^{\otimes k-1}\otimes I\otimes V^{\otimes n-k-1}=$$

$$\langle\, x^\beta f_\eta x^\gamma \mid \eta\in\overline{S},\ f_\eta=x^\eta-\sum c_{\eta\varsigma}x^\varsigma,\ \beta\eta\gamma\leqslant\alpha\,\rangle$$

and similarly define $J^n_{<\alpha}$. We can prove by induction in α that the monomials $(x^\beta \mid \beta\in S^{(n)},\ \beta\leqslant\alpha)$ are linearly independent modulo $J^n_{\leqslant\alpha}$. Indeed, it suffices to show that $x^\beta y_\eta x^\gamma - x^{\beta'}y_{\eta'}x^{\gamma'}\in J^n_{<\alpha}$ for any $\beta\eta\gamma=\alpha=\beta'\eta'\gamma'$. This is clear if the subwords η and η' do not intersect each other. Otherwise, it follows from the condition in degree 3. □

Another proof: Consider a sequence $i_1,\ i_2,\ i_3,\ \dots$ as above and let

$$\pi=\cdots\pi^{i_3,i_3+1}\pi^{i_2,i_2+1}\pi^{i_1,i_1+1}=\pi'\pi^{i_1,i_1+1}$$

be the corresponding product. It is enough to prove that $\pi\pi^{j,j+1}=\pi$ for any j. We use induction in α: assuming that $\pi\pi^{j,j+1}$ agrees with π on $\mathbb{T}^n(V)_{<\alpha}$ for any j and any π as above we have to show that $\pi\pi^{j,j+1}x^\alpha=\pi x^\alpha$. If $\pi^{j,j+1}x^\alpha=x^\alpha$ then there is nothing to prove, so we can assume that $\pi^{j,j+1}x^\alpha\in\mathbb{T}^n(V)_{<\alpha}$. Then by the induction hypothesis

$$\pi\pi^{j,j+1}x^\alpha=\pi'\pi^{i_1,i_1+1}\pi^{j,j+1}x^\alpha=\pi'\pi^{j,j+1}x^\alpha. \tag{2.2}$$

We claim that it suffices to consider the case when $\pi^{i_1,i_1+1}x^\alpha\in\mathbb{T}^n(V)_{<\alpha}$. Indeed, otherwise $\pi x^\alpha=\pi' x^\alpha$ and using (2.2) we can replace π by π'. Decomposing further π' as $\pi'=\pi''\pi^{i_2,i_2+1}$ we can go on until after several steps we obtain $i_t=j$. Now applying π to the equation

$$\cdots\pi^{j,j+1}\pi^{i_1,i_1+1}\pi^{j,j+1}x^\alpha=\cdots\pi^{i_1,i_1+1}\pi^{j,j+1}\pi^{i_1,i_1+1}x^\alpha$$

(that follows from (2.1)) and using the induction assumption we obtain $\pi\pi^{j,j+1}x^\alpha$ on the left and $\pi\pi^{i_1,i_1+1}x^\alpha=\pi x^\alpha$ on the right. □

Example 1. Let SP be a *skew-polynomial algebra*, i.e., a quadratic algebra with generators $x_1,\dots,x_n$ and relations

$$x_jx_i=q_{ij}x_ix_j,\ i<j,$$

where $(q_{ij}, 1\leqslant i<j\leqslant n)$ is a collection of nonzero constants. Then $x_1,\dots,x_n$ is the set of PBW-generators. The corresponding PBW-basis of SP consists of all monomials of the form $x_1^{m_1}\dots x_n^{m_n}$, where $m_1,\dots,m_n\geqslant 0$. For example, every basis of a vector space V gives PBW-generators of the symmetric algebra $\mathbb{S}(V)$.

Example 2. The classical Poincaré–Birkhoff–Witt theorem can be deduced from Theorem 2.1 as follows. Let $\mathfrak{g}$ be a finite-dimensional Lie algebra. Consider the graded version $\widetilde{U}(\mathfrak{g})$ of the universal enveloping algebra generated in degree one by $\mathfrak{g}$ together with an additional central generator t and with relations $xy-yx=t[x,y]$ for $x,\ y\in\mathfrak{g}$. If $x_1,\dots,x_m$ is a basis of $\mathfrak{g}$ then the elements $(t,x_1,\dots,x_m)$ are

PBW-generators of $\widetilde{U}(\mathfrak{g})$ and the corresponding PBW-basis consists of the monomials $(t^k x_1^{k_1} \cdots x_m^{k_m})$, where k, $k_i \geqslant 0$. Indeed, the PBW condition in degree 3 is easily seen to be equivalent to the Jacobi identity. Using the isomorphism $U(\mathfrak{g}) = \widetilde{U}(\mathfrak{g})/(t-1)$ we obtain that the monomials $(x_1^{k_1} \cdots x_m^{k_m})$ form a basis in $U(\mathfrak{g})$. (See also Theorem 2.1 of chapter 5 and Remark 2 in section 2 of chapter 6.)

3. PBW-bases and Koszulness

The following result is due to S. Priddy [**104**].

THEOREM 3.1. *A quadratic PBW-algebra is Koszul.*

Proof: Define a multiindex-valued filtration on the algebra A by the rule $F_\alpha A_n = \langle x^\beta \mid \beta \leqslant \alpha \rangle$, where length $\alpha = n$. Since the elements $(x_1, \ldots, x_m)$ generate a PBW-basis of A, the associated graded algebra $\mathrm{gr}^F A = \bigoplus_\alpha F_\alpha / F_{\alpha'}$ (where α' precedes α in the multiindex order) is quadratic. Thus, $\mathrm{gr}^F A$ is a quadratic monomial algebra, so it is Koszul by Corollary 4.3 of chapter 2. Now the assertion follows easily from the spectral sequence $E_1 = \mathrm{Tor}^{\mathrm{gr}^F A}(\Bbbk, \Bbbk) \implies \mathrm{Tor}^A(\Bbbk, \Bbbk)$ associated with the corresponding multiindex-valued filtration of the bar-complex computing the Tor (cf. proof of Theorem 7.1). Another proof will be given in section 5. □

Example. Here is a (minimal in the numbers of generators and relations for algebras over an algebraically closed field) example due to J. Backelin [**18**] of a Koszul algebra that has no PBW-basis:

$$\begin{cases} x^2 + \ \ yz = 0 \\ x^2 + azy = 0, \qquad a \neq 0,\, 1. \end{cases}$$

The proof of Koszulness is as follows. One can compute the Hilbert series $h_A(z) = \sum_n (\dim A_n) z^n$ using the noncommutative Gröbner bases technique [**26, 36**]. This computation shows that $h_A(z)^{-1}$ is a polynomial of degree 2. Hence, A is Koszul by Proposition 2.3 of chapter 2. We will see in section 10 that the algebra of this example has a so-called $\mathbb{Z}$-PBW-basis which also implies Koszulness.

Remark. The set $\mathcal{PBW}(V)$ of quadratic PBW-algebras with a fixed space of generators V is a Zariski constructible subset in the Grassmannian variety of quadratic algebras $\mathbb{G}(V^{\otimes 2})$. Indeed, let us fix a basis $x = (x_1, \ldots, x_m)$ of V and consider the Schubert stratification $\mathbb{G}(V^{\otimes 2}) = \bigcup_{S \subset [1,m]^{\times 2}} \mathbb{G}_S(V^{\otimes 2}; x^{\otimes 2})$ defined in Lemma 1.1. Let $\mathcal{PBW}(V; x; S) \subset \mathcal{PBW}(V)$ denote the set of quadratic algebras having $x_1, \ldots, x_m$ as PBW-generators with the corresponding set S. Then $\mathcal{PBW}(V; x; S)$ is a closed algebraic subvariety of $\mathbb{G}_S(V^{\otimes 2}; x^{\otimes 2})$, preserved by the action of the upper triangular subgroup $B(m) \subset \mathrm{GL}(m)$. Consider the finite disjoint union $\mathcal{PBW}(V; x) = \coprod_S \mathcal{PBW}(V; x; S)$. Then the set $\mathcal{PBW}(V)$ can be presented as the image of the algebraic morphism

$$\mathrm{GL}(m) \times_{B(m)} \mathcal{PBW}(V; x) \longrightarrow \mathbb{G}(V^{\otimes 2}),$$

hence it is constructible. On the other hand, it turns out that *the set of all Koszul algebras with m generators is not constructible for $m \geqslant 3$* (see section 6 of chapter 6).

4. PBW-bases and operations on quadratic algebras

As we have shown in chapter 3, a number of natural operations on quadratic algebras preserves Koszulness. In this section we will check that these operations also preserve the class of PBW-algebras.

THEOREM 4.1. *If $(x_1, \ldots, x_m)$ are PBW-generators of a quadratic algebra A then the elements $(x_m^*, \ldots, x_1^*)$ of the dual basis are PBW-generators of the dual quadratic algebra $A^!$.*

Proof: Let A^0 denote the monomial algebra with the generators $(x_1, \ldots, x_m)$ and relations $x_i x_j = 0$ for $(i,j) \in \overline{S}$. Note that $(x_1, \ldots, x_m)$ are PBW-generators of A iff $\dim A_3 = \dim A_3^0$. The analogous monomial algebra $A^{!0}$ for the quadratic dual algebra $A^!$ with the set of generators $(x_m^*, \ldots, x_1^*)$ corresponds to the complementary subset $S^! = \overline{S}$. Hence, $A^{!0}$ coincides with the quadratic dual to A^0. For any quadratic algebra $A = \{V, I\}$ we have $A_3 = V^{\otimes 3}/(I \otimes V + V \otimes I)$ and $A_3^{!*} = I \otimes V \cap V \otimes I$. Therefore, the dimension of $A_3^!$ is given by the formula

$$\dim A_3^! = (\dim A_1)^3 - 2(\dim A_1)(\dim A_2) + (\dim A_3).$$

Applying this to A and A^0 we derive that $\dim A_3 = \dim A_3^0$ iff $\dim A_3^! = \dim A_3^{!0}$. Another proof can be derived from the proof of Proposition 5.1. □

Next, we are going to show that the class of PBW-algebras is closed under all binary operations on quadratic algebras considered in chapter 3.

PROPOSITION 4.2. *Let $(x_1, \ldots, x_m)$ and $(y_1, \ldots, y_l)$ be PBW-generators of quadratic algebras A and B, respectively. Then the basis $(x_1, \ldots, x_m, y_1, \ldots, y_l)$ of the vector space $A_1 \oplus B_1$ is a set of PBW-generators in each of the algebras $A \sqcap B$, $A \sqcup B$, and $A \otimes^q B$, where $q \in \mathbb{P}^1_{\Bbbk}$. Also, the basis*

$$(x_1 \otimes y_1, \ldots, x_1 \otimes y_l, x_2 \otimes y_1, \ldots, x_m \otimes y_1, \ldots, x_m \otimes y_l)$$

of the vector space $A_1 \otimes B_1$ is a set of PBW-generators in each of the algebras $A \circ B$ and $A \bullet B$.

Proof: This follows from the fact that the set of PBW-monomials in each of these algebras (except for $A \bullet B$) coincides with the explicit basis constructed from the PBW-bases of A and B. For example, $S_{A \otimes B} = S_A \cup S_B \cup \{(i,j) \mid x_i \in A_1,\ y_j \in B_1\}$, hence, $S^{(n)}_{A \otimes B} = \bigcup_{k+l=n} S^{(k)}_A \times S^{(l)}_B$. For $A \bullet B$ we use Theorem 4.1. □

The PBW-property is also preserved by the Veronese power construction.

PROPOSITION 4.3. *Let $(x_1, \ldots, x_m)$ be PBW-generators of a quadratic algebra A. Then the elements of the PBW-basis $(x^\alpha)_{\alpha \in S^{(d)}}$ of A_d taken in lexicographical order are PBW-generators of the Veronese subalgebra $A^{(d)}$.*

Proof: This is an immediate consequence of the fact that the set of multiindices $S^{(dn)}$ can be described as follows:

$$S^{(dn)} = \{(\alpha_1, \ldots, \alpha_n) \in (S^d)^n \ :\ (\alpha_i, \alpha_{i+1}) \in S^{(2d)} \text{ for all } i = 1, \ldots, n-1\}.$$

□

Remark. A more interesting interaction between the Veronese construction and the PBW-property is connected with noncommutative Gröbner bases (see [**36**]). Namely, let A be a one-generated algebra. Assume that the Gröbner basis of the ideal of relations with respect to generators $(x_1, \ldots, x_m)$ in A_1 is finite. Then for all sufficiently large d the set of normal monomials in A_d ordered lexicographically is a set of PBW-generators of $A^{(d)}$. More precisely, it suffices to take d bigger than the degree of all elements in the Gröbner basis.

5. PBW-bases and distributing bases

The following result explains the relation between PBW-bases and the criterion of Koszulness in terms of distributive lattices.

PROPOSITION 5.1. *A basis* $x_1, \dots, x_m$ *of a vector space* V *generates a PBW-basis of a quadratic algebra* $A = \{V, I\}$ *iff for any* $n \geqslant 0$ *there exists a basis* $(y_{i_1 \dots i_n})$ *of the vector space* $V^{\otimes n}$ *distributing the collection of subspaces* $V^{\otimes k-1} \otimes I \otimes V^{\otimes n-k-1} \subset V^{\otimes n}$, *such that* $(y_{i_1 \dots i_n})$ *differs from the monomial basis* $(x_{i_1} \otimes \cdots \otimes x_{i_n})$ *by the action of an upper triangular matrix:*

$$y_{i_1 \dots i_n} = x_{i_1} \otimes \cdots \otimes x_{i_n} - \sum_{(j_1 \dots j_n) < (i_1 \dots i_n)} c_{i_1 \dots i_n}^{j_1 \dots j_n} x_{j_1} \otimes \cdots \otimes x_{j_n}.$$

Proof: "If". Apply Lemma 1.1 to the subspace $V^{\otimes k-1} \otimes I \otimes V^{\otimes n-k-1} \subset V^{\otimes n}$, where $V^{\otimes n}$ is equipped with the lexicographically ordered basis $(x_{i_1} \otimes \cdots \otimes x_{i_n})$. We have two bases of this subspace: one consisting of some subset of the basis $(y_{i_1 \dots i_n})$ and the other consisting of the elements

$$x_{i_1} \otimes \cdots \otimes (x_{i_k} \otimes x_{i_{j+1}} - \pi(x_{i_k} \otimes x_{i_{k+1}})) \otimes \cdots \otimes x_{i_n}$$

with $(i_k, i_{k+1}) \in \bar{S}$. Using the uniqueness part of Lemma 1.1 we derive that $y_{i_1 \dots i_n}$ belongs to $V^{\otimes k-1} \otimes I \otimes V^{\otimes n-k-1}$ iff $(i_k, i_{k+1}) \in \bar{S}$. Therefore, the elements

$$(y_{i_1 \dots i_n} \mid \exists k : (i_k, i_{k+1}) \in \bar{S})$$

form a basis of $\sum_{k=1}^{n-1} V^{\otimes k-1} \otimes I \otimes V^{\otimes n-k-1}$. It remains to observe that $A_n = V^{\otimes n} / \sum V^{\otimes k-1} \otimes I \otimes V^{\otimes n-k-1}$ and to apply Lemma 1.1 again.
"Only if". Consider the dual quadratic algebra $A^!$. It has a PBW-basis generated by $(x_m^*, \dots, x_1^*)$ with the corresponding set $S^! = \bar{S}$. Let $\pi_h^! \colon V^{*\otimes h} \longrightarrow V^{*\otimes h}$ be the projection onto the subspace $\langle x^\alpha, \alpha \in \bar{S}^{(h)} \rangle$ along the relations subspace $J_h^!$. Then the dual operator $p_h = \pi_h^{!*} \colon V^{\otimes h} \longrightarrow V^{\otimes h}$ is a projection onto $\bigcap_t V^{\otimes t-1} \otimes I \otimes V^{\otimes h-t-1}$. For every multiindex $\alpha = (i_1, \dots, i_n)$ consider the partition of the segment $[1, n]$ into the union $[1, k_1] \sqcup \cdots \sqcup [k_s + 1, n]$ of the maximal subsegments $[k, l]$ with the property that for all $i \in [k, l-1]$ one has $(i, i+1) \in \bar{S}$. It is easy to see that the elements

$$y_\alpha = (p_{k_1} \otimes p_{k_2 - k_1} \otimes \cdots \otimes p_{n-k_s})(x^\alpha)$$

form the required distributing basis of $V^{\otimes n}$. □

Remark. The above criterion looks similar to the Backelin's criterion of Koszulness in terms of distributivity of the collection $(V^{\otimes k-1} \otimes I \otimes V^{\otimes n-k-1})$ in $V^{\otimes n}$ for all n. The crucial difference is that for the PBW-condition we are requiring the existence of an *upper triangular* distributing basis. By Theorem 2.1 it suffices to check that this condition is satisfied for $n = 3$. On the other hand, in Backelin's criterion the distributivity condition is empty for $n = 3$. The first nontrivial Koszulness condition occurs in degree 4 and no finite degree condition is sufficient: for any $n \geqslant 3$ there exists a quadratic algebra with 3 generators which is n-Koszul, but not $(n+1)$-Koszul (see section 4 of chapter 2 and section 6 of chapter 6).

6. Hilbert series of PBW-algebras

Clearly, the Hilbert series of a PBW-algebra A is equal to that of the corresponding monomial algebra A^0 and depends only on the subset $S = S_I \subset [1,m]^2$. It is convenient to view S as the set of edges of an oriented graph G_S with the set of vertices $[1,m]$. Then $\dim A_n = |S^{(n)}|$ is equal to the number of *paths* of length $n-1$ in G_S (where $n \geqslant 1$). Here by a path of length n in G_S we mean a sequence of edges of the form $v_0 \to v_1 \to v_2 \to \ldots \to v_{n-1} \to v_n$. Thus, the study of Hilbert series of PBW-algebras reduces to the study of generating series of the numbers of paths in oriented graphs. Below we present some basic results about these generating series (see also [**119**]).

Let Γ be a finite oriented graph with the set of vertices $[1,m]$ (we allow Γ to have two or more edges sharing the same source and target). With Γ we associate an $m \times m$ *adjacency matrix* $M^\Gamma = (m_{ij})$, where m_{ij} is the number of edges going from i to j. Note that in the case of the graphs G_S above the adjacency matrix consists of 0's and 1's. Let $a_n(\Gamma)$ be the number of paths of length $n-1$ in Γ, where $n \geqslant 1$. For example, $a_1(\Gamma) = m$, $a_2(\Gamma)$ is the number of edges in Γ. We also set $a_0(\Gamma) = 1$ and form a generating series

$$h_\Gamma(z) = \sum_{n \geqslant 0} a_n(\Gamma) z^n.$$

The following result gives a way to compute this series.

PROPOSITION 6.1. *Let E denote the $m \times m$ matrix with all entries equal to 1. Then*

$$h_\Gamma(z) = \frac{\det(1 + z(E - M^\Gamma))}{\det(1 - zM^\Gamma)}.$$

Proof: Let us set $M = M^\Gamma$. It is easy to see that

$$a_n(\Gamma) = \operatorname{tr}(M^{n-1}E)$$

for $n \geqslant 1$. Therefore,

$$h_\Gamma(z) = 1 + \operatorname{tr}(zE + z^2 ME + z^3 M^2 E + \ldots) = 1 + \operatorname{tr}((1 - zM)^{-1} zE).$$

Since $\det(1+T) = 1 + \operatorname{tr}(T)$ for any rank-one matrix T we can rewrite the above expression as

$$h_\Gamma(z) = \det(1 + (1 - zM)^{-1} zE) = \det((1 - zM)^{-1}(1 + z(E - M))).$$

□

For any $S \subset [1,m]^2$ let M_S denote the $m \times m$ matrix with 1's at all entries belonging to S and 0's at all other entries.

COROLLARY 6.2. *The Hilbert series of a PBW-algebra A is rational. More precisely,*

$$h_A(z) = \frac{\det(1 + zM_{\overline{S}})}{\det(1 - zM_S)}.$$

Now we are going to show that the rate of growth of the sequence $a_n(\Gamma)$ is determined by the configuration of (oriented) cycles in Γ. A *cycle* of length n is a path of the form $v_1 \to v_2 \to \ldots \to v_n \to v_1$ in Γ such that all vertices $v_1, \ldots, v_n$ are distinct (thus, a loop $v \to v$ is a cycle of length 1). We say that two cycles intersect if they pass through the same vertex. Let γ_1 and γ_2 be a pair of non-intersecting cycles. We say that γ_2 is *adjacent* to γ_1 and write $\gamma_1 \to \gamma_2$ if there exists an edge $v_1 \to v_2$ such that γ_1 passes through v_1 and γ_2 passes through v_2. A *chain* of cycles of length d is a sequence $\gamma_0 \to \gamma_1 \to \ldots \to \gamma_d$.

PROPOSITION 6.3. *(i) One has $a_n(\Gamma) = 0$ for $n \gg 0$ iff Γ has no cycles.*
(ii) If Γ has a pair of distinct intersecting cycles then $a_n(\Gamma)$ grows exponentially, i.e., there exists a constant $c > 1$ such that $a_n(\Gamma) \geqslant c^n$ for $n \gg 0$.
(iii) If Γ has no intersecting cycles then $a_n(\Gamma)$ grows polynomially. More precisely, if d is the maximal length of a chain of cycles in Γ then there exist constants $c_1 > c_2 > 0$ such that

$$c_1 n^d \geqslant a_n(\Gamma) \geqslant c_2 n^d$$

for $n \gg 0$.

Proof: (i) The proof is straightforward.
(ii) Let γ_1 and γ_2 be cycles of lengths $n_1 \geqslant n_2$ passing through a common vertex v. Let us consider only paths that go along γ_1 and γ_2 sometimes switching from one cycle to another when passing through v. Counting the number of such paths we immediately get an estimate

$$a_n \geqslant 2^{\lfloor \frac{n}{n_1} \rfloor}.$$

(iii) First, we observe that in this case for every pair of adjacent distinct cycles $\gamma_1 \to \gamma_2$ there exists a unique edge $v_1 \to v_2$ such that γ_1 passes through v_1 and γ_2 passes through v_2. Indeed, if there were two such edges then we would be able to construct a new cycle intersecting γ_1 and γ_2. Now let us fix a chain of cycles $\gamma_0 \to \gamma_1 \to \ldots \to \gamma_d$ and a vertex v_0 on γ_0. Let us denote by p_n the number of paths of length n that start at v_0, go around γ_0 several times then switch to γ_1, etc. and finally end on γ_d. It suffices to prove that there exists $c_1 > c_2 > 0$ such that

$$c_1 n^d \geqslant p_n \geqslant c_2 n^d$$

for $n \gg 0$. It is clear that if we decrease the length of one of the cycles by 1 the number of paths p_n can only increase. Therefore, it is enough to consider the case when all cycles $\gamma_1, \ldots, \gamma_d$ have the same length s. Let t be the minimal length of a path satisfying the above restrictions. Then

$$p_n = \#\{(n_0, \ldots, n_d) \in \mathbb{Z}^{d+1}_{\geqslant 0} \mid n_0 + \ldots n_d = \lfloor \frac{n-t}{s} \rfloor\} = \binom{\lfloor \frac{n-t}{s} \rfloor + d}{d},$$

so our assertion follows. □

7. Filtrations on quadratic algebras

In this section we generalize the PBW-theorem to the case of filtrations with values in an arbitrary graded ordered semigroup.

Definition. A *graded ordered semigroup* Γ is a (noncommutative) semigroup with a unit e equipped with a homomorphism $g\colon \Gamma \longrightarrow \mathbb{N}$ and a collection of total orders on the fibers $\Gamma_n = g^{-1}(n)$ satisfying the following condition:

$$\alpha < \beta \implies \alpha\gamma < \beta\gamma, \ \gamma\alpha < \gamma\beta \qquad \text{for } \alpha, \beta \in \Gamma_n, \ \gamma \in \Gamma_k$$

(the strict inequality here is essential for the definition of the associated graded object). In addition we assume that $g^{-1}(0) = \{e\}$.

A Γ*-valued filtration* on a graded algebra A is a collection of subspaces ($F_\alpha = F_\alpha A_n \subset A_n$, $\alpha \in \Gamma_n$) for every $n \geqslant 0$ such that

(i) $\alpha \leqslant \beta$ implies $F_\alpha \subset F_\beta$;
(ii) $F_{\gamma_n} = A_n$ for the maximal element $\gamma_n \in \Gamma_n$;
(iii) $F_\alpha \cdot F_\beta \subset F_{\alpha\beta}$ for arbitrary $\alpha, \beta \in \Gamma$.

The *associated* Γ*-graded algebra* for a Γ-filtered algebra A is defined in the usual way: $\mathrm{gr}^F A = \bigoplus_{\alpha\in\Gamma} F_\alpha/F_{\alpha'}$, where α' precedes α in the total order (if α is minimal in Γ_n then we take $F_{\alpha'} = 0$). Note that the algebra $\mathrm{gr}^F A$ is one-generated iff the filtration F is *one-generated*, i.e., $F_\alpha A_n = \sum_{i_1\cdots i_n \leqslant \alpha} F_{i_1}V \cdots F_{i_n}V$, where $i_s \in \Gamma_1$. A one-generated filtration is determined by its restriction to A_1.

If A is a quadratic algebra equipped with a one-generated Γ-valued filtration F then Koszulness of $\mathrm{gr}^F A$ implies Koszulness of A. In fact, the following stronger version of this assertion holds.

THEOREM 7.1. *Let A be a quadratic algebra equipped with a one-generated Γ-valued filtration. Assume that the associated Γ-graded algebra $\mathrm{gr}^F A$ satisfies the following conditions:*

(1) *$\mathrm{gr}^F A$ has no defining relations of degree 3;*
(2) *the quadratic part $\mathrm{q}(\mathrm{gr}^F A)$ of $\mathrm{gr}^F A$ is Koszul.*

Then $\mathrm{gr}^F A$ is quadratic (and hence Koszul by (2)), and A is Koszul.

Proof: First, let us check that $\mathrm{gr}^F A$ is quadratic. We are going to prove by induction in n that $\mathrm{gr}^F A$ has no defining relations in degrees from 3 to n. Assume that this is true for $n-1$ and consider the following fragment $C_\bullet(A)$ of the Koszul complex of A:

$$A_{n-3} \otimes A_3^{!*} \longrightarrow A_{n-2} \otimes I \longrightarrow A_{n-1} \otimes V \longrightarrow A_n \longrightarrow 0.$$

Note that it is equipped with the induced filtration with values in Γ_n. Condition (1) implies that $\mathrm{gr}^F A_3^{!*} = (\mathrm{gr}^F A)_3^{!*}$, hence the associated Γ_n-graded complex coincides with $C_\bullet(\mathrm{gr}^F A)$. We have $H_0 C_\bullet(\mathrm{gr}^F A) = 0$ since $\mathrm{gr}^F A$ is one-generated, and $H_1 C_\bullet(A) = 0$ since A is quadratic. We also claim that $H_2 C_\bullet(\mathrm{gr}^F A) = 0$. Indeed, by the induction assumption the homomorphism $\mathrm{q}(\mathrm{gr}^F A) \to \mathrm{gr}^F A$ is an isomorphism in degrees $< n$. Therefore, our claim follows from the assumption that $\mathrm{q}(\mathrm{gr}^F A)$ is Koszul. Now the spectral sequence associated with the filtered complex $C_\bullet(A)$ immediately shows that $H_1 C_\bullet(\mathrm{gr}^F A) = 0$. Hence, $\mathrm{gr}^F A$ has no relations in degree n. This proves that $\mathrm{gr}^F A$ is quadratic. To verify Koszulness of A we consider the induced Γ-valued filtration on the bar-complex of A:

$$F_\alpha(A_+^{\otimes i}) = \sum_{\beta_1\cdots\beta_i \leqslant \alpha} F_{\beta_1} \otimes \cdots \otimes F_{\beta_i}.$$

The associated Γ-graded complex is isomorphic to the bar-complex of $\mathrm{gr}^F A$, so we get a spectral sequence $E_1 = \mathrm{Tor}^{\mathrm{gr}^F A}(\Bbbk, \Bbbk) \Longrightarrow \mathrm{Tor}^A(\Bbbk, \Bbbk)$. Since $\mathrm{Tor}_{ij}^{\mathrm{gr}^F A}(\Bbbk, \Bbbk) = 0$ for $i \neq j$, the same is true for $\mathrm{Tor}_{ij}^A(\Bbbk, \Bbbk)$. □

Remark 1. Note that to prove only that $\mathrm{gr}^F A$ is quadratic one can replace (2) with the weaker assumption that the Koszul complex of $\mathrm{q}(\mathrm{gr}^F A)$ is exact in homological degree 3 (cf. Remark 1 in section 7 of chapter 5 and Remark 1 in section 2 of chapter 6).

Example 1. To derive the PBW-Theorem 2.1 from Theorem 7.1 we take Γ to be the free noncommutative semigroup generated by the symbols $1, \ldots, m \in \Gamma_1$ and consider the Γ-filtration generated by the filtration $\langle x_1 \rangle \subset \langle x_1, x_2 \rangle \subset \cdots \subset$

$\langle x_1, \ldots, x_m \rangle = A_1$ associated with the basis $x_1, \ldots, x_m$. It is clear that the PBW-monomials form a basis in A_n iff the associated Γ-graded algebra $\mathrm{gr}^F A$ has no defining relations in degree k, where $3 \leqslant k \leqslant n$. In this case the quadratic part of the associated Γ-graded algebra is always Koszul since it is a monomial algebra (see Corollary 4.3 of chapter 2).

Example 2. Let $f\colon A \longrightarrow C$ be a homomorphism of quadratic algebras such that $C_1 f(A_1) \subset f(A_1) C_1$. Assume that A and $C/f(A_1)C$ are Koszul algebras and the left action of A on C is free in degrees $\leqslant 3$, i.e., $\mathrm{Ext}_A^{ij}(C, \Bbbk) = 0$ for $i > 0$ and $j \leqslant 3$. Then C is a Koszul algebra and a free left A-module. Indeed, this follows from Theorem 7.1 applied to the filtration of C with values in the free semigroup with two generators $\alpha < \gamma$ (and the lexicographical order $\alpha\gamma < \gamma\alpha$) defined by $F_\alpha = f(A_1)$ and $F_\gamma = C_1$. Indeed, the only nonzero components of $\mathrm{gr}^F C$ are $\mathrm{gr}^F_{\alpha^l \gamma^k} C = f(A_l) C_k / f(A_{l+1}) C_{k-1}$, hence $\mathrm{gr}^F C$ is isomorphic to the one-sided product $A \otimes^0 (C/f(A_1)C)$ in degrees $\leqslant 3$. Therefore, the conditions of Theorem 7.1 are satisfied and we conclude that $\mathrm{gr}^F C \simeq A \otimes^0 (C/f(A_1)C)$. For example, if C is a quadratic algebra and $t \in C_1$ is such an element that $Ct \subset tC$, C/tC is Koszul, and the left multiplication by t is injective on C_1 and C_2, then t is not a left zero divisor in C. For a central element t, this statement is equivalent to the nonhomogeneous PBW-Theorem 2.1 of chapter 5.

Example 3. The twisted tensor product $A \otimes^R B$ of two quadratic algebras A and B with respect to an operator $R\colon B_1 \otimes A_1 \longrightarrow A_1 \otimes B_1$ is defined as the quadratic algebra with the generators space $A_1 \oplus B_1$ and the relations space $I_A + I_B + \langle b \otimes a - R(b \otimes a) \rangle$. Applying Example 2 to the natural morphism $A \longrightarrow A \otimes^R B$ we find that the multiplication map of vector spaces $A \otimes B \longrightarrow A \otimes^R B$ is an isomorphism iff the condition in degree 3

$$R^{23} R^{12} (B_1 \otimes I_A) \subset I_A \otimes B_1$$
$$R^{12} R^{23} (I_B \otimes A_1) \subset A_1 \otimes I_B$$

is satisfied. It is easy to see that the Koszulness condition is not needed here.

Example 4. Consider the case when $\Gamma = \mathbb{N}^m$, $\gamma(a_1, \ldots, a_m) = a_1 + \cdots + a_m$, and the Γ-filtration on A is generated by the complete flag in A_1 associated with a basis $x_1, \ldots, x_m$. Then the quadratic part B of the associated Γ-graded algebra is defined by relations of the following form: for any i there may be a relation $x_i^2 = 0$ and for any $i \neq j$ one has the following four possibilities:

(1) $x_i x_j = q_{ij} x_j x_i$ with some $q_{ij} \in \Bbbk^*$;
(2) either $x_i x_j = 0$, or $x_j x_i = 0$;
(3) $x_i x_j = x_j x_i = 0$;
(4) no relation.

Let $y_i = x_i^*$ be the generators of $B^!$. It is shown by Fröberg in [**52**] that such an algebra B is Koszul iff for any two monomials $x^\alpha \in B$ and $y^\beta \in B^!$ there exists an index i and a multiindex γ such that either $x^\alpha = c x^\gamma x_i$ and $y_i y^\beta \neq 0$ or $y^\beta = c y_i y^\gamma$ and $x^\alpha x_i \neq 0$, where $c \in \Bbbk^*$. We give a proof of this fact in the following three cases when this condition is satisfied automatically:
(i) there are no relations of type (1) (B is a monomial algebra);
(ii) there are no relations of types (2) and (4) (B is a quotient of a skew-polynomial algebra by monomial relations);
(iii) there are no relations of types (2) and (3).

In case (i) Koszulness of B follows from Corollary 4.3 of chapter 2. In case (ii) Koszulness of B will be proved in Theorem 8.1 below. Finally, case (iii) is dual to case (ii).

The simplest example of an algebra of the above type for which Fröberg's condition is not satisfied (and hence the algebra is not Koszul) is the algebra with generators x_1, x_2, x_3 and relations $x_1x_2 = x_2x_1,\ x_1x_3 = x_3x_1 = 0$.

The following statement generalizes the results of sections 3 and 5 to arbitrary (one-generated) Γ-valued filtrations.

PROPOSITION 7.2. *Let $A = \{V, I\}$ be a quadratic algebra equipped with a one-generated Γ-valued filtration F. Then the following two conditions are equivalent:*

(a) *the algebra $\mathrm{gr}^F A$ is Koszul;*

(b) *for every $n \geqslant 0$ the collection of all the subspaces $V^{\otimes k-1} \otimes I \otimes V^{\otimes n-k-1} \subset V^{\otimes n}$ together with the subspaces*

$$F_\alpha(V^{\otimes n}) = \sum_{i_1 \cdots i_n \leqslant \alpha} F_{i_1}V \otimes \cdots \otimes F_{i_n}V, \qquad \alpha \in \Gamma_n,$$

in $V^{\otimes n}$ is distributive.

Proof: This follows from the Backelin's criterion of Koszulness. For the part (b) $\Longrightarrow$ (a) one has to check first that the algebra $\mathrm{gr}^F A$ is quadratic. For the implication (a) $\Longrightarrow$ (b) one can apply Corollary 7.3 of chapter 1, (b) $\Longrightarrow$ (a) (the condition $\mathrm{gr}^F(X_i + X_j) = \mathrm{gr}^F X_i + \mathrm{gr}^F X_j$ needed in this result follows from the absence of cubic defining relations in $\mathrm{gr}^F A$). □

Remark 2. Actually, the above argument proving (a) $\Longrightarrow$ (b) uses only Koszulness of $q(\mathrm{gr}^F A)$ and the absence of relations of degree 3. Hence, it provides another proof of Theorem 7.1.

For a graded ordered semigroup Γ, let Γ° denote the same graded semigroup with the opposite order. Let F be a one-generated Γ-valued filtration on a quadratic algebra A. Let us define the dual Γ°-valued filtration F° on the quadratic dual algebra $A^!$ by the rule $F^\circ_i A^*_1 = (F_{i'}A_1)^\perp$, where i' is the element preceding $i \in \Gamma_1$ in the order of Γ.

COROLLARY 7.3. *Let F be a one-generated Γ-valued filtration on a quadratic algebra A. Then the associated Γ-graded algebras $\mathrm{gr}^F A$ and $\mathrm{gr}^{F^\circ} A^!$ are Koszul simultaneously.*

Proof: One can either use Proposition 7.2, or observe that $\mathrm{q}(\mathrm{gr}^{F^\circ} A^!) \simeq (\mathrm{q}\mathrm{gr}^F A)^!$ and apply Theorem 7.1 and the argument from the proof of Theorem 4.1. □

8. Commutative PBW-bases

In this section we discuss the notion of a commutative PBW-basis for a commutative algebra. This is not the same as a noncommutative PBW-basis for such an algebra (although there is a relation between the two notions, see below). Note that both notions fit into the context of section 7 but the corresponding semigroups are different: in the latter case it is the free noncommutative semigroup with m generators, and in the former case one has to replace it with the free commutative semigroup $\mathbb{N}^m$. On the other hand, commutative PBW-bases correspond to quadratic (commutative) Gröbner bases for the ideals of relations [**35, 26**]. Commutative algebras admitting such a basis are called G-quadratic (see [**38**]).

The definition is completely parallel to the noncommutative case, with the only difference that noncommutative monomials are replaced by the commutative ones $x^a = x_1^{a_1} \dots x_m^{a_m}$. Let $A = \Bbbk[x_1, \dots, x_m]/J_c$ be a commutative quadratic algebra with a fixed set of generators. Let us order commutative multiindices by the inverse lexicographical order: $(a_1, \dots, a_m) < (b_1, \dots, b_m)$ iff there is such k that $a_1 = b_1$, $\dots$, $a_{k-1} = b_{k-1}$ and $a_k > b_k$. Note that we can also identify the set of commutative multiindices of degree n with $\mathrm{Sym}^n[1, m]$, the set of unordered n-tuples of elements of $[1, m]$. Applying the construction of Lemma 1.1 to the subspace of quadratic relations $I_c \subset \mathbb{S}^2(V)$ we obtain the corresponding subset $S \subset \mathrm{Sym}^2[1, m]$. Let $S^{(n)} \subset \mathrm{Sym}^n[1, m]$ be the set of all multiindices which are not divisible by any $a \in \bar{S}$. It is easy to see that the commutative PBW-monomials $(x^b, b \in \bigcup_{n \geqslant 0} S^{(n)})$ generate A as a vector space. If they are also linearly independent then we say that they form a *commutative PBW-basis* of A. If A admits such a basis then it is called *G-quadratic* (or *commutative PBW-algebra*).

The commutative PBW-Theorem states that *if the commutative PBW-monomials form a basis in A_3 then the same is true in any degree n*. Also, *any commutative G-quadratic algebra is Koszul* (see [**72**]). To derive these statements from Theorem 7.1, let us consider the $\mathbb{N}^m$-valued filtration $F_a = \langle x^b \mid b \leqslant a \rangle$ on A (see Example 4 of section 7). The quadratic part of the associated $\mathbb{N}^m$-graded algebra is isomorphic to the commutative monomial algebra $\Bbbk[x_1, \dots, x_m]/(x^a = 0 \mid a \in \bar{S})$. It remains to use the following result due to Fröberg [**52**] (our proof differs from the proof given in [**52**]).

THEOREM 8.1. *Let SP be a skew-polynomial algebra, i.e., a quadratic algebra with generators $x_1, \dots, x_N$ and relations*

$$x_j x_i = q_{ij} x_i x_j, \ i < j,$$

where $(q_{ij}, 1 \leqslant i < j \leqslant N)$ is a collection of nonzero constants. Then every quotient-algebra of SP by monomial quadratic relations with respect to generators $x_1, \dots, x_N$ is Koszul.

Proof: Let A be such a quotient. We are going to prove a stronger statement: all A-modules A/I, where I is a left ideal generated by some subset of x_i's, have free linear resolutions as left A-modules (in fact, I is a two-sided ideal but this is not important for us). We use double induction. The first induction is in cohomological degree: we want to prove that for every n and every ideal I as above one has $\mathrm{Tor}^A_{i,j}(\Bbbk, A/I) = 0$ for $i \leqslant n$ and $j > i$. For $n = 0$ this is clear. Now let us assume that this is true for some n. To deduce the assertion for $n+1$ we use the induction in the number of generators of I. For $I = 0$ the assertion is clearly true. Assume that our assertion that $\mathrm{Tor}^A_{i,j}(\Bbbk, A/I) = 0$ for $i \leqslant n+1$ and $j > i$ holds for every I generated by $\leqslant m$ elements. Given an ideal I generated by $m+1$ elements $x_{i_1}, \dots, x_{i_{m+1}}$ we can consider the left ideal $J \subset I$ generated by the m elements $x_{i_2}, \dots, x_{i_{m+1}}$. Then the exact sequence of A-modules

$$0 \longrightarrow I/J \longrightarrow A/J \longrightarrow A/I \longrightarrow 0$$

leads to a long exact sequence

$$\dots \longrightarrow \mathrm{Tor}^A_{n+1}(\Bbbk, A/J) \longrightarrow \mathrm{Tor}^A_{n+1}(\Bbbk, A/I) \longrightarrow \mathrm{Tor}^A_n(\Bbbk, I/J) \longrightarrow \dots.$$

Since $\mathrm{Tor}^A_{n+1}(\Bbbk, A/J)$ is concentrated in the internal degree $n+1$ by the induction assumption, it suffices to prove that $\mathrm{Tor}^A_n(\Bbbk, I/J)$ has internal degree $\leqslant n+1$. But the A-module I/J has one generator $\overline{x}_{i_1}$ of degree 1 (the image of x_{i_1}) and the relations $x_i\overline{x}_{i_1} = 0$, where either $i \in \{i_2, \dots, i_{m+1}\}$ or $x_i x_{i_1} = 0$ in A. Therefore, $I/J \simeq A/J'(-1)$ for some ideal J' of the same type as above. It remains to apply the assumption of the external induction to the space $\operatorname{Tor}_n^A(\Bbbk, A/J')$. □

Remark 1. The collection of left ideals considered in the above proof is an example of a *Koszul filtration*. For commutative algebras this notion was introduced in [**40**]. Piontkovskii observed in [**93**] that one can also consider a similar notion for noncommutative algebras.

Remark 2. Commutative Gröbner bases of the ideals of relations between a finite number of generators are always finite (see [**35**]). Thus, the commutative version of Remark from section 4 states that for every commutative graded algebra A generated by A_1 the Veronese subalgebra $A^{(d)}$ is G-quadratic for all sufficiently large d. In particular, we obtain another proof of the fact that $A^{(d)}$ is Koszul for all $d \gg 0$ (see Corollary 3.4 of chapter 3, [**16**], [**46**]).

The natural question is whether there is any relation between commutative and noncommutative PBW-bases of commutative algebras. It can be easily seen that if a commutative algebra A has a PBW-basis in the noncommutative sense (with respect to the lexicographical order) then it also has a commutative PBW-basis with the same generators (with respect to the inverse lexicographical order on commutative monomials). Now assume that A has a commutative PBW-basis. Let $S \subset \operatorname{Sym}^2[1,m]$ be the corresponding subset of commutative monomials (such that $|S| = \dim A_2$). Let us consider also the subset $\widetilde{S} \subset [1,m]^2$ obtained by looking at noncommutative quadratic monomials in A. Then $\widetilde{S}$ consists of all pairs (i,j) such that $(i,j) \leqslant (j,i)$ and $p(i,j) \in S$, where p is the natural projection $[1,m]^2 \longrightarrow \operatorname{Sym}^2[1,m]$. Our commutative PBW-generators define a noncommutative PBW-basis iff the relation $(i,j) \in \widetilde{S}$ is transitive:

$$(i,j) \in \widetilde{S} \text{ and } (j,k) \in \widetilde{S} \implies (i,k) \in \widetilde{S}.$$

In fact, commutative PBW-generators often do not generate a noncommutative PBW-basis. Furthermore, below we will give examples of commutative quadratic algebras that possess commutative PBW-bases but no noncommutative PBW-bases.

One can view the subset $S \subset \operatorname{Sym}^2[1,m]$ defining a commutative quadratic monomial algebra A as the set of effective divisors of degree 2 on the set $[1,m]$. Here by *divisors* on $[1,m]$ we mean formal linear combinations $\sum_{i=1}^m n_i(i)$ where $n_i \in \mathbb{Z}$. The degree of such a divisor is $\sum n_i$. *Effective divisors* are those with $n_i \geqslant 0$. For a pair of divisors D_1 and D_2 we write $D_1 \leqslant D_2$ if $D_2 - D_1$ is effective. The *support* of a divisor $\sum_{i=1}^m n_i(i)$ is the set of i such that $n_i \neq 0$. We have $\dim A_n = |S^{(n)}|$, where $S^{(n)}$ is the set of effective divisors D on $[1,m]$ of degree n such that for every effective divisor E of degree 2 with $E \leqslant D$ one has $E \in S$.

The next result gives a formula for the Hilbert series of A. Let us say that a subset $H \subset [1,m]$ is *S-complete* if for every $i,j \in H$ with $i \neq j$ one has $(i)+(j) \in S$. We denote by $\mathcal{H}(S)$ the set of all S-complete subsets of $[1,m]$. For every subset $H \subset [1,m]$ we denote by $L(H)$ the number of $i \in H$ such that $2(i) \in S$.

PROPOSITION 8.2. *For every commutative quadratic monomial algebra A one has*

$$h_A(z) = \sum_{H \in \mathcal{H}(S)} \frac{z^{|H|}}{(1-z)^{L(H)}}.$$

Proof: It is clear that the support of a divisor in $S^{(n)}$ is an S-complete subset of $[1, m]$. Let $H \subset [1, m]$ be an S-complete subset, $K \subset H$ be the subset of i such that $2(i) \in S$ (recall that $|K| = L(H)$). Then divisors in $S^{(n)}$ that have H as a support are exactly all divisors of the form

$$\sum_{i \in H} (i) + \sum_{j \in K} n_j(j),$$

where $n_j \geqslant 0$, $\sum_{j \in K} n_j = n - |H|$. This immediately gives the required formula. □

Example 1. Here is the promised example (found using a computer) of a commutative monomial quadratic algebra with the Hilbert series that differs from that of any noncommutative monomial algebra. In particular, this algebra has no noncommutative PBW-bases. Namely, consider the algebra

$$A = k[x_1, \ldots, x_7]/(x_1^2, \ldots, x_7^2, x_1x_4, x_2x_5, x_3x_6, x_2x_7, x_4x_7, x_6x_7).$$

Its Hilbert series is $h_A(z) = 1 + 7z + 15z^2 + 11z^3 + z^4$. Assume that there is a noncommutative monomial algebra with the same Hilbert series. This would mean that the corresponding oriented graph has 7 vertices, 15 edges, 11 oriented paths of length 2, 1 path of length 3 and no paths of length > 3 (see section 6). This graph cannot have loops or oriented cycles (because the algebra is finite-dimensional). Therefore, the unique path of length 3 has to pass through 4 distinct vertices v_1, v_2, v_3, v_4. It follows easily that each of the remaining 3 vertices w_1, w_2, w_3 is joined with the set $\{v_1, \ldots, v_4\}$ by $\leqslant 2$ edges. There may be also $\leqslant 6$ edges in the subgraph with the vertices $\{v_1, \ldots, v_4\}$ and $\leqslant 3$ edges in the subgraph $\{w_1, w_2, w_3\}$. To obtain the total number of 15 edges all of these estimates have to be attained. This implies that there is a unique oriented graph (without oriented cycles) that has 7 vertices, 15 edges, and 1 path of length 3. However, this graph has 10 paths of length 2 instead of 11.

Example 2. As we have seen above, Hilbert series of commutative and noncommutative PBW-algebras are given by explicit rational functions determined by the combinatorial structure of quadratic relations. It would be interesting to determine how "big" is the subset of Hilbert series of PBW-algebras in the set of all Hilbert series of Koszul algebras. Among 83 cases of various Hilbert and Poincaré series of commutative quadratic algebras with four generators listed in [**107**] we found two examples of Koszul algebras whose Hilbert series differ from that of any commutative or noncommutative PBW-algebra. These algebras are

$$A' = \Bbbk[x, y, z, u]/(x^2 + xy, x^2 + zu, y^2, z^2, xz + yu, xu)$$
$$A'' = \Bbbk[x, y, z, u]/(xy + yz, xy + z^2 + yu, yu + zu, y^2, xz)$$

with the Hilbert series

$$h_{A'}(z) = 1 + 4z + 4z^2 + z^3 + z^4 + z^5 + z^6 + z^7 + \cdots$$
$$h_{A''}(z) = 1 + 4z + 5z^2 + 4z^3 + 5z^4 + 6z^5 + 7z^6 + 8z^7 + \cdots.$$

Example 3. Even though not every Koszul commutative algebra is G-quadratic, there are situations when these two notions coincide. For example, Backelin showed in [**15**] that a commutative quadratic algebra A is Koszul provided $\dim A_2 \leqslant 2$. Later Conca [**38**] proved that (if characteristic is $\neq 2$ and the field is algebraic closed) with one exception every such algebra is G-quadratic (the exception is $\Bbbk[x,y,z]/(x^2,xy,y^2-xz,yz)$). Also, it is shown in [**38**] that a generic commutative quadratic algebra with $\dim A_2 < \dim A_1$ is G-quadratic.

Example 4. It is an interesting question for which projective varieties $X \subset \mathbb{P}^n$ the homogeneous coordinate ring R_X is G-quadratic. For example, it is known to be true in the following cases:
(1) X is a set of $\leqslant 2n$ points in a general linear position in $\mathbb{P}^n$ (see Thm. 3.1 of [**39**]);
(2) X is a canonical curve (nonhyperelliptic, nontrigonal, not a plane quintic) - see [**50**] and [**39**], Thm. 5.1. In fact, in [**50**] even a noncommutative PBW-basis is constructed.

On the other hand, it is known that if X is a generic complete n-dimensional intersection of quadrics in $\mathbb{P}^{e+n}$ and $n < \frac{(e-1)(e-2)}{6} - 1$ then R_X is not G-quadratic (see [**46**], Cor.20). For example, this is the case for a generic complete intersection of 5 quadrics in $\mathbb{P}^5$.

9. ℤ-algebras

The notion of *ℤ-algebra* (introduced in [**25, 23**]) is a convenient generalization of the notion of graded algebra. Perhaps, a better term would be $(\mathbb{Z},\leqslant)$-algebra, i.e., an algebra over the partially ordered set $(\mathbb{Z},\leqslant)$. Since we never explicitly mention algebras over $\mathbb{Z}$, hopefully, using the term ℤ-algebra should not lead to confusion.

Definition 1. A (positively graded) *ℤ-algebra* $\mathcal{A}$ over the ground field $\Bbbk$ is defined by a collection of finite-dimensional vector spaces $(\mathcal{A}_{\sigma\tau}, \sigma,\tau\in\mathbb{Z})$ such that $\mathcal{A}_{\sigma\tau}=0$ for $\sigma<\tau$ and $\mathcal{A}_{\sigma\sigma}=\Bbbk$, equipped with associative product maps

$$\mathcal{A}_{\rho\sigma}\otimes\mathcal{A}_{\sigma\tau}\longrightarrow\mathcal{A}_{\rho\tau},\qquad \rho,\sigma,\tau\in\mathbb{Z}.$$

An *$\mathcal{A}$-module* $\mathcal{M}$ is a collection of vector spaces $(\mathcal{M}_\tau,\tau\in\mathbb{Z})$ equipped with action maps $\mathcal{A}_{\sigma\tau}\otimes\mathcal{M}_\tau\longrightarrow\mathcal{M}_\sigma$ that are compatible with the multiplication in $\mathcal{A}$. The abelian category of $\mathcal{A}$-modules is defined in the obvious way.

For a ℤ-algebra $\mathcal{A}$ the relation spaces $\mathcal{I}_{\rho\tau}$ are defined from the exact sequences

$$0\longrightarrow\mathcal{I}_{\rho\tau}\longrightarrow\mathcal{A}_{\rho,\rho-1}\otimes\mathcal{A}_{\rho-1,\rho-2}\otimes\cdots\otimes\mathcal{A}_{\tau+2,\tau+1}\otimes\mathcal{A}_{\tau+1,\tau}\stackrel{m_{\rho\tau}}{\longrightarrow}\mathcal{A}_{\rho\tau},$$

where the maps $m_{\rho\tau}$ are induced by the product in $\mathcal{A}$.

Definition 2. A ℤ-algebra $\mathcal{A}$ is called *quadratic* if the product maps $m_{\rho\tau}$ are surjective and the relation spaces $\mathcal{I}_{\rho\tau}$ are generated by the quadratic relation spaces $\mathcal{I}_{\sigma+1,\sigma-1}$:

$$\bigoplus_{\rho-1\geqslant\sigma\geqslant\tau+1}\mathcal{A}_{\rho,\rho-1}\otimes\cdots\otimes\mathcal{A}_{\sigma+2,\sigma+1}\otimes\mathcal{I}_{\sigma+1,\sigma-1}\otimes\mathcal{A}_{\sigma-1,\sigma-2}\otimes\cdots\otimes\mathcal{A}_{\tau+1,\tau}\longrightarrow\mathcal{I}_{\rho\tau}\longrightarrow 0.$$

The *dual quadratic ℤ-algebra* $\mathcal{A}^!$ has the generator spaces $\mathcal{A}^!_{\sigma+1,\sigma}=\mathcal{A}^*_{\sigma+1,\sigma}$ and the quadratic relation spaces $\mathcal{I}^!_{\sigma+1,\sigma-1}=\mathcal{I}^\perp_{\sigma+1,\sigma-1}\subset\mathcal{A}^*_{\sigma+1,\sigma}\otimes\mathcal{A}^*_{\sigma,\sigma-1}$.

For a ℤ-algebra $\mathcal{A}$ and for $\sigma\in\mathbb{Z}$ let $\Bbbk_\sigma$ denote the irreducible $\mathcal{A}$-module defined by $(\Bbbk_\sigma)_\tau=0$ for $\sigma\neq\tau$ and $(\Bbbk_\sigma)_\sigma=\Bbbk$. We have the following analogue of Proposition 3.1 of chapter 1.

PROPOSITION 9.1. *One has* $\mathrm{Ext}^i_{\mathcal{A}}(\Bbbk_\sigma, \Bbbk_\tau) = 0$ *for* $i > \sigma - \tau$ *and the diagonal* $\mathbb{Z}$*-algebra* $\mathcal{B}_{\sigma\tau} = \mathrm{Ext}^{\sigma-\tau}_{\mathcal{A}}(\Bbbk_\sigma, \Bbbk_\tau)$ *is the quadratic* $\mathbb{Z}$*-algebra dual to the quadratic part of* $\mathcal{A}$. □

Generalizing Backelin's Koszulness criterion to $\mathbb{Z}$-algebras we can make the following

DEFINITION–PROPOSITION 9.2. *A* $\mathbb{Z}$*-algebra* $\mathcal{A}$ *is called Koszul if the following equivalent conditions hold:*

(1) $\mathrm{Ext}^i_{\mathcal{A}}(\Bbbk_\sigma, \Bbbk_\tau) = 0$ *for* $i \neq \sigma - \tau$;

(2) $\mathcal{A}$ *is quadratic and for any* $\rho \geqslant \tau \in \mathbb{Z}$ *the collection of subspaces*

$$\mathcal{A}_{\rho,\rho-1} \otimes \cdots \otimes \mathcal{I}_{\sigma+1,\sigma-1} \otimes \cdots \otimes \mathcal{A}_{\tau+1,\tau} \subset \mathcal{A}_{\rho,\rho-1} \otimes \cdots \otimes \mathcal{A}_{\tau+1,\tau}, \quad \rho - 1 \geqslant \sigma \geqslant \tau + 1$$

is distributive.

In particular, quadratic dual $\mathbb{Z}$*-algebras* $\mathcal{A}$ *and* $\mathcal{A}^!$ *are Koszul simultaneously, and for a Koszul algebra* $\mathcal{A}$ *one has* $\mathrm{Ext}^*_{\mathcal{A}}(\Bbbk_\sigma, \Bbbk_\tau) \simeq \mathcal{A}^!_{\sigma\tau}$. □

Example. There is a natural embedding of the category of graded algebras into the category of $\mathbb{Z}$-algebras. Namely, for a graded algebra A we define the associated $\mathbb{Z}$-algebra $A^{\mathbb{Z}}$ by the rule $A^{\mathbb{Z}}_{\sigma\tau} = A_{\sigma-\tau}$. Then the category of $A^{\mathbb{Z}}$-modules coincides with the category of graded A-modules. A graded algebra A is quadratic (resp. Koszul) iff the corresponding $\mathbb{Z}$-algebra $A^{\mathbb{Z}}$ is of the same type.

Remark. For $n > 0$ one can define an $[n]$*-algebra* $\mathcal{A}$ as a collection of vector spaces $(\mathcal{A}_{\sigma\tau}, n \geqslant \sigma, \tau \geqslant 1)$ and of product maps of the same form as above. Similarly, replacing the partially ordered set $(\mathbb{Z}, \leqslant)$ with $([1,n], \leqslant)$ one can define all the notions above for $[n]$-algebras.

Now for a graded algebra A we can define the associated $[n]$-algebra by the rule $A^{[n]}_{\sigma\tau} = A_{\sigma-\tau}$ for $n \geqslant \sigma, \tau \geqslant 1$. It is easy to see that the $[n]$-algebra $A^{[n]}$ is Koszul iff A is n-Koszul.

On the other hand, with an $[n]$-algebra $\mathcal{A}$ one can associate the graded algebra $\mathcal{A}^{\oplus}$ by the rule $\mathcal{A}^{\oplus}_i = \bigoplus_{\sigma-\tau=i} \mathcal{A}_{\sigma\tau}$. Since $A^{\oplus}_0 = \Bbbk^n$ the notion of Koszulness still makes sense for $A^{\oplus}$ (see Example 2 of section 9 in chapter 2). Note that the category of $\mathcal{A}$-modules is equivalent to the category of *nongraded* $\mathcal{A}^{\oplus}$-modules. As for the category of graded $\mathcal{A}^{\oplus}$-modules, it splits into the direct sum of $\mathbb{Z}$ copies of the category of $\mathcal{A}$-modules [**126**]. It follows easily that *an* $[n]$*-algebra* $\mathcal{A}$ *is Koszul iff the graded algebra* $\mathcal{A}^{\oplus}$ *is Koszul.*

Finally, by modifying the degree zero component of a graded algebra B with $B_0 \simeq \Bbbk^{\oplus n}$ we get the graded algebra $B^{\Bbbk} = \Bbbk \oplus (\bigoplus_{i>0} B_i)$, and this operation preserves Koszulness.

Considering the composition of the three operations above, $A \longmapsto A^{[n]} \longmapsto A^{[n],\oplus,\Bbbk}$, we get a transformation on the class of graded algebras that kills the homological information in internal degrees $> n$, while preserving it in internal degrees $\leqslant n$. This kind of construction can be used instead of the operation $A \longmapsto A \circ C$, where C is the exterior algebra with n generators, to produce counterexamples with Hilbert series of non-Koszul algebras (see section 5 of chapter 3).

10. $\mathbb{Z}$-PBW-bases

While the Koszulness condition does not change when passing from a graded algebra to the corresponding $\mathbb{Z}$-algebra, the notion of PBW-basis becomes more general.

Definition. Let $\mathcal{A}$ be a quadratic $\mathbb{Z}$-algebra. Choose a family of bases

$$\mathcal{X}^{\sigma+1,\sigma} = \{x_1^{\sigma+1,\sigma}, \dots, x_{m_{\sigma+1,\sigma}}^{\sigma+1,\sigma}\} \subset \mathcal{A}_{\sigma+1,\sigma} \qquad (m_{\sigma+1,\sigma} = \dim \mathcal{A}_{\sigma+1,\sigma})$$

in the generator spaces $\mathcal{A}_{\sigma+1,\sigma}$. Applying Lemma 1.1 to the space $\mathcal{A}_{\sigma+1,\sigma} \otimes \mathcal{A}_{\sigma,\sigma-1}$ with the basis $x_i^{\sigma+1,\sigma} \otimes x_j^{\sigma,\sigma-1}$ in the lexicographical order and the subspace of quadratic relations $\mathcal{I}_{\sigma+1,\sigma-1}$, we get the corresponding subset of indices $S^{\sigma+1,\sigma-1} \subset [1, m_{\sigma+1,\sigma}] \times [1, m_{\sigma,\sigma-1}]$. We say that the family of bases $\mathcal{X}^{\sigma+1,\sigma}$ is a set of PBW-generators for a quadratic $\mathbb{Z}$-algebra $\mathcal{A}$ if the monomials of the form

$$x_{i_{\rho,\rho-1}}^{\rho,\rho-1} \cdots x_{i_{\tau+1,\tau}}^{\tau+1,\tau}, \quad \text{where } (i_{\sigma+1,\sigma}, i_{\sigma,\sigma-1}) \in S^{\sigma+1,\sigma-1} \text{ for all } \rho - 1 \geqslant \sigma \geqslant \tau + 1,$$

form a basis of $\mathcal{A}_{\rho\tau}$ for all $\rho \geqslant \tau$.

Proposition 10.1. *For any quadratic $\mathbb{Z}$-algebra $\mathcal{A}$, the PBW condition above for $\rho - \tau = 3$ implies the same property for all (ρ, τ). Furthermore, a family of bases $\mathcal{X}^{\sigma+1,\sigma}$ is a set of PBW-generators for $\mathcal{A}$ iff for any $\rho \geqslant \tau$ there exists a basis $(y_{i_{\rho,\rho-1},\dots,i_{\tau+1,\tau}})$ of the space $\mathcal{A}_{\rho,\rho-1} \otimes \cdots \otimes \mathcal{A}_{\tau+1,\tau}$ distributing the lattice generated by the subspaces $\mathcal{A}_{\rho,\rho-1} \otimes \cdots \otimes \mathcal{I}_{\sigma+1,\sigma-1} \otimes \cdots \otimes \mathcal{A}_{\tau+1,\tau}$, such that $(y_{i_{\rho,\rho-1},\dots,i_{\tau+1,\tau}})$ differs from the basis $x_{i_{\rho,\rho-1}}^{\rho,\rho-1} \otimes \cdots \otimes x_{i_{\tau+1,\tau}}^{\tau+1,\tau}$ by the action of an upper triangular matrix. A $\mathbb{Z}$-algebra admitting a PBW-basis is Koszul. If $\mathcal{A}$ is a quadratic $\mathbb{Z}$-algebra admitting a PBW-basis then the same is true for the dual quadratic $\mathbb{Z}$-algebra $\mathcal{A}^!$.*

Proof: The proof is similar to that of Theorem 2.1 and Proposition 5.1. □

Let A be a quadratic graded algebra and let $A^{\mathbb{Z}}$ be the corresponding $\mathbb{Z}$-algebra. By definition, a *$\mathbb{Z}$-PBW basis* for A is a PBW-basis for $A^{\mathbb{Z}}$. We say that A is a $\mathbb{Z}$-PBW algebra if it admits a $\mathbb{Z}$-PBW basis. Thus, we obtain the following hierarchy of quadratic graded algebras:

$$\textit{quadratic} \supset \textit{Koszul} \supset \mathbb{Z} - \textit{PBW} \supset \textit{PBW}.$$

The example below shows that the third inclusion here is strict. We already know that the first inclusion is strict (see Example 1 in section 2 of chapter 2). One can show that the second inclusion is strict using generic algebras (see Remark 3 in section 4 of chapter 6). More refined counterexamples are provided by elliptic Sklyanin algebras described in the next section.

Example. Consider the quadratic algebra mentioned in section 3:

$$\begin{cases} x^2 + \ \ yz = 0 \\ x^2 + azy = 0, \qquad a \neq 0, 1. \end{cases}$$

It is Koszul and has no PBW-basis. However, we claim that it has a $\mathbb{Z}$-PBW-basis. Indeed, consider the family of bases $\{x^{\sigma+1,\sigma}, y^{\sigma+1,\sigma}, z^{\sigma+1,\sigma}\}$ defined by

$$\begin{cases} x = x^{\sigma+1,\sigma} \\ y = y^{\sigma+1,\sigma} + b_\sigma x^{\sigma+1,\sigma} \\ z = z^{\sigma+1,\sigma} + c_\sigma x^{\sigma+1,\sigma} \end{cases}$$

where $c_{\sigma+1} = -1/b_\sigma$, $b_{\sigma+1} = -1/(ac_\sigma)$. One can check that it is a set of PBW-generators for $A^{\mathbb{Z}}$.

Remark. It is an open problem whether the set of quadratic algebras with a given generator space admitting a $\mathbb{Z}$-PBW-basis is Zariski constructible (as is the case for usual PBW-algebras, see Remark in section 3).

11. Three-dimensional Sklyanin algebras

Let E be an elliptic curve, and let $(\mathcal{L}_\sigma)_{\sigma\in\mathbb{Z}}$ be an arithmetic progression of line bundles of degree 3 on E. This means that all the line bundles $\mathcal{L}_{\sigma+1}\otimes\mathcal{L}_\sigma^{-1}$ are isomorphic to a fixed degree 0 line bundle (and $\deg\mathcal{L}_\sigma = 3$). Following [**31**] we associate with such a progression a $\mathbb{Z}$-algebra $\mathcal{A}$ with $\mathcal{A}_{\sigma+1,\sigma}\simeq H^0(E,\mathcal{L}_\sigma)$ and $\mathcal{A}_{\sigma+2,\sigma}\simeq H^0(E,\mathcal{L}_{\sigma+1}\otimes\mathcal{L}_\sigma)$, where the multiplication $\mathcal{A}_{\sigma+2,\sigma+1}\otimes\mathcal{A}_{\sigma+1,\sigma}\longrightarrow\mathcal{A}_{\sigma+2,\sigma}$ is given by the product of sections. Note that over an algebraically closed field one can always find a graded algebra A with $A^{\mathbb{Z}}\simeq\mathcal{A}$. Indeed, one can choose a cubic root of the line bundle $\mathcal{L}_{\sigma+1}/\mathcal{L}_\sigma$ and consider the translation $t_x : E\longrightarrow E$ associated with the corresponding point $x\in E$. Then one has an isomorphism $t_x^*\mathcal{L}_\sigma\simeq\mathcal{L}_{\sigma+1}$ for all $\sigma\in\mathbb{Z}$. This immediately leads to the description of $\mathcal{A}$ in the form $A^{\mathbb{Z}}$. Note that in the case $\mathcal{L}_{\sigma+1}\simeq\mathcal{L}_\sigma$ the algebra A is isomorphic to $\Bbbk[x,y,z]$.

THEOREM 11.1 ([**31**], Thm. 7.4). *The $\mathbb{Z}$-algebra $\mathcal{A}$ is Koszul and the corresponding graded algebra A has the Hilbert series $h_A(z)=(1-z)^{-3}$.*

Proof: The proof is somewhat similar to that of Theorem 7.7 from chapter 2. Let $(\mathcal{M}_\sigma)$ be a sequence of line bundles such that $\mathcal{M}_{\sigma+1}\simeq\mathcal{M}_\sigma\otimes\mathcal{L}_\sigma$. We can associate with such a sequence a $\mathbb{Z}$-algebra $\mathcal{B}$ by setting $\mathcal{B}_{\sigma,\tau}=\mathrm{Hom}(\mathcal{M}_\tau,\mathcal{M}_\sigma)$ and using the natural composition as a product. Note that the quadratic part of $\mathcal{B}$ is naturally isomorphic to $\mathcal{A}$, so we have a homomorphism $\mathcal{B}\longrightarrow\mathcal{A}$. We are going to check that $\mathcal{A}$ is Koszul by applying the $\mathbb{Z}$-algebra version of Corollary 5.7 of chapter 2.

Note that the natural maps $H^0(E,\mathcal{L}_\sigma)\otimes\mathcal{O}_E\longrightarrow\mathcal{L}_\sigma$ are surjective, so we can define rank-2 vector bundles V_σ from the exact sequences

$$0\longrightarrow V_\sigma\longrightarrow H^0(\mathcal{L}_\sigma)\otimes\mathcal{O}_E\longrightarrow\mathcal{L}_\sigma\longrightarrow 0. \tag{11.1}$$

It is easy to see that the vector bundle $\mathcal{L}_{\sigma+1}\otimes V_\sigma$ is generated by global sections and that there is an exact sequence

(11.2)
$$0\longrightarrow\mathcal{L}_{\sigma+2}^{-1}\longrightarrow H^0(\mathcal{L}_{\sigma+1}\otimes V_\sigma)\otimes\mathcal{O}_E\longrightarrow\mathcal{L}_{\sigma+1}\otimes V_\sigma\longrightarrow 0$$

(here we use an isomorphism $\mathcal{L}_{\sigma+2}\simeq\mathcal{L}_{\sigma+1}^2\otimes\mathcal{L}_\sigma^{-1}$). Tensoring (11.1) with $\mathcal{M}_{\sigma+n}\otimes\mathcal{M}_{\sigma+1}^{-1}$, where $n\geqslant 1$, and taking global sections we get exact sequences

$$0\longrightarrow H^0(\mathcal{M}_{\sigma+n}\otimes\mathcal{M}_{\sigma+1}^{-1}\otimes V_\sigma)\longrightarrow\mathcal{B}_{\sigma+n,\sigma+1}\otimes H^0(\mathcal{L}_\sigma)\longrightarrow\mathcal{B}_{\sigma+n,\sigma}\longrightarrow 0.$$

Similarly, from (11.2) we get exact sequences

$$0\longrightarrow\mathcal{B}_{\sigma+n,\sigma+3}\longrightarrow\mathcal{B}_{\sigma+n,\sigma+2}\otimes H^0(\mathcal{L}_{\sigma+1}\otimes V_\sigma)\xrightarrow{f_n}H^0(\mathcal{M}_{\sigma+n}\otimes\mathcal{M}_{\sigma+1}^{-1}\otimes V_\sigma),$$

where the maps f_n are surjective for $n\neq 3$. Note that we can combine the above sequences into complexes

$$0\longrightarrow\mathcal{B}_{\sigma+n,\sigma+3}\longrightarrow\mathcal{B}_{\sigma+n,\sigma+2}\otimes H^0(\mathcal{L}_{\sigma+1}\otimes V_\sigma)\longrightarrow\mathcal{B}_{\sigma+n,\sigma+1}\otimes H^0(\mathcal{L}_\sigma)\longrightarrow\mathcal{B}_{\sigma+n,\sigma}\longrightarrow 0,$$

exact for $n\neq 3$ and with the only cohomology at the term $\mathcal{B}_{\sigma+3,\sigma+1}\otimes H^0(\mathcal{L}_\sigma)$ for $n=3$. It is easy to see that these complexes are grading components of complexes of projective $\mathcal{B}$-modules. Therefore, Koszulness of $\mathcal{A}$ follows from the $\mathbb{Z}$-algebra version of Corollary 5.7 of chapter 2. The above argument also shows that the quadratic dual algebra to A has the Hilbert series $1+3z+3z^2+z^3$ which implies the formula for h_A by Corollary 2.2 of chapter 2. □

PROPOSITION 11.2. *The above $\mathbb{Z}$-algebra $\mathcal{A}$ does not admit a PBW-basis (in the lexicographical order) unless $\mathcal{L}_\sigma \simeq \mathcal{L}_{\sigma+1}$.*

Proof: Assume the contrary, and let

$$J_{\sigma+1,\sigma} = \langle x_1^{\sigma+1,\sigma}\rangle \subset K_{\sigma+1,\sigma} = \langle x_1^{\sigma+1,\sigma}, x_2^{\sigma+1,\sigma}\rangle \subset V_{\sigma+1,\sigma} = H^0(E, \mathcal{L}_\sigma)$$

be the corresponding complete flags in the generator spaces $V_{\sigma+1,\sigma} = \mathcal{A}_{\sigma+1,\sigma}$ of $\mathcal{A}$. Let $S^{\sigma+2,\sigma} \subset \{1,2,3\}^2$ be the corresponding 6-element subsets and $\overline{S}^{\sigma+2,\sigma}$ be their 3-element complements.

The PBW property means that for each σ there is exactly one triple of indices $(a,b,c) \in \{1,2,3\}^3$ such that $(a,b) \in \overline{S}^{\sigma+3,\sigma+1}$ and $(b,c) \in \overline{S}^{\sigma+2,\sigma}$. Note that $\mathcal{A}$ has no zero divisors, hence $\overline{S}^{\sigma+2,\sigma} \subset \{2,3\} \times \{1,2,3\}$ for all σ and it follows that $(2,1)$, $(3,1) \in \overline{S}^{\sigma+2,\sigma}$. Also, if $(3,3) \in \overline{S}^{\sigma+2,\sigma}$ then $(3,3) \notin \overline{S}^{\tau+2,\tau}$ for $\tau = \sigma - 1$ and $\tau = \sigma + 1$. Next, we claim that if $(3,3) \notin \overline{S}^{\tau+2,\tau}$ then for both $\rho = \tau$ and $\rho = \tau + 1$ one has $K_{\rho+1,\rho} = H^0(\mathcal{L}_\rho(-x_\rho)) \subset H^0(\mathcal{L}_\rho)$ for some points $x_\rho \in E$. Indeed, this is just a reformulation of the fact that degree 1 zero divisors in $\mathcal{A}^!$ correspond to points of E. Therefore, for all σ we have $K_{\sigma+1,\sigma} = H^0(\mathcal{L}_\sigma(-x_\sigma))$ with some $x_\sigma \in E$. On the other hand, the fact that the dimension of the image of the natural map

$$H^0(\mathcal{L}_{\sigma+1}(-x_{\sigma+1})) \otimes H^0(\mathcal{L}_\sigma) \longrightarrow H^0(\mathcal{L}_{\sigma+1} \otimes \mathcal{L}_\sigma)$$

is equal to 5 implies that $\overline{S}^{\sigma+2,\sigma} \cap \{1,2\} \times \{1,2,3\} = \{(2,1)\}$. Hence, the third element of $\overline{S}^{\sigma+2,\sigma}$ is either $(3,2)$ or $(3,3)$. Applying the PBW condition again, we obtain that $\overline{S}^{\sigma+2,\sigma} = \{(2,1),(3,1),(3,2)\}$ for all σ.

In particular, we see that $V_{\sigma+2,\sigma+1}K_{\sigma+1,\sigma} = K_{\sigma+2,\sigma+1}V_{\sigma+1,\sigma}$ in $\mathcal{A}_{\sigma+2,\sigma}$. Therefore, the subspaces $H^0(\mathcal{L}_{\sigma+1} \otimes \mathcal{L}_\sigma(-x_\tau)) \subset H^0(\mathcal{L}_{\sigma+1} \otimes \mathcal{L}_\sigma)$ coincide for $\tau = \sigma$ and $\tau = \sigma+1$. This implies that all the points x_σ coincide with each other: $x_\sigma = x \in E$.

Let us denote by $Z_{\sigma+1,\sigma}$ the divisor of zeros of a nonzero section from $J_{\sigma+1,\sigma}$. The conditions $(2,1)$, $(3,1) \in \overline{S}^{\sigma+2,\sigma}$ give the following inclusions between subspaces of $\mathcal{A}_{\sigma+2,\sigma}$:

$$K_{\sigma+2,\sigma+1}J_{\sigma+1,\sigma} \subset J_{\sigma+2,\sigma+1}V_{\sigma+1,\sigma},$$
$$V_{\sigma+2,\sigma+1}J_{\sigma+1,\sigma} \subset K_{\sigma+2,\sigma+1}V_{\sigma+1,\sigma}.$$

Hence, $Z_{\sigma+2,\sigma+1} \subset Z_{\sigma+1,\sigma}+(x)$ and $(x) \subset Z_{\sigma+1,\sigma}$. It follows easily that $Z_{\sigma+2,\sigma+1} = Z_{\sigma+1,\sigma}$, and therefore

$$\mathcal{L}_{\sigma+1} \simeq \mathcal{O}_E(Z_{\sigma+2,\sigma+1}) = \mathcal{O}_E(Z_{\sigma+1,\sigma}) \simeq \mathcal{L}_\sigma.$$

□

CHAPTER 5

Nonhomogeneous Quadratic Algebras

In this chapter we consider algebras defined by nonhomogeneous quadratic relations. Note that in principle nonhomogeneous quadratic relations can imply nontrivial linear relations as well as other quadratic relations. It is easy to write a necessary (cubic) self-consistence condition generalizing the Jacobi identity for Lie brackets (see section 1). In section 2 we prove the analogue of the Poincaré–Birkhoff–Witt theorem stating that if the algebra defined by the quadratic parts of the relations is Koszul then it coincides with the associated graded algebra of the original nonhomogenous algebra provided the analogue of the Jacobi identity is satisfied. In section 5 we consider some examples of this generalized Jacobi identity. In section 4 we consider the analogue of quadratic duality in the nonhomogeneous case. Following [**99**] we show that dual objects to nonhomogeneous quadratic algebras are quadratic curved-DG-algebras (*CDG-algebras*). This notion appeared also in the works of A. Schwarz [**110**] and A. Connes [**41**]. Interactions of nonhomogeneous quadratic duality with various homology and cohomology functors are studied in sections 6 and 7. In section 8 we consider homology of the completed cobar-complex of a DG-algebra (or a DG-module). We show that for a Koszul DG-algebra this homology can be viewed as the derived functor of the A_+-adic completion of the corresponding Koszul quadratic-linear algebra A.

1. Jacobi identity

For simplicity, we start with the case of an augmented algebra.

Definition 1. A *quadratic-linear algebra*, or an *augmented nonhomogeneous quadratic algebra* is an augmented algebra $A = \Bbbk \oplus A_+$ with a fixed space of generators $V \subset A_+$ such that the defining relations are quadratic-linear, i.e., the ideal of relations $J = \ker(\mathbb{T}(V) \longrightarrow A)$ in $\mathbb{T}(V)$ is generated by the subspace $J_2 = J \cap (V + V^{\otimes 2})$. A *morphism* between quadratic-linear algebras (A, V) and (A', V') is a homomorphism of algebras $A \to A'$ sending V to V'.

Example 1. Let $\mathfrak{g}$ be a finite-dimensional Lie algebra. Then the universal enveloping algebra $U\mathfrak{g}$ can be viewed as a quadratic-linear algebra with the space of generators $\mathfrak{g}$ and the defining relations $X \otimes Y - Y \otimes X - [X, Y]$, where $X, Y \in \mathfrak{g}$.

Definition 2. A *nonhomogeneous quadratic algebra* is an algebra A with a fixed generating subspace $\widetilde{V} \subset A$ such that $1_A \in \widetilde{V}$ (where 1_A is the unit in A), and the ideal of relations $J \subset \widetilde{\mathbb{T}}(\widetilde{V}) = \mathbb{T}(\widetilde{V})/(1_{\mathbb{T}} - 1_A)$ in $\widetilde{\mathbb{T}}(\widetilde{V})$ is generated by the subspace $J_2 = J \cap \widetilde{\mathbb{T}}_2(\widetilde{V})$, where $\widetilde{\mathbb{T}}_i(\widetilde{V})$ is the image of $\mathbb{T}^i(\widetilde{V})$ under the projection $\mathbb{T}(\widetilde{V}) \to \widetilde{\mathbb{T}}(\widetilde{V})$. A *morphism* between nonhomogeneous quadratic algebras $(A, \widetilde{V})$ and $(A', \widetilde{V}')$ is a homomorphism of algebras $A \to A'$ sending $\widetilde{V}$ to $\widetilde{V}'$.

Example 2. Let $0 \to \Bbbk \cdot \kappa \to \widetilde{\mathfrak{g}} \to \mathfrak{g} \to 0$ be a central extension of a finite-dimensional Lie algebra. Then the quotient $U\widetilde{\mathfrak{g}}/(\kappa - 1)$ can be viewed as a nonhomogeneous quadratic algebra with the space of generators $\mathfrak{g}$ and the defining relations $X \otimes Y - Y \otimes X - [X, Y] - c(X, Y)$, where $X, Y \in \mathfrak{g}$, and $c(X, Y)$ is a 2-cocycle describing the extension.

Example 3. A finite-dimensional algebra A can be viewed as a nonhomogeneous quadratic algebra with the space of generators $\widetilde{V} = A$ and the defining relations $a \otimes a' - aa'$ for $a, a' \in A$.

For a vector space $\widetilde{V}$ with a fixed vector $1_A \in \widetilde{V}$ consider the modified tensor algebra $\widetilde{\mathbb{T}}(\widetilde{V})$ defined above. A subspace $J_2 \subset \widetilde{\mathbb{T}}_2(\widetilde{V})$ appears as the space of quadratic relations of a nonhomogeneous quadratic algebra A with generators $\widetilde{V} \subset A$ and unit $1_A \in \widetilde{V}$ iff $J_2 \cap \widetilde{\mathbb{T}}_1(\widetilde{V}) = 0$ and the ideal $J \subset \widetilde{\mathbb{T}}(\widetilde{V})$ generated by J_2 intersects $\widetilde{\mathbb{T}}_2(\widetilde{V})$ exactly at J_2. For any subspace $J_2 \subset \widetilde{\mathbb{T}}_2(\widetilde{V})$ we can consider the quotient algebra $A = \widetilde{\mathbb{T}}(\widetilde{V})/(J_2)$. Note that the natural map $\widetilde{V} \to A$ is not necessarily an embedding. However, its image generates A as an algebra, and we can consider the corresponding filtration $F_n A = \widetilde{V}^n$ (n-tuple product in A). On the other hand, we have the quadratic algebra $A^{(0)} = \{V, I\}$ with $V = \widetilde{V}/\langle 1_A \rangle$ and $I = J_2 \bmod \widetilde{\mathbb{T}}_1(\widetilde{V})$. In this situation A is a nonhomogeneous quadratic algebra with generators $\widetilde{V}$ and relations J_2 iff the quadratic part of the associated graded algebra $\operatorname{gr}^F A$ is isomorphic to $A^{(0)}$. For example, if A is an algebra equipped with a filtration $\Bbbk = F_0 A \subset F_1 A \subset \cdots \subset A$ such that $\operatorname{gr}^F A$ is quadratic then A is a nonhomogeneous quadratic algebra with $\widetilde{V} = F_1 A$.

Let us choose a complementary subspace $V \subset \widetilde{V}$ to the line spanned by 1_A. Then we have a natural isomorphism $\widetilde{\mathbb{T}}(\widetilde{V}) \simeq \mathbb{T}(V)$, so that $\widetilde{\mathbb{T}}_2(\widetilde{V}) \simeq V^{\otimes 2} \oplus V \oplus \Bbbk$. Every subspace $J_2 \subset \widetilde{\mathbb{T}}_2(\widetilde{V})$ such that $J_2 \cap \widetilde{\mathbb{T}}_1(\widetilde{V}) = 0$ is the graph of a linear map $\psi \colon I \longrightarrow V \oplus \Bbbk$, i.e.,

$$J_2 = \{a + \psi(a) \mid a \in I\}. \tag{1.1}$$

Let us write $\psi = (\varphi, \theta)$, where $\varphi \colon I \longrightarrow V$, $\theta \colon I \longrightarrow \Bbbk$.

Proposition 1.1. *If $J_2 \subset \widetilde{\mathbb{T}}_2(\widetilde{V})$ is the space of quadratic relations of a nonhomogeneous quadratic algebra generated by $\widetilde{V}$ then the map $\psi = (\varphi, \theta)$ satisfies the following condition:*

$$(\psi^{12} - \psi^{23})(V \otimes I \cap I \otimes V) \subset \{a + \psi(a) \mid a \in I\}, \tag{1.2}$$

where $V \otimes I \cap I \otimes V \subset V^{\otimes 3}$, $\psi^{12} = \psi \otimes \mathrm{id} : I \otimes V \to V^{\otimes 2} \oplus V$ (resp., $\psi^{23} = \mathrm{id} \otimes \psi : V \otimes I \to V^{\otimes 2} \oplus V$). It can be rewritten as the system of equations

$$(\varphi^{12} - \varphi^{23})(V \otimes I \cap I \otimes V) \subset I \tag{1.3}$$

$$(\varphi \circ (\varphi^{12} - \varphi^{23}) - (\theta^{12} - \theta^{23}))(V \otimes I \cap I \otimes V) = 0 \tag{1.4}$$

$$(\theta \circ (\varphi^{12} - \varphi^{23}))(V \otimes I \cap I \otimes V) = 0. \tag{1.5}$$

Proof: For $x \in V \otimes I \cap I \otimes V$ we have $x + \psi^{12}(x) \in J$ and $x + \psi^{23}(x) \in J$. Hence, $\psi^{12}(x) - \psi^{23}(x) \in J \cap \widetilde{\mathbb{T}}_2(\widetilde{V}) = J_2$. □

For a quadratic-linear algebra we have $\theta = 0$ and equations (1.4) and (1.5) reduce to

$$\varphi \circ (\varphi^{12} - \varphi^{23}) = 0. \tag{1.6}$$

In the nonaugmented case the definition of φ and θ depends on the choice of a complementary subspace $V \subset \widetilde{V}$. If $V' \subset \widetilde{V}$ is another such subspace then there is a map $\alpha\colon V \longrightarrow \Bbbk$ such that $V' = \{v - \alpha(v) \mid v \in V\}$. The corresponding maps (φ', θ') are related to (φ, θ) by

$$\varphi' = \varphi + \alpha \otimes \mathrm{id} + \mathrm{id} \otimes \alpha, \qquad \theta' = \theta + \alpha \circ \varphi + \alpha \otimes \alpha. \tag{1.7}$$

2. Nonhomogeneous PBW-theorem

The following theorem gives a sufficient criterion for the subspace $J_2 \subset V^{\otimes 2} \oplus V \oplus \Bbbk$ to be the space of quadratic relations of a nonhomogeneous quadratic algebra A. In addition, it describes the associated graded algebra of A with respect to the natural filtration (see also [**99, 33**]). We keep the notation of Proposition 1.1.

THEOREM 2.1. *Let $A^{(0)} = \{V, I\}$ be a Koszul algebra, and let $\psi\colon I \longrightarrow V \oplus \Bbbk$ be a map satisfying* (1.2)*. Then the corresponding quotient algebra $A = \mathbb{T}(V)/(J_2)$ is a nonhomogeneous quadratic algebra with $\mathrm{gr}^F A \simeq A^{(0)}$.*

Proof: Let us define recursively $J_n := \widetilde{V} J_{n-1} + J_{n-1}\widetilde{V} \subset \widetilde{\mathbb{T}}_n(\widetilde{V})$, where $n \geqslant 3$. Then the ideal generated by J_2 in $\widetilde{\mathbb{T}}(\widetilde{V})$ is $J = \bigcup_{n \geqslant 2} J_n$. It is easy to see that $\mathrm{gr}^F A \simeq A^{(0)}$ iff $J \cap \widetilde{\mathbb{T}}_n(\widetilde{V}) = J_n$ for all n. Clearly, it suffices to check that $J_n \cap \widetilde{\mathbb{T}}_{n-1}(\widetilde{V}) = J_{n-1}$. Note that condition (1.2) is equivalent to $J_3 \cap \widetilde{\mathbb{T}}_2(\widetilde{V}) = J_2$.

Set $R_i = V^{\otimes i-1} \otimes I \otimes V^{\otimes n-i-1} \subset V^{\otimes n}$ for $i = 1, \dots, n-1$. For any element $x \in J_n$ we have

$$x = \sum_{i=1}^{n-1} (r_i + \psi^{i,i+1} r_i) \mod J_{n-1},$$

where $r_i \in R_i$. If $x \in J_n \cap \widetilde{\mathbb{T}}_{n-1}(\widetilde{V})$ then we have $r_1 + \dots + r_{n-1} = 0$. We need to show that $\sum_{i=1}^{n-1} \psi^{i,i+1} r_i \in J_{n-1}$ in this case.

We use the characterization of Koszul algebras in terms of distributive lattices (see Theorem 4.1 of chapter 2). Distributivity of the collection $(R_1, \dots, R_{n-1})$ implies that

$$R_i \cap (R_{i+1} + \dots + R_{n-1}) = R_i \cap R_{i+1} + \dots + R_i \cap R_{n-1}.$$

Hence, any vector $(r_1, \dots, r_{n-1}) \in \bigoplus_{i=1}^{n-1} R_i$ such that $\sum r_i = 0$ in $V^{\otimes n}$ can be presented as a sum of vectors of the same type with only two nonzero components $r_i = -r_j$ (cf. Proposition 7.2 of chapter 1, (a) $\Longrightarrow$ (c*)). It remains to show that $\psi^{i,i+1} r - \psi^{j,j+1} r \in J_{n-1}$ for any element $r \in R_i \cap R_j$. If $|i - j| = 1$ then this follows from (1.2). Otherwise, we have $\psi^{i,i+1} \psi^{j,j+1} = \psi^{j-1,j} \psi^{i,i+1}$, and hence

$$\psi^{i,i+1} r - \psi^{j,j+1} r = (\mathrm{id} + \psi^{j,j+1})\psi^{i,i+1} r - (\mathrm{id} + \psi^{i,i+1})\psi^{j,j+1} r \in J_{n-1}.$$

□

Another proof will be given in section 7 (see Proposition 7.2).

Example. Without the Koszulness assumption the statement of the above theorem is wrong. For example, consider two nonhomogeneous quadratic algebras generated by x, y and z with defining relations

$$\begin{cases} xy = 1 \\ x^2 + y^2 = 0 \end{cases} \qquad \text{or} \qquad \begin{cases} xy = x + y \\ x^2 + y^2 = 0. \end{cases}$$

In both cases by an easy calculation in degree 4 one can deduce the relation $xy = yx$. On the other hand, we have $I \otimes V \cap V \otimes I = 0$, so the condition (1.2) is trivially satisfied.

Theorem 2.1 justifies to some extent the following

Definition. A nonhomogeneous quadratic algebra A is said to be *Koszul* if the corresponding quadratic algebra $A^{(0)}$ (obtained by taking homogeneous parts of quadratic relations) is Koszul.

3. Nonhomogeneous quadratic modules

In this section we prove analogues of Proposition 1.1 and Theorem 2.1 for modules over nonhomogeneous quadratic algebras. [1]

Definition 1(M). A *nonhomogeneous quadratic module* over a nonhomogeneous quadratic algebra $A \supset \widetilde{V} \supset \Bbbk \cdot 1_A$ is a left A-module M together with a fixed generating subspace $H \subset M$ such that the submodule of relations $L = \ker(A \otimes H \longrightarrow M)$ is generated as an A-module by the subspace $L_1 = L \cap (\widetilde{V} \otimes H) \subset A \otimes H$.

A subspace $L_1 \subset \widetilde{V} \otimes H$ appears as the relation space of a nonhomogeneous quadratic module generated by H iff the submodule $L \subset A \otimes H$ generated by L_1 intersects $\widetilde{V} \otimes H$ exactly at L_1. For any subspace $L_1 \subset \widetilde{V} \otimes H$ we can consider the quotient A-module $M = A \otimes H / A \cdot L_1$ equipped with the filtration $F_n M = F_n A \cdot H$. On the other hand, we have the quadratic $A^{(0)}$-module $M^{(0)} = \langle H, K \rangle_{A^{(0)}}$ where $K = L_1 \bmod \Bbbk \otimes H \subset V \otimes H$. In this situation M is a nonhomogeneous quadratic module with generators H and relations L_1 iff the quadratic part of the associated graded module $\operatorname{gr}^F M$ is isomorphic to $M^{(0)}$. In particular, if $\operatorname{gr}^F A$ is a quadratic algebra and $F_0 M \subset F_1 M \subset \cdots \subset M$ is a filtered A-module such that the $\operatorname{gr}^F A$-module $\operatorname{gr}^F M$ is quadratic, then M is a nonhomogeneous quadratic A-module with $H = F_0 M$.

As before, we choose a complementary hyperplane $V \subset \widetilde{V}$ to the line $\Bbbk \cdot 1_A \subset \widetilde{V}$. Then we have $\widetilde{V} \simeq V \oplus \Bbbk$ and $\widetilde{V} \otimes H \simeq H \otimes V \oplus H$. The subspace $L_1 \subset \widetilde{V} \otimes H$ is the graph of a linear map $\mu \colon K \longrightarrow H$:

$$L_1 = \{c + \mu(c) \mid c \in K\}. \tag{3.1}$$

Proposition 3.1.(M). *Assume that $J_2 \subset \widetilde{\mathbb{T}}_2(\widetilde{V})$ is the space of relations of a nonhomogeneous quadratic algebra A generated by $\widetilde{V}$ and $L_1 \subset \widetilde{V} \otimes H$ is the space of relations of a nonhomogeneous quadratic A-module generated by H. Then the maps ψ and μ defined by (1.1) and (3.1) satisfy the condition*

$$(\psi^{12} - \mu^{23})(V \otimes K \cap I \otimes H) \subset \{c + \mu(c) \mid c \in K\}, \tag{3.2}$$

where $V \otimes K \cap I \otimes H \subset V \otimes V \otimes H$, $\psi^{12} = \psi \otimes \mathrm{id} : I \otimes H \to V \otimes H \oplus H$ (resp., $\mu^{23} = \mathrm{id} \otimes \mu : V \otimes K \to V \otimes H$). Setting $\psi = (\varphi, \theta)$, where $\varphi : I \longrightarrow V$ and $\theta : I \longrightarrow \Bbbk$, we can rewrite this condition as the system of equations

$$(\varphi^{12} - \mu^{23})(V \otimes K \cap I \otimes H) \subset K \tag{3.3}$$

$$(\mu \circ (\varphi^{12} - \mu^{23}) - \theta^{12})(V \otimes K \cap I \otimes H) = 0. \tag{3.4}$$

[1]We are grateful to A. Braverman for pointing out to us the possibility of considering such analogues.

Proof: For $y \in V \otimes K \cap I \otimes H$ the image of $y + \psi^{12}(y)$ is equal to zero in $A \otimes H$, while the image of $y + \mu^{23}(y)$ belongs to $L \subset A \otimes H$. Hence, $\psi^{12}(y) - \mu^{23}(y) \in L \cap \widetilde{V} \otimes H = L_1$. □

If we change the complementary hyperplane $V \subset \widetilde{V}$ to $V' = \{v - \alpha(v) \mid v \in V\}$, where $\alpha : V \longrightarrow \Bbbk$, then the map μ gets replaced by

$$\mu' = \mu + \alpha \otimes \mathrm{id}\,. \tag{3.5}$$

THEOREM 3.2.(M). *Let A be a nonhomogeneous Koszul algebra with $A^{(0)} = \{V, I\}$, and let $M^{(0)} = \langle H, K\rangle_{A^{(0)}}$ be a Koszul $A^{(0)}$-module. Then for any map $\mu\colon K \longrightarrow H$ satisfying* (3.2) *the corresponding quotient module $M = A \otimes H/A \cdot L_1$ is a nonhomogeneous quadratic A-module with* $\mathrm{gr}^F M = M^{(0)}$.

Proof: Consider the natural map $\widetilde{\mathbb{T}}(\widetilde{V}) \otimes H \longrightarrow M$ and denote by $P \subset \widetilde{\mathbb{T}}(\widetilde{V}) \otimes H$ its kernel. We have $P = \bigcup_{n \geqslant 1} P_n$, where $P_1 = L_1$ and $P_n = \widetilde{V} P_{n-1} + J_n \otimes H$ for $n \geqslant 2$ (with $J_n \subset \widetilde{\mathbb{T}}_n(\widetilde{V})$ defined in the proof of Theorem 2.1). To prove the desired isomorphism $\mathrm{gr}^F M \simeq M^{(0)}$ we have to check that $P \cap (\widetilde{\mathbb{T}}_n(\widetilde{V}) \otimes H) = P_n$ for all n. Note that condition (3.2) is equivalent to $P_2 \cap \mathbb{T}_1(\widetilde{V}) \otimes H = P_1$. The rest of the proof is parallel to the corresponding part of the proof of Theorem 2.1: one has to use conditions (1.2) and (3.2) and distributivity of the collection of subspaces $(S_1, \dots, S_n)$, where $S_i = R_i \otimes H$ for $i = 1, \dots, n-1$ and $S_n = V^{\otimes n-1} \otimes K \subset V^{\otimes n} \otimes H$. □

Similarly to the case of algebras we make the following definition.

Definition 2(M). A nonhomogeneous quadratic module M over a nonhomogeneous Koszul algebra A is called *Koszul* if the quadratic module $M^{(0)}$ over the Koszul algebra $A^{(0)}$ is Koszul.

4. Nonhomogeneous quadratic duality

Let A be a nonhomogeneous quadratic algebra with the corresponding quadratic algebra $A^{(0)} = \{V, I\}$. Consider the dual quadratic algebra $B = A^{(0)!}$. Recall that we have $B_2 \simeq I^*$. Define $\varphi^* : B_1 \longrightarrow B_2$ and $\theta_B \in B_2$ as dual to the maps $\varphi\colon I \longrightarrow V$ and $\theta\colon I \longrightarrow \Bbbk$. The relation dual to (1.3) has the form

$$(\varphi^{*12} - \varphi^{*23})(I^{\perp}) \subset V \otimes I^{\perp} + I^{\perp} \otimes V.$$

This means that φ^* extends to an odd derivation of degree $+1$ on B, i.e., a map $d_B : B \to B$ of degree $+1$ such that

$$d_B(xy) = d_B(x)y + (-1)^{\tilde{x}} x d_B(y), \qquad x \in B_{\tilde{x}},\ \ y \in B_{\tilde{y}}.$$

Then (1.4) and (1.5) are equivalent to

$$d_B^2(x) = [\theta_B, x] \text{ for all } x \in B, \qquad d_B(\theta_B) = 0$$

(it suffices to check the first relation for $x \in B_1$). In the augmented case we have $\theta_B = 0$ and $d_B^2 = 0$. The transformation (1.7) corresponds to

$$d'_B(x) = d_B(x) + [\alpha, x], \qquad \theta'_B = \theta_B + d_B(\alpha) + \alpha^2 \tag{4.1}$$

for $\alpha \in B_1$, where $[\ ,\ ]$ denotes the supercommutator.

Now let M be a nonhomogeneous quadratic A-module with the corresponding quadratic $A^{(0)}$-module $M^{(0)} = \langle H, K\rangle_{A^{(0)}}$. Let $N = M^{(0)!}$ be the dual quadratic

B-module. Recall that we have $N_0 = H^*$ and $N_1 \simeq K^*$. Let $\mu^*\colon N_0 \longrightarrow N_1$ be the map dual to μ. The relation dual to (3.3) has the form

$$(\varphi^{*12} - \mu^{*23})(K^\perp) \subset V \otimes K^\perp + I^\perp \otimes H.$$

This means that μ^* can be extended to an odd derivation of the B-module N compatible with the derivation d_B of B, i.e., a map $d_N : N \to N$ of degree $+1$ such that

$$d_N(xu) = d_B(x)u + (-1)^{\tilde{x}} x d_N(u), \qquad x \in B_{\tilde{x}},\ \ u \in N_{\tilde{u}}.$$

The equation (3.4) is equivalent to

$$d_N^2(u) = \theta_B u \text{ for all } u \in N.$$

In particular, in the augmented case we have $d_N^2 = 0$. The transformation (3.5) corresponds to

$$d'_N(u) = d_N(u) + \alpha u.$$

Definition 1.(cf. [**99**], [**110**]) A *curved DG-algebra* (*CDG-algebra*) $\boldsymbol{B}$ is a triple $\boldsymbol{B} = (B, d_B, \theta_B)$, where $B = \bigoplus_{i=0}^\infty B_i$ is a graded algebra, $d_B\colon B_i \longrightarrow B_{i+1}$ is an odd derivation of degree $+1$, and $\theta_B \in B_2$ is an element such that $d_B^2(x) = [\theta_B, x]$ and $d_B(\theta_B) = 0$.

A *morphism* of CDG-algebras $\boldsymbol{g}\colon \boldsymbol{C} \longrightarrow \boldsymbol{B}$ is given by a pair $\boldsymbol{g} = (g, \alpha)$, where $g\colon C \longrightarrow B$ is a morphism of graded algebras and $\alpha \in B_1$ is an element such that $g(d_C x) = d_B g(x) + [\alpha, g(x)]$ and $g(\theta_C) = \theta_B + d_B\alpha + \alpha^2$.

A CDG-algebra is called *quadratic* (resp., *Koszul*) if the underlying graded algebra is quadratic (resp., Koszul).

Note that the transformation $\boldsymbol{B} = (B, d_B, \theta_B) \longmapsto \boldsymbol{B}' = (B, d'_B, \theta'_B)$ given by (4.1) leads to an isomorphism $\boldsymbol{\alpha} = (\mathrm{id}, \alpha)\colon \boldsymbol{B}' \longrightarrow \boldsymbol{B}$ in the category of CDG-algebras.

A nonnegatively graded *DG-algebra* $\boldsymbol{B} = (B, d_B)$ (see Appendix) can be viewed as a CDG-algebra with $\theta_B = 0$. *Morphisms* of such DG-algebras are given by morphisms of the corresponding CDG-algebras (g, α) with $\alpha = 0$.

PROPOSITION 4.1. *(i) The above construction defines an anti-equivalence of the category of nonhomogeneous quadratic (resp., quadratic-linear) algebras with a full subcategory of the category of quadratic CDG-algebras (resp., DG-algebras).*
(ii) We have an anti-equivalence of the category of nonhomogeneous (resp. quadratic-linear) Koszul algebras with the category of Koszul CDG-algebras (resp. DG-algebras).

Proof: Given a nonhomogeneous quadratic algebra $A \supset \widetilde{V}$, we choose a complementary hyperplane $V \subset \widetilde{V}$ to the line $\Bbbk$ and define the corresponding CDG-algebra (B, d_B, θ_B) as above. A morphism $f\colon A' \longrightarrow A''$ of nonhomogeneous quadratic algebras is determined by its restriction $p\colon \widetilde{V}' \longrightarrow \widetilde{V}''$ to the generating subspace $\widetilde{V}' \subset A'$. Let $V' \subset \widetilde{V}'$ and $V'' \subset \widetilde{V}''$ be complementary subspaces to the unit lines. Then we can write $p|_{V'} = q - \alpha$, where $q \in \mathrm{Hom}(V', V'')$, and $\alpha \in V'^*$. It is easy to check that the dual map $q^*\colon V''^* \longrightarrow V'^*$ extends to a homomorphism of the dual quadratic algebras $g\colon B'' \longrightarrow B'$ and the pair $\boldsymbol{g} = (g, \alpha)$ defines a morphism of the corresponding CDG-algebras. This functor is fully faithful since a map between generating subspaces extends a morphism of nonhomogeneous quadratic algebras iff it preserves the quadratic relations. Part (ii) follows from Theorem 2.1. (See also [**99**].) □

We will describe the class of quadratic CDG-algebras appearing in part (i) of the above proposition in Proposition 7.2.

Definition 2(M). A *(left) curved DG-module (CDG-module)* $\boldsymbol{N} = (N, d_N)$ over a CDG-algebra $\boldsymbol{B} = (B, d_B, \theta_B)$ is a graded left B-module N equipped with an odd derivation $d_N\colon N_i \longrightarrow N_{i+1}$ compatible with d_B and such that $d_N^2(u) = \theta_B u$. A *morphism* of CDG-modules $(N, d_N) \to (N', d_{N'})$ is a degree 0 morphism $f : N \to N'$ of B-modules such that $d_{N'} f = f d_N$. A CDG-module $\boldsymbol{N}$ over a quadratic (resp., Koszul) CDG-algebra $\boldsymbol{B}$ is called *quadratic* (resp., *Koszul*) if such is the underlying graded B-module N.

Note that a *DG-module* over a nonnegatively graded DG-algebra $\boldsymbol{B}$ (see Appendix) is the same as a CDG-module over the algebra $\boldsymbol{B}$ viewed as a CDG-algebra. We say that a DG-module is quadratic (resp., Koszul) if the underlying graded module over B is quadratic (resp., Koszul).

It is easy to see that the categories of CDG-modules over isomorphic CDG-algebras are naturally equivalent. More generally, for any morphism of CDG-algebras $\boldsymbol{g} = (g, \alpha)\colon \boldsymbol{C} \longrightarrow \boldsymbol{B}$ and a CDG-module $\boldsymbol{N}$ over $\boldsymbol{B}$ the *pull-back of $\boldsymbol{N}$ along $\boldsymbol{g}$* is a CDG-module over $\boldsymbol{C}$ defined by $\boldsymbol{g}^*\boldsymbol{N} = (g^*N, d_N^\alpha)$, where g^*N is the same graded vector space N equipped with a C-module structure via g and $d_N^\alpha(u) = d_N(u) + \alpha u$.

PROPOSITION 4.2.(M). *(i) For any nonhomogeneous quadratic algebra A the category of nonhomogeneous quadratic A-modules is anti-equivalent to a full subcategory in the category of quadratic CDG-modules over the dual quadratic CDG-algebra $\boldsymbol{B}$.*
(ii) If in addition A is Koszul then the category of nonhomogeneous Koszul A-modules is anti-equivalent to the category of Koszul CDG-modules over $\boldsymbol{B}$. □

If A is a quadratic-linear algebra then one can replace in the above proposition CDG-modules by DG-modules over the dual quadratic DG-algebra $\boldsymbol{B}$.

Example. The trivial module $\Bbbk$ over a quadratic-linear algebra A is dual to the free DG-module $\boldsymbol{N} = \boldsymbol{B}$ with $d_N = d_B$. Note that there is no natural way to define a CDG-module structure on the free B-module $N = B$ for a CDG-algebra $\boldsymbol{B}$ with $\theta_B \neq 0$. On the other hand, for any CDG-algebra $\boldsymbol{B}$ one can consider the trivial CDG-module $N = N_0 = \Bbbk$ with $d_N = 0$. It is dual to the free A-module $M = A$.

Note that the opposite algebra A^{op} to a nonhomogeneous quadratic algebra A is also a nonhomogeneous quadratic algebra with the same generating space $\widetilde{V}$. A right A-module R with a fixed generating subspace $H \subset R$ is called a *right nonhomogeneous quadratic module* if the opposite module R^{op} over A^{op} is a left nonhomogeneous quadratic A^{op}-module with respect to generators H. The CDG-algebra dual to A^{op} is $\boldsymbol{B}^{\mathrm{op}} = (B^{\mathrm{op}}, d_{B^{\mathrm{op}}}, \theta_{B^{\mathrm{op}}})$, where (contrary to our conventions in chapter 1) we equip B^{op} with the multiplication that incorporates the sign rule $x^{\mathrm{op}}y^{\mathrm{op}} = (-1)^{\tilde{x}\tilde{y}}(yx)^{\mathrm{op}}$, and set $d_{B^{\mathrm{op}}}(x^{\mathrm{op}}) = (d_B(x))^{\mathrm{op}}$ and $\theta_{B^{\mathrm{op}}} = -(\theta_B)^{\mathrm{op}}$. We define a *right CDG-module $\boldsymbol{S}$* over $\boldsymbol{B}$ as a right graded B-module S equipped with a differential d_S of degree $+1$ such that $d_S(sx) = d_S(s)x + (-1)^{\tilde{s}} s d_B(x)$ and $d_S^2(s) = -s\theta_B$. There is a natural equivalence between the categories of right CDG-modules over $\boldsymbol{B}$ and left CDG-modules over $\boldsymbol{B}^{\mathrm{op}}$ given by the rule $\boldsymbol{S} \longmapsto \boldsymbol{S}^{\mathrm{op}}$, where $x^{\mathrm{op}}s^{\mathrm{op}} = (-1)^{\tilde{x}\tilde{s}}(sx)^{\mathrm{op}}$ and $d_{S^{\mathrm{op}}}(s^{\mathrm{op}}) = (d_S(s))^{\mathrm{op}}$. The right module version of nonhomogeneous quadratic duality associates with a right nonhomogeneous A-module $R = H \otimes A/\{c + \mu(c) \mid c \in K\} \cdot A$ the right quadratic CDG-module $\boldsymbol{S} = (S, d_S)$ over $\boldsymbol{B}$ such that $S = R^{(0)!}$ and $d_{S,0} = \mu^*$.

5. Examples

Here we describe examples of solutions of our equations (1.2) and (3.2) and the corresponding nonhomogeneous quadratic duality. In particular, Example 5 includes the case of Lie superalgebras, while Example 6 is related to quantum groups.

1. Any finite-dimensional algebra A can be viewed as a nonhomogeneous quadratic algebra with $\widetilde{V} = A$. In this case $I = V \otimes V$, $A^{(0)} = \Bbbk \oplus V$, and (1.2) is just the associativity condition for the map $\psi\colon V \otimes V \longrightarrow V \oplus \Bbbk$. Similarly, an augmented algebra A can be viewed as a quadratic-linear algebra with $V = A_+$. The dual DG-algebra is the normalized cobar-complex $\mathcal{COB}^\bullet(A) = \bigoplus_n A_+^{*\otimes n}$ (see section 1 of chapter 1). We can view a finite-dimensional A-module M as a nonhomogeneous quadratic A-module with $H = M$. The dual DG-module over $\mathcal{COB}^\bullet(A)$ can be identified with the cobar-complex $\mathcal{COB}^\bullet(A, M) = \bigoplus_n A_+^{*\otimes n} \otimes M^*$.

2. More generally, we can view a finite-dimensional module M over any nonhomogeneous quadratic algebra $A \supset \widetilde{V}$ as a nonhomogeneous quadratic module with $H = M$. The corresponding graded $A^{(0)}$-module is trivial: $M^{(0)} = M_0^{(0)} = H$. In this case $K = V \otimes H$ and (3.4) is just the compatibility of the action map $\mu\colon V \otimes H \longrightarrow H$ with quadratic relations in A. As the dual object we get a CDG-module structure on the free B-module $N = B \otimes M^*$. The differential d_N can be described explicitly as follows. The vector space M^* has a natural structure of a right A-module, so we can view the tensor product $B \otimes M^*$ as a right $B \otimes A$-module. Let $e_A \in V^* \otimes V \subset B \otimes A$ be the identity element. We can consider e_A as an operator on $B \otimes M^*$. Then it is easy to verify that $d_N = d_B \otimes \mathrm{id} + e_A$ (for $d_B = 0$ such complexes appear in section 3 of chapter 2).

Assume in addition that A is quadratic-linear, so that $d_N^2 = 0$. The complex $K_\bullet(A, M) = B^* \otimes M$ dual to $\boldsymbol{N}$ is sometimes called the *nonhomogeneous Koszul complex* of a module M over a quadratic-linear algebra A (its definition can be extended to the case when M is infinite-dimensional). If A is Koszul then the homology of $K_\bullet(A, M)$ is isomorphic to $\mathrm{Tor}_*^A(\Bbbk, M)$ (as follows from Proposition 6.1 below). In particular, for $M = A$ we get a free resolution of the right A-module $\Bbbk$ called the *nonhomogeneous Koszul resolution*. More general nonhomogeneous Koszul complexes are considered in [**104**].

3. Let $A^{(0)}$ be the symmetric algebra $\mathbb{S}(V)$, so that $I = \bigwedge^2 V \subset V \otimes V$. Then (1.3) is satisfied automatically and (1.6) is equivalent to the Jacobi identity for the skew-symmetric bracket given by $\varphi\colon \bigwedge^2 V \longrightarrow V$. Furthermore, any element $\theta \in B_2$ is central, so (1.4) is equivalent to (1.6), so φ is a Lie algebra structure on V. Assume first that $\theta = 0$, i.e., A is quadratic-linear. In this case A is isomorphic to the universal enveloping algebra $U(V)$. The dual DG-algebra is the standard cohomological complex $C^\bullet(V)$. The standard complexes $C^\bullet(V, M)$ for V-modules M appear in the nonhomogeneous duality for modules (see the previous example). In the general case θ is a 2-cocycle on V and A is the enveloping algebra of the corresponding central extension $0 \longrightarrow \Bbbk\varkappa \longrightarrow \widehat{V} \longrightarrow V \longrightarrow 0$, i.e.,

$A = U(\widehat{V})/(\varkappa - 1)$. Note that transformation (1.7) corresponds to adding to θ the coboundary 2-cocycle $d\alpha$.

4. Let $A(0) = \bigwedge(V)$ be the exterior algebra, so that $I = \mathbb{S}^2 V \subset V \otimes V$. Assume first that $\operatorname{char} \Bbbk \neq 2$. Equation (1.3) implies that $\varphi(\xi^{\otimes 2}) \otimes \xi - \xi \otimes \varphi(\xi^{\otimes 2}) = 0$ for $\xi \in V$. Therefore, $\varphi(\xi^2) = \lambda(\xi)\xi$ for some linear function $\lambda \colon V \longrightarrow \Bbbk$. If φ has this form then equations (1.4) and (1.5) are trivially satisfied. Using transformation (1.7) with $\alpha = -\lambda/2$ we can make $\varphi = 0$. Therefore, the algebra A is the Clifford algebra associated with the quadratic form θ on the vector space V. In characteristic 2 we get a structure of a restricted Lie algebra on V, i.e., a Lie bracket together with a quadratic map $V \longrightarrow V : x \longmapsto x^2$ such that $[x, y] = (x + y)^2 - x^2 - y^2$ and $[x, [x, y]] = [x^2, y]$. In this case the construction from **2** provides the well-known Koszul resolution for 2-restricted Lie algebras (see [**104**]). In the nonaugmented case we also get a cocycle c on V together with a quadratic map $\tilde{c} \colon V \longrightarrow \Bbbk$ such that $c(x, y) = \tilde{c}(x + y) - \tilde{c}(x) - \tilde{c}(y)$ and $c(x, [x, y]) = c(x^2, y)$.

5. Let $V = V_0 \oplus V_1$ be a super vector space and let $I \subset V \otimes V$ be the space of quadratic relations of the free commutative superalgebra generated by V: $I \simeq \bigwedge^2 V_0 \oplus V_0 \otimes V_1 \oplus \mathbb{S}^2 V_1$, where the embedding of $V_0 \otimes V_1$ into $V \otimes V$ is given by $v_0 \otimes v_1 \longmapsto v_0 \otimes v_1 - v_1 \otimes v_0$. Let $(\varphi, \theta) \colon I \longrightarrow V \oplus \Bbbk$ be a map satisfying equations (1.3), (1.4) and (1.5). It is easy to see that if this map preserves parity then our equations just mean that φ defines a Lie superalgebra structure and θ is an even 2-cocycle (in the characteristic 2 or 3 as well). Let us denote by φ_c^{ab} and θ^{ab} the components of φ and θ with respect to the above decompositions of V and I (so that $\varphi_c^{00} : \bigwedge^2 V_0 \to V_c$, $\varphi_c^{01} : V_0 \otimes V_1 \to V_c$, etc.). The linear condition (1.3) is equivalent to the system

$$\begin{aligned} &\varphi_1^{00}(x \wedge y) \otimes \xi - \xi \otimes \varphi_1^{00}(x \wedge y) \in \mathbb{S}^2 V_1, \\ &\qquad \varphi_0^{01}(x\xi) \otimes \xi + \xi \otimes \varphi_0^{01}(x\xi) \in \langle\, v_0 \otimes v_1 - v_1 \otimes v_0 \mid v_s \in V_s \,\rangle, \\ &\qquad \varphi_1^{11}\xi^2 \otimes \xi - \xi \otimes \varphi_1^{11}\xi^2 \in \mathbb{S}^2 V_1, \end{aligned}$$

where x, $y \in V_0$, $\xi \in V_1$, and $x\xi = x \otimes \xi - \xi \otimes x$. Now assume that $\operatorname{char} \Bbbk \neq 2$ and $\dim V_1 > 1$. Then it follows that $\varphi_1^{00} = 0$, $\varphi_0^{01} = 0$, and $\varphi_1^{11}(\xi^2) = \lambda(\xi)\xi$ as in section **4**. Using transformation (1.7) we can also make $\varphi_1^{11} = 0$. Then it follows from (1.4) that $\theta^{01} = 0$, and therefore A is the enveloping algebra of an even central extension of a Lie superalgebra.

Note that in the remaining cases $\dim V_1 = 1$ or $\operatorname{char} \Bbbk = 2$ the subspace $I \subset V^{\otimes 2}$ is determined by the subspace $V_1 \subset V$. Consider first the case $\dim V_1 = 1$ and $\operatorname{char} \Bbbk \neq 2$. Then $I = \bigwedge^2 V \oplus V_1^{\otimes 2}$. In this case the only nontrivial linear condition is that $\varphi(x \otimes \xi - \xi \otimes x) \in V_1$ for any $x \in V$ and $\xi \in V_1$. Let $([\ \ ], m)$ and (c, f) be the components of φ and θ with respect to the decomposition of I. Our conditions mean that $[\ \ ]$ is a Lie algebra structure on V, $V_1 \subset V$ is an ideal, c is a 2-cocycle, and the following equations are satisfied:

$$[x, m(\xi^2)] = 2m(\xi[x, \xi]) - 2c(x, \xi)\xi, \qquad c(x, m(\xi^2)) = 2f(\xi[x, \xi]),$$

where $x \in V$, $\xi \in V_1$. The transformation (1.7) has the following form:

$$\begin{gathered} m'(\xi^2) = m(\xi^2) + 2\alpha(\xi)\xi, \qquad c'(x, y) = c(x, y) + \alpha([x, y]), \\ f'(\xi^2) = f(\xi^2) + \alpha(m(\xi^2)) + \alpha(\xi)^2. \end{gathered}$$

Consider the central extension $\widehat{V}$ of V defined by the cocycle c, and let $U = U(\widehat{V})/(\kappa - 1)$ be the corresponding enveloping algebra. Then our algebra A is isomorphic to $U/(z)$, where $z = \xi^2 + m(\xi^2) + f(\xi^2)$ is a quadratic element in U. The above equations are equivalent to the condition that z is normal, i.e., $Uz = zU$. It is easy to see that any normal quadratic element of U with the symbol of the form $\xi^2 \in \mathbb{S}^2(V)$ for some $\xi \in V$ arises in this way.

If $\operatorname{char} \Bbbk = 2$ then the linear conditions are trivial. Our data consist of the Lie algebra structure on V together with a restriction map $x \longmapsto x^2$ (as in section **4**) defined for $x \in V_1$ and the appropriate 2-cocycle.

6. Let $(\Gamma, +)$ be an abelian monoid, $\Phi \subset \Gamma$ be a finite subset. Consider the algebra A with generators x_α numbered by $\alpha \in \Phi$ and defining relations

$$x_\alpha x_\beta = q_{\alpha,\beta} x_\beta x_\alpha + c_{\alpha,\beta} x_{\alpha+\beta},$$

where we set $x_\alpha = 0$ for $\alpha \notin \Phi$ and assume that $q_{\beta,\alpha} q_{\alpha,\beta} = 1$, $q_{\alpha,\alpha} = 1$, $c_{\beta,\alpha} = -q_{\beta,\alpha} c_{\alpha,\beta}$ and $c_{\alpha,\alpha} = 0$. It is easy to see that the analogue of the Jacobi identity in this case reduces to the following equations:

$$(q_{\alpha,\gamma} q_{\beta,\gamma} - q_{\alpha+\beta,\gamma}) c_{\alpha,\beta} = 0,$$

$$q_{\gamma,\alpha} c_{\alpha,\beta} c_{\alpha+\beta,\gamma} + q_{\alpha,\beta} c_{\beta,\gamma} c_{\beta+\gamma,\alpha} + q_{\beta,\gamma} c_{\gamma,\alpha} c_{\gamma+\alpha,\beta} = 0,$$

where $\alpha, \beta, \gamma \in \Phi$. Note that $A^{(0)}$ is a PBW-algebra (see Example 1 of section 2 of chapter 4), hence it is Koszul. Therefore, Theorem 2.1 implies that $\operatorname{gr}^F A \simeq A^{(0)}$ provided the above equations are satisfied. For example, this is the case for the quantum universal enveloping algebra $U_q\mathfrak{n} \subset U_q sl(N)$, where as generators we take the root vectors (see [**108**]).

7. Let A be a nonhomogeneous quadratic algebra with $A^{(0)} = \{V, I\}$, and let $E \subset V$ be a subspace. We want to study nonhomogeneous quadratic A-modules M such that the associated quadratic module $M^{(0)}$ is generated by a given vector space $H \neq 0$ with defining relations $K = E \otimes H \subset V \otimes H$, i.e., $M^{(0)} \simeq (A^{(0)}/A^{(0)}E) \otimes H$. Let $\boldsymbol{B}$ be the dual quadratic CDG-algebra to A, and let $\boldsymbol{N}$ be the quadratic CDG-module corresponding to such M. By definition, we have $N = B/BE^\perp \otimes H^*$, where $E^\perp \subset B_1$ is the orthogonal complement to E. Let us assume that $E^\perp$ is left-normal in B, i.e., $E^\perp B \subset BE^\perp$. Then the quotient module $\overline{B} = B/BE^\perp$ is a quadratic algebra. Now for any $x \in E^\perp$ and $u \in H$ one has

$$0 = d_N(xu) = d_B(x)u - x d_N(u) = d_B(x)u.$$

This implies that $d_B(E^\perp) = 0$. Thus, in order to have any CDG-module structures on N one should have $d_B(E^\perp) = 0$. Assume that this is the case. Then we can equip $\overline{B}$ with the CDG-algebra structure, such that the differential $d_{\overline{B}}$ is induced by d_B and the element $\theta_{\overline{B}} \in \overline{B}_2$ is the image of $\theta_B \in B_2$. Let us call the obtained CDG-algebra $\overline{\boldsymbol{B}}$. The CDG-module $\boldsymbol{N}$ over $\boldsymbol{B}$ is the pull-back of a CDG-module over $\overline{\boldsymbol{B}}$. Dualizing the relation $d_B(E^\perp) = 0$ we get $\varphi(E^{\otimes 2} \cap I) \subset E$. It follows easily that there exists a nonhomogeneous quadratic algebra $\overline{A}$ dual to $\overline{\boldsymbol{B}}$ together with a homomorphism $\overline{A} \longrightarrow A$ and an $\overline{A}$-module structure on H. The original A-module M is isomorphic to the induced module $M = A \otimes_{\overline{A}} H$.

6. Nonhomogeneous duality and cohomology

Now we are going to show that nonhomogeneous quadratic duality provides a way to compute the cohomology of nonhomogeneous Koszul modules over Koszul quadratic-linear algebras.

PROPOSITION 6.1. *Let $A \supset A_+ \supset V$ be a Koszul quadratic-linear algebra and let $M \supset H$ be a nonhomogeneous Koszul A-module. Let also $\boldsymbol{B}$ be the quadratic dual DG-algebra to A and $\boldsymbol{N}$ the quadratic dual DG-module to M. Then there is a natural quasi-isomorphism of DG-algebras*

$$\mathcal{COB}^\bullet(A) \longrightarrow \boldsymbol{B} \tag{6.1}$$

and a compatible quasi-isomorphism of DG-modules

$$\mathcal{COB}^\bullet(A, M) \longrightarrow \boldsymbol{N}. \tag{6.2}$$

Passing to cohomology we obtain an isomorphism of algebras

$$\mathrm{EXT}^*_A(\Bbbk, \Bbbk) \simeq H^*(\boldsymbol{B}) := H^*_{d_B}(B)$$

and a compatible isomorphism of modules

$$\mathrm{EXT}^*_A(M, \Bbbk) \simeq H^*(\boldsymbol{N}) := H^*_{d_N}(N).$$

Proof: We have $\mathcal{COB}^\bullet(A) = \mathcal{B}ar_\bullet(A)^\vee$ and (6.1) will be dual to a natural morphism of DG-coalgebras

$$\boldsymbol{B}^* \longrightarrow \mathcal{B}ar_\bullet(A) \tag{6.3}$$

constructed as follows. Since $\mathcal{B}ar_\bullet(A)$ is the cofree coalgebra cogenerated by $A_+ = \mathcal{B}ar_1(A)$, the natural embedding $B_1^* = V \hookrightarrow A_+$ extends uniquely to a morphism of coalgebras $B^* \longrightarrow \mathcal{B}ar_\bullet(A)$. This morphism commutes with differentials due to commutativity of the diagram

$$\begin{CD} I @>>> A_+ \otimes A_+ \\ @V{\varphi}VV @VV{m}V \\ V @>>> A_+ \end{CD}$$

where $I \subset V^{\otimes 2}$ is the space of quadratic relations in $A^{(0)}$, $m : A_+ \otimes A_+ \longrightarrow A_+$ is the multiplication map. Similarly, since $\mathcal{B}ar_\bullet(A, M)$ is the cofree comodule over $\mathcal{B}ar_\bullet(A)$ cogenerated by $M = \mathcal{B}ar_0(A, M)$, the identical embedding $N_0^* = H \longrightarrow M$ extends uniquely to a comodule homomorphism $N^* \longrightarrow \mathcal{B}ar_\bullet(A, M)$. Moreover, we get a morphism of DG-comodules

$$\boldsymbol{N}^* \longrightarrow \mathcal{B}ar_\bullet(A, M) \tag{6.4}$$

as follows from commutativity of the diagram

$$\begin{CD} K @>>> A_+ \otimes M \\ @V{\mu}VV @VV{a}V \\ H @>>> A_+ \end{CD}$$

where $K \subset V \otimes H$ is the space of quadratic relations in $M^{(0)}$, $a : A_+ \otimes M \longrightarrow M$ is the action map. We define (6.2) to be dual to (6.4). It remains to show that (6.3) and (6.4) are quasi-isomorphisms. The increasing filtrations F on the algebra

A and A-module M induce an increasing filtration on the complex $\mathcal{B}ar_\bullet(A,M)$ by the rule

$$F_j(A_+^{\otimes i}\otimes M)=\sum_{j_1+\cdots+j_i+k=n} F_{j_1}A_+\otimes\cdots\otimes F_{j_i}A_+\otimes F_kM.$$

Set $F_{j-1}N_j^*=0$ and $F_jN_j^*=N_j^*$. Then (6.4) is compatible with filtrations and the induced morphism of associated graded complexes is $N^*\longrightarrow \mathcal{B}ar_\bullet(\mathrm{gr}^F A,\ \mathrm{gr}^F M)$, where N^* is equipped with a zero differential. Since the algebra $A^{(0)}=\mathrm{gr}^F A$ and the module $M^{(0)}=\mathrm{gr}^F M$ are Koszul, the latter map is a quasi-isomorphism (see Proposition 3.1 of chapter 1). It follows that (6.4) is also a quasi-isomorphism. The same argument works for (6.3). □

Remark 1. Let us define a *Koszul quadratic-linear structure* on an augmented algebra A as a choice of a generator subspace $V\subset A_+\subset A$ with respect to which A becomes nonhomogeneous Koszul. In this situation we say that V is a *Koszul generator subspace.* The above result allows us to associate (functorially) with an augmented algebra admitting a Koszul quadratic-linear structure a quasi-isomorphism class of DG-algebras. Namely, we claim that the DG-algebra $\boldsymbol{B}$ above does not depend on the choice of a Koszul generator subspace $V\subset A$. Indeed, if $V'\subset V''\subset A_+$ are two Koszul generating subspaces for A then the identity map $(A,V')\longrightarrow(A,V'')$ is a morphism of quadratic-linear algebras, and therefore it induces a morphism of the dual DG-algebras $\boldsymbol{B}''\longrightarrow\boldsymbol{B}'$. By the above proposition, it is a quasi-isomorphism. Also, it is easy to see that for any two Koszul generating subspaces V', $V''\subset A_+$ there exists a third one V such that $V'+V''\subset V$ (this follows from Proposition 2.2 of chapter 3). Analogously, the quasi-isomorphism class of the DG-module $\boldsymbol{N}$ is an invariant of an augmented algebra A and an A-module M admitting nonhomogeneous Koszul structures.

The question of *existence* of a nonhomogeneous Koszul structure on a given algebra A or module M seems to be very nontrivial. Let us only observe that it should be viewed as a *finiteness* question: indeed, on any finite-dimensional algebra A (resp., module M) there is an obvious nonhomogeneous Koszul structure given by $V=A$ (resp., $H=M$), see Examples 1 and 2 in section 5.

Remark 2. Conversely, one can ask whether the quasi-isomorphism class of the DG-algebra $\boldsymbol{B}$ (resp., DG-module $\boldsymbol{N}$) determines the isomorphism class of an augmented algebra A (resp., A-module M). Here is a simple counterexample. Consider the quadratic-linear algebra $A=\Bbbk[x]/(x^2+x)$. Then we have $A^{(0)}=\Bbbk[x]/x^2$, $B=\Bbbk[\xi]$, and $d_B(\xi)=\xi^2$. Clearly, $H^*_{d_B}(B)=\Bbbk$ and the morphisms of quadratic-linear algebras $\Bbbk\longrightarrow A$ and $A\longrightarrow\Bbbk$ induce quasi-isomorphisms of the dual DG-algebras $\Bbbk\longrightarrow\boldsymbol{B}\longrightarrow\Bbbk$. For the A-module $M=\Bbbk u$, where x acts by $xu=u$, we have $H^*_{d_N}(N)=0$. (See also Example in section 8.)

As we will show in section 8, the quasi-isomorphism class of the dual DG-algebra to A (resp., dual DG-module to M) determines the A_+-adic completion of A (resp., M) and its higher derived functors.

7. Bar construction for CDG-algebras and modules

For a CDG-algebra $\boldsymbol{B}=(B,d_B,\theta_B)$ let $\mathcal{B}ar(B)=\bigoplus_{i=0}^{\infty}B_+^{\otimes i}$ denote the bar construction of the graded algebra B (see section 1 of chapter 1). Recall that it has a natural bigrading $\mathcal{B}ar_{ij}(B)=\langle b_1\otimes\cdots\otimes b_i \mid j_1+\cdots+j_i=j\rangle$, where $b_t\in B_{j_t}$. The CDG-algebra structure on B provides the coalgebra $\mathcal{B}ar(B)$ with three differentials depending on the multiplication in B, the differential d_B and the element θ_B, respectively:

$$\partial(b_1 \otimes \cdots \otimes b_i) = \sum_{t=1}^{i-1} (-1)^{j_1+\cdots+j_t+t} b_1 \otimes \cdots \otimes b_s b_{t+1} \otimes \cdots \otimes b_i,$$

$$d(b_1 \otimes \cdots \otimes b_i) = \sum_{t=1}^{i} (-1)^{j_1+\cdots+j_{t-1}+t-1} b_1 \otimes \cdots \otimes d_B(b_t) \otimes \cdots \otimes b_i,$$

$$\delta(b_1 \otimes \cdots \otimes b_i) = \sum_{t=1}^{i+1} (-1)^{j_1+\cdots+j_{t-1}+t-1} b_1 \otimes \cdots \otimes b_{t-1} \otimes \theta_B \otimes b_t \otimes \cdots \otimes b_i.$$

The first differential ∂ coincides with the differential on $\mathcal{B}ar(B)$ defined in section 1 of chapter 1 up to the signs reflecting the grading of B (these signs do not affect the homology spaces).

Now let $\mathcal{B}ar^\bullet(\boldsymbol{B}) = \mathcal{B}ar^\bullet(B, d_B, \theta_B)$ denote the total complex with the differential $D = \partial + d + \delta$, where $\mathcal{B}ar^n(\boldsymbol{B}) = \bigoplus_{j-i=n} \mathcal{B}ar_{ij}(B)$. It follows from the CDG-algebra axioms that $D^2 = 0$, so $\mathcal{B}ar^\bullet(\boldsymbol{B})$ is a DG-coalgebra. Let $\overline{\mathcal{C}ob}_\bullet(\boldsymbol{B})$ be the dual DG-algebra with $\overline{\mathcal{C}ob}_m(\boldsymbol{B}) = \mathcal{B}ar^m(\boldsymbol{B})^\vee$, and let $\mathcal{C}ob_\bullet(\boldsymbol{B}) \subset \overline{\mathcal{C}ob}_\bullet(\boldsymbol{B})$ be the DG-subalgebra given by $\mathcal{C}ob_m(\boldsymbol{B}) = \mathcal{B}ar^m(\boldsymbol{B})^* = \bigoplus_{j-i=m} \mathcal{B}ar_{ij}(\boldsymbol{B})^*$. This DG-subalgebra looks as follows:

$$\cdots \longrightarrow \mathcal{C}ob_3(\boldsymbol{B}) \longrightarrow \mathcal{C}ob_2(\boldsymbol{B}) \longrightarrow \bigoplus_{j_1, j_2} B_1^{*\otimes j_1} \otimes B_2^* \otimes B_1^{*\otimes j_2} \longrightarrow \bigoplus_{j=0}^{\infty} B_1^{*\otimes j}.$$

Note that as a graded algebra $\mathcal{C}ob(\boldsymbol{B})$ can be identified (up to a linear change in the grading) with the algebra $\mathcal{C}ob(B)$ defined in section 1 of chapter 1. However, the completions $\mathcal{COB}(B)$ and $\overline{\mathcal{C}ob}(\boldsymbol{B})$ are entirely different even as vector spaces.

The next result and Proposition 7 below provide a generalization of Proposition 3.1 of chapter 1 to quadratic CDG-algebras.

PROPOSITION 7.1. *Let $\boldsymbol{B} = (B, d_B, \theta_B)$ be a quadratic CDG-algebra, $\{V, I\} = B^!$ the quadratic dual algebra to B. Consider the maps $\varphi = d^*_{B,1} \colon I \longrightarrow V$ and $\theta = \theta_B \colon I \longrightarrow \Bbbk$ (cf. section 4). Then there is a natural isomorphism $H_0\mathcal{C}ob_\bullet(\boldsymbol{B}) = \mathbb{T}(V)/(J_2)$, where $J_2 = \{a + \varphi(a) + \theta(a) \mid a \in I\}$.*

The proof is straightforward and is left for the reader.

There is a natural increasing filtration $F_k\mathcal{C}ob_m(\boldsymbol{B}) = \bigoplus_{j-i=m}^{j\leqslant k} \mathcal{B}ar_{ij}(\boldsymbol{B})^*$ on the complex $\mathcal{C}ob_\bullet(\boldsymbol{B})$. It induces the standard filtration $F_kA = \widetilde{V}^k$ on $A = H_0\mathcal{C}ob_\bullet(\boldsymbol{B})$. The associated graded complex $\operatorname{gr}^F\mathcal{C}ob_\bullet(\boldsymbol{B})$ can be identified with the usual cobar-complex of the graded algebra B, so its homology is isomorphic to $\operatorname{Ext}^*_B(\Bbbk, \Bbbk)$. Thus, we get a spectral sequence

$$'E^1_{p,q} = \operatorname{Ext}_B^{-q,p}(k,k) \Longrightarrow \operatorname{gr}^F_p H_{p+q}\mathcal{C}ob_\bullet(\boldsymbol{B}) \tag{7.1}$$

with the differentials $d^r_{p,q} \colon E^r_{p,q} \longrightarrow E^r_{p-r,q+r-1}$. Now we can characterize the class of quadratic CDG-algebras dual to nonhomogeneous quadratic algebras.

PROPOSITION 7.2. *(i) A quadratic CDG-algebra $\boldsymbol{B}$ corresponds to a nonhomogeneous quadratic algebra A iff the spectral sequence (7.1) degenerates at $'E^1_{-i,i}$ for $i = 0$, 1, and 2 (i.e., $'E^1_{-i,i} = {}'E^\infty_{-i,i}$ for $i = 0$, 1, and 2). Furthermore, one has $\operatorname{gr}^F A = A^{(0)}$ iff this spectral sequence degenerates at $'E^1_{-i,i}$ for all $i \geqslant 0$.*

(ii) If B is Koszul then (7.1) degenerates at $'E^1$ and $H_{\neq 0}\mathcal{C}ob_\bullet(\boldsymbol{B}) = 0$, so the natural map $\mathcal{C}ob_\bullet(\boldsymbol{B}) \to A$ is a quasi-isomorphism.

Proof: Note that if $\boldsymbol{B}$ corresponds to some nonhomogeneous quadratic algebra then this algebra has to be $A = H_0\mathcal{C}ob_\bullet(\boldsymbol{B})$ (by Proposition 7.1). Now (i) follows from the identifications $'E^1_{-i,i} = \mathrm{Ext}^{ii}_B(\Bbbk,\Bbbk) = A^{(0)}_i$ and $'E^\infty_{-i,i} = \mathrm{gr}^F_i H_0\mathcal{C}ob_\bullet(\boldsymbol{B}) = \mathrm{gr}^F_i A$. For (ii) we observe that if B is a Koszul algebra then all terms in $'E^1$ with $p+q\neq 0$ vanish, so the sequence degenerates. □

Combining parts (i) and (ii) of the above proposition we obtain a new proof of the nonhomogeneous PBW-Theorem 2.1.

Remark 1. The above argument shows that the assumption of Koszulness of $A^{(0)} = B^!$ in Theorem 2.1 can be replaced by the weaker condition $\mathrm{Ext}^{j-1,j}_B(\Bbbk,\Bbbk) = 0$ for $j \geqslant 0$. On the other hand, our first proof of Theorem 2.1 uses the distributivity condition which is easily seen to be equivalent to the vanishing of $\mathrm{Ext}^{3,j}_{A^{(0)}}(\Bbbk,\Bbbk)$ for $j > 3$ (the same condition is also used in the proof of the PBW-theorem for filtrations in section 7 of chapter 4 and in the proof of Theorem 2.1 in chapter 6). These conditions on $A^{(0)}$ and on B are actually dual to each other (see Corollary 3.3 of chapter 2).

Now let $\boldsymbol{S}$ (resp., $\boldsymbol{N}$) be a right (resp., left) CDG-module over a CDG-algebra $\boldsymbol{B}$. Consider the bar construction $\mathcal{B}ar(S,B,N) = \bigoplus_{i=0}^\infty S\otimes B_+^{\otimes i}\otimes N$ (see section 1 of chapter 1) with the standard bigrading $\mathcal{B}ar_{ij}(S,B,N) = \langle s\otimes b_1\otimes\cdots\otimes b_i\otimes n \mid k+j_1+\cdots+j_i+l = j\rangle$, where $s\in S_k$, $b_t\in B_{j_t}$, and $n\in N_l$. Generalizing the above construction of $\mathcal{B}ar(\boldsymbol{B})$ we define the following differentials on $\mathcal{B}ar(S,B,N)$:

$$\partial(s\otimes b\otimes n) = sb\otimes n + (-1)^k s\otimes\partial(b)\otimes n + (-1)^{k+j_1+\cdots+j_i+i} s\otimes bn,$$

$$d(s\otimes b\otimes n) = d_S(s)\otimes b\otimes n + (-1)^k s\otimes d(b)\otimes n + (-1)^{k+j_1+\cdots+j_i+i} s\otimes b\otimes d_N(n),$$

$$\delta(s\otimes b\otimes n) = s\otimes\delta(b)\otimes n,$$

where $b = b_1\otimes\cdots\otimes b_i\in B_{j_1}\otimes\cdots\otimes B_{j_i}$ and $sb = sb_1\otimes\cdots\otimes b_i$, $bn = b_1\otimes\cdots\otimes b_i n$. Let $\mathcal{B}ar^\bullet(\boldsymbol{S},\boldsymbol{B},\boldsymbol{N})$ denote the total complex with the differential $D = \partial + d + \delta$, where $\mathcal{B}ar^n(\boldsymbol{S},\boldsymbol{B},\boldsymbol{N}) = \bigoplus_{j-i=n}\mathcal{B}ar_{ij}(S,B,N)$. Let $\overline{\mathcal{C}ob}_\bullet(\boldsymbol{S},\boldsymbol{B},\boldsymbol{N})$ denote the dual complex with $\overline{\mathcal{C}ob}_m(\boldsymbol{S},\boldsymbol{B},\boldsymbol{N}) = \mathcal{B}ar^m(\boldsymbol{S},\boldsymbol{B},\boldsymbol{N})^\vee$, and let $\mathcal{C}ob_\bullet(\boldsymbol{S},\boldsymbol{B},\boldsymbol{N})\subset\overline{\mathcal{C}ob}_\bullet(\boldsymbol{S},\boldsymbol{B},\boldsymbol{N})$ be the subcomplex given by $\mathcal{C}ob_m(\boldsymbol{S},\boldsymbol{B},\boldsymbol{N}) = \bigoplus_{j-i=m}\mathcal{B}ar_{ij}(S,B,N)^*$. Note that we have $\mathcal{B}ar^\bullet(\boldsymbol{B}) = \mathcal{B}ar^\bullet(\Bbbk,\boldsymbol{B},\Bbbk)$. Let us also set

$$\mathcal{B}ar^\bullet(\boldsymbol{B},\boldsymbol{N}) = \mathcal{B}ar^\bullet(\Bbbk,\boldsymbol{B},\boldsymbol{N}),$$

$$\overline{\mathcal{C}ob}_\bullet(\boldsymbol{B},\boldsymbol{N}) = \overline{\mathcal{C}ob}_\bullet(\Bbbk,\boldsymbol{B},\boldsymbol{N}),$$

$$\mathcal{C}ob_\bullet(\boldsymbol{B},\boldsymbol{N}) = \mathcal{C}ob_\bullet(\Bbbk,\boldsymbol{B},\boldsymbol{N}),$$

where $\Bbbk$ denotes the trivial right CDG-module over $\boldsymbol{B}$ (see Example 2 in section 5). Similarly, we set $\mathcal{B}ar^\bullet(\boldsymbol{S},\boldsymbol{B}) = \mathcal{B}ar^\bullet(\boldsymbol{S},\boldsymbol{B},\Bbbk)$, etc. For example, the complex $\mathcal{C}ob_\bullet(\boldsymbol{B},\boldsymbol{N})$ has the form

$$\cdots\longrightarrow\mathcal{C}ob_2(\boldsymbol{B},\boldsymbol{N})$$

$$\longrightarrow\bigoplus_{j_1,j_2} B_1^{*\otimes j_1}\otimes B_2^*\otimes B_1^{*\otimes j_2}\otimes N_0^*\oplus\bigoplus_j B_1^{*\otimes j}\otimes N_1^*\longrightarrow\bigoplus_{j=0}^\infty B_1^{*\otimes j}\otimes N_0^*.$$

PROPOSITION 7.3.(M). *Let* $\boldsymbol{B}$ *be a quadratic CDG-algebra. Set* $\{V, I\} = B^!$, $A = H_0\mathcal{C}ob_\bullet(\boldsymbol{B})$ *(see Proposition 7.1). Let* $\boldsymbol{N} = (N, d_N)$ *be a quadratic CDG-module over* $\boldsymbol{B}$ *and let* $\langle H, K\rangle_{\{V,I\}} = N^{!_B}$ *be the quadratic dual module to* N. *Consider the map* $\mu = d^*_{N,0}\colon K \longrightarrow H$. *Then there is an isomorphism* $H_0\mathcal{C}ob_\bullet(\boldsymbol{B}, \boldsymbol{N}) = A \otimes H/(A \cdot L_1)$, *where* $L_1 = \{c + \mu(c) \mid c \in K\}$. □

The increasing filtration $F_k\mathcal{C}ob_m(\boldsymbol{B}, \boldsymbol{N}) = \bigoplus_{j-i=m}^{j\leqslant k} \mathcal{B}ar_{ij}(\boldsymbol{B}, \boldsymbol{N})^*$ on the complex $\mathcal{C}ob_\bullet(\boldsymbol{B}, \boldsymbol{N})$ induces the standard filtration $F_kM = \widetilde{V}^kH$ on $M = H_0\mathcal{C}ob_\bullet(\boldsymbol{B}, \boldsymbol{N})$. The homology of the associated graded complex is isomorphic to $\operatorname{Ext}^*_B(N, \Bbbk)$, hence we get a spectral sequence

$$'E^1_{p,q} = \operatorname{Ext}_B^{-q,p}(N, \Bbbk) \Longrightarrow \operatorname{gr}^F_p H_{p+q}\mathcal{C}ob_\bullet(\boldsymbol{B}, \boldsymbol{N}). \tag{7.2}$$

We have the following version of Proposition 7.2 for CDG-modules.

PROPOSITION 7.4.(M). *(i) Let* A *be a nonhomogeneous quadratic algebra and* $\boldsymbol{B}$ *be the dual CDG-algebra. Then a quadratic CDG-module* $\boldsymbol{N}$ *over* $\boldsymbol{B}$ *corresponds to a nonhomogeneous quadratic* A*-module* M *iff the spectral sequence* (7.2) *degenerates at* $'E^1_{-i,i}$ *for* $i = 0$ *and* 1. *One has* $\operatorname{gr}^F M = M^{(0)}$ *iff this spectral sequence degenerates at* $'E^1_{-i,i}$ *for all* $i \geqslant 0$.
(ii) If B *and* N *are Koszul then* (7.2) *degenerates at* $'E^1$ *and* $H_{\neq 0}\mathcal{C}ob_\bullet(\boldsymbol{B}, \boldsymbol{N}) = 0$, *so the natural map* $\mathcal{C}ob_\bullet(\boldsymbol{B}, \boldsymbol{N}) \to M$ *is a quasi-isomorphism.*

Note that the above proposition provides a new proof of Theorem 3.2. We leave it for the reader to prove the analogue of Proposition 7.4 for right CDG-modules.

Remark 2(M). Using the above spectral sequence one can show that the condition of Koszulness of $M^{(0)}$ in Theorem 3.2 can be weakened to

$$\operatorname{Ext}_B^{j-1,j}(N, \Bbbk) = 0 \text{ for all } j. \tag{7.3}$$

In fact, for any quadratic CDG-module $\boldsymbol{N}$ over a quadratic CDG-algebra $\boldsymbol{B}$ such that (7.3) holds, we get a $\widetilde{\mathbb{T}}(\widetilde{V})/(J_2)$-module M with the relations given by d_N and an isomorphism of $A^{(0)}$-modules $\operatorname{gr}^F M \simeq M^{(0)}$. On the other hand, the distributivity condition used in the proof of Theorem 3.2 is equivalent to the vanishing of $\operatorname{Ext}^{3,j}_{A^{(0)}}(\Bbbk, \Bbbk)$ for $j > 3$ and of $\operatorname{Ext}^{2,j}_{A^{(0)}}(M^{(0)}, \Bbbk)$ for $j > 2$. It follows from Corollary 3.5 of chapter 2 that this pair of conditions on $A^{(0)}$ and $M^{(0)}$ implies (7.3). Conversely, (7.3) implies that $\operatorname{Ext}^{2,j}_{A^{(0)}}(M^{(0)}, \Bbbk) = 0$ for $j > 2$ but not the vanishing of $\operatorname{Ext}^{3,j}_{A^{(0)}}(\Bbbk, \Bbbk)$. Thus, the approach via quadratic duality gives a slightly sharper result.

Next, we will give a description of the Tor-spaces between nonhomogeneous Koszul modules in terms of the cobar-complex associated with the dual CDG-modules.

PROPOSITION 7.5. *Let* $\boldsymbol{B}$ *be a quadratic CDG-algebra and let* $\boldsymbol{S}$ *(resp.* $\boldsymbol{N}$*) be a right (resp., left) quadratic CDG-module over* $\boldsymbol{B}$. *Set* $A = H_0\mathcal{C}ob_\bullet(\boldsymbol{B})$ *and let* $R = H_0\mathcal{C}ob_\bullet(\boldsymbol{S}, \boldsymbol{B})$ *and* $M = H_0\mathcal{C}ob_\bullet(\boldsymbol{B}, \boldsymbol{N})$ *be the corresponding right and left modules over* A, *respectively. Then one has* $H_0\mathcal{C}ob_\bullet(\boldsymbol{S}, \boldsymbol{B}, \boldsymbol{N}) \simeq R \otimes_A M$. *Moreover, if an algebra* B *and modules* S *and* N *are Koszul then there are natural isomorphisms*

$$H_k\mathcal{C}ob_\bullet(\boldsymbol{S}, \boldsymbol{B}, \boldsymbol{N}) \simeq \operatorname{Tor}^A_k(R, M)$$

for all k.

Proof: Both statements follow immediately from the next lemma applied to the DG-algebra $\mathcal{C}ob_\bullet(\boldsymbol{B})$ and the DG-modules $\mathcal{C}ob_\bullet(\boldsymbol{S},\boldsymbol{B})$ and $\mathcal{C}ob_\bullet(\boldsymbol{B},\boldsymbol{N})$ over it. □

LEMMA 7.6. *Let $E_\bullet = (\cdots \longrightarrow E_2 \longrightarrow E_1 \longrightarrow E_0)$ be a DG-algebra, $P_\bullet = (\cdots \longrightarrow P_2 \longrightarrow P_1 \longrightarrow P_0)$ a right DG-module over $E_\bullet$ and $Q_\bullet = (\cdots \longrightarrow Q_2 \longrightarrow Q_1 \longrightarrow Q_0)$ a left DG-module over $E_\bullet$. Let $P_\bullet \otimes_{E_\bullet} Q_\bullet$ be the tensor product complex (with the underlying graded vector space $P \otimes_E Q$). Then one has $H_0(P_\bullet \otimes_{E_\bullet} Q_\bullet) \simeq H_0P_\bullet \otimes_{H_0E_\bullet} H_0Q_\bullet$. Assume in addition that P (resp., Q) is a free right (resp., left) module over E and $H_nP_\bullet = H_nQ_\bullet = 0$ for $n \neq 0$. Then there are natural isomorphisms*

$$H_k(P_\bullet \otimes_{E_\bullet} Q_\bullet) \simeq \operatorname{Tor}_k^{H_0E_\bullet}(H_0P_\bullet,\, H_0Q_\bullet)$$

for all k.

Proof: The statement about H_0 is straightforward. To compare higher homology spaces with Tor-spaces let us consider the bar-complex $\mathcal{B}ar_\bullet(P_\bullet, E_\bullet, Q_\bullet)$ equipped with the bigrading $\mathcal{B}ar_{ij}(P_\bullet, E_\bullet, Q_\bullet) = \langle p \otimes e_1 \otimes \cdots \otimes e_i \otimes q \mid k + j_1 + \cdots + j_i + l = j \rangle$ (where $p \in P_k,\ e_t \in E_{j_t},\ q \in Q_l$), the total differential D defined as above, and the total (lower) grading $i + j$. We have two spectral sequences

$$'E^1_{p,q} = \operatorname{Tor}^E_{qp}(P, Q) \implies H_{p+q}\mathcal{B}ar_\bullet(P_\bullet, E_\bullet, Q_\bullet),$$
$$''E^2_{p,q} = \operatorname{Tor}^{H_*E_\bullet}_{pq}(H_*P_\bullet,\, H_*Q_\bullet) \implies H_{p+q}\mathcal{B}ar_\bullet(P_\bullet, E_\bullet, Q_\bullet).$$

The first sequence shows that the natural morphism of complexes

$$\mathcal{B}ar_\bullet(P_\bullet, E_\bullet, Q_\bullet) \longrightarrow P_\bullet \otimes_{E_\bullet} Q_\bullet$$

is a quasi-isomorphism. The second sequence implies that

$$H_k\mathcal{B}ar_\bullet(P_\bullet, E_\bullet, Q_\bullet) \simeq \operatorname{Tor}_k^{H_0E_\bullet}(H_0P_\bullet,\, H_0Q_\bullet).$$

Combining these two facts we get the result. □

In the case when graded modules $R^{(0)}$ and $M^{(0)}$ are relatively Koszul over the graded algebra $A^{(0)}$ (see section 10 of chapter 2) one can simplify the answer for $\operatorname{Tor}^A_*(R, M)$ given in Proposition 7.1. Let us equip the tensor product $S \otimes_B N$ with the differential $d_\otimes(s \otimes n) = d_S(s) \otimes n + (-1)^{\tilde{s}} s \otimes d_N(n)$. Using the identities $d_S^2(s) = -s\theta_B$ and $d_N^2(n) = \theta_B n$ one can easily check that $d_\otimes^2 = 0$. Let us denote the obtained complex $\boldsymbol{S} \otimes_{\boldsymbol{B}} \boldsymbol{N}$.

COROLLARY 7.7. *Let R (resp., M) be a right (resp., left) nonhomogeneous Koszul module over a nonhomogeneous Koszul algebra A. Assume that the corresponding graded modules $R^{(0)}$ and $M^{(0)}$ over $A^{(0)}$ are relatively Koszul. Let $\boldsymbol{B}$ be the CDG-algebra dual to A, $\boldsymbol{S}$ (resp., $\boldsymbol{N}$) the right (resp., left) CDG-module over $\boldsymbol{B}$ dual to R (resp., M). Define the increasing filtration F on the vector space $\operatorname{Tor}^A_0(R, M) = R \otimes_A M$ by $F_j(R \otimes_A M) = \sum_{k+l=j} \operatorname{im}(F_kR \otimes F_lM)$. Then there are natural isomorphisms*

$$\operatorname{gr}^F_j(R \otimes_A M) \simeq (R^{(0)} \otimes_{A^{(0)}} M^{(0)})_j \qquad \textit{for } j > 0,$$
$$\operatorname{Tor}^A_i(R, M) \simeq H^i(\boldsymbol{S} \otimes_{\boldsymbol{B}} \boldsymbol{N})^* \qquad \textit{for } i > 0,$$
$$F_0(R \otimes_A M) \simeq H^0(\boldsymbol{S} \otimes_{\boldsymbol{B}} \boldsymbol{N})^*.$$

Proof: Consider the spectral sequence

$$'E^1_{p,q} = \operatorname{Tor}^B_{-q,p}(S, N)^* \implies \operatorname{gr}^F_p H_{p+q}\mathcal{C}ob_\bullet(\boldsymbol{S}, \boldsymbol{B}, \boldsymbol{N})$$

associated with the natural filtration $F_k\mathcal{C}ob_m(\boldsymbol{B},\boldsymbol{N},\boldsymbol{S}) = \bigoplus_{j-i=m}^{j\leqslant k} \mathcal{B}ar_{ij}(\boldsymbol{B},\boldsymbol{N},\boldsymbol{S})^*$ of the complex $\mathcal{C}ob_\bullet(\boldsymbol{B},\boldsymbol{N},\boldsymbol{S})$. By the definition of relative Koszulness, we have $'E_1^{p,q} = 0$ unless $q = 0$ or $p+q = 0$. Now it is clear from the form of the spectral sequence that it degenerates at $E^1_{p,q}$ for $q \neq 0$ and at $E^2_{p,q}$ for $q = 0$. Moreover, the complex $'E^1_{\bullet,0}$ can be identified with $(\boldsymbol{S}\otimes_{\boldsymbol{B}}\boldsymbol{N})^*$. It follows that the natural morphism of complexes $(\boldsymbol{S}\otimes_{\boldsymbol{B}}\boldsymbol{N})^* \longrightarrow \mathcal{C}ob_\bullet(\boldsymbol{S},\boldsymbol{B},\boldsymbol{N})$ induces an isomorphism on the homology H_i with $i > 0$ and identifies $H_0(\boldsymbol{S}\otimes_{\boldsymbol{B}}\boldsymbol{N})^*$ with $F_0H_0\mathcal{C}ob_\bullet(\boldsymbol{S},\boldsymbol{B},\boldsymbol{N})$. According to Proposition 7.5 the homology of the cobar-complex is isomorphic to $\mathrm{Tor}^A_*(R,M)$. This immediately implies the isomorphism for $\mathrm{Tor}^A_i(R,M)$ for $i > 0$. To describe $\mathrm{gr}^F_j(R\otimes_A M)$ we use in addition Proposition 10.1 of chapter 2. □

8. Homology of completed cobar-complexes

First, let us observe that if $\boldsymbol{B} = (B, d_B, \theta_B)$ is a CDG-algebra with $\theta_B \neq 0$ then the complex $\overline{\mathcal{C}ob}_\bullet(\boldsymbol{B})$ (resp., complex $\overline{\mathcal{C}ob}_\bullet(\boldsymbol{B},\boldsymbol{N})$ for a CDG-module $\boldsymbol{N}$) is acyclic. Indeed, this follows easily from the fact that the unit in the DG-algebra $\overline{\mathcal{C}ob}_\bullet(\boldsymbol{B})$ is a coboundary.

Now let $\boldsymbol{B} = (B, d_B)$ be a (nonnegatively graded) DG-algebra. We can view its cobar-construction as a bicomplex, where $\mathcal{C}ob^{ij}(\boldsymbol{B}) = \mathcal{B}ar_{ij}(\boldsymbol{B})^*$. We already used one spectral sequence (7.1) associated with this bicomplex. Now let us consider the second spectral sequence associated with it. In other words, we want to look at the spectral sequence associated with the decreasing filtration $G^l\mathcal{C}ob(\boldsymbol{B}) = \bigoplus^{i\geqslant l}\mathcal{B}ar_{ij}(\boldsymbol{B})^*$. It is easy to see that this spectral sequence converges to the homology of the completed cobar-complex:

$$''E^2_{p,q} = \mathrm{Ext}^{-p,q}_{H^*(\boldsymbol{B})}(\Bbbk,\Bbbk) \implies \mathrm{gr}_G^{-p} H_{p+q}\overline{\mathcal{C}ob}_\bullet(\boldsymbol{B}).$$

Indeed, we have $H_n\overline{\mathcal{C}ob}_\bullet(\boldsymbol{B}) = H^n\mathcal{B}ar^\bullet(\boldsymbol{B})^\vee$ and the dual spectral sequence for the increasing filtration of $\mathcal{B}ar^\bullet(\boldsymbol{B})$ clearly converges. Analogously, for a DG-module $\boldsymbol{N}$ over $\boldsymbol{B}$ we have a spectral sequence

$$''E^2_{p,q} = \mathrm{Ext}^{-p,q}_{H^*(\boldsymbol{B})}(H^*(\boldsymbol{N}),\Bbbk) \implies \mathrm{gr}^p_G H_{p+q}\overline{\mathcal{C}ob}_\bullet(\boldsymbol{B},\boldsymbol{N})$$

associated with the decreasing filtration $G^l\mathcal{C}ob(\boldsymbol{B},\boldsymbol{N}) = \bigoplus^{i\geqslant l}\mathcal{B}ar^{ij}(\boldsymbol{B},\boldsymbol{N})^*$. Using these spectral sequences we immediately derive the following result.

Proposition 8.1. *A quasi-isomorphism of nonnegatively graded DG-algebras* $f : \boldsymbol{B}' \longrightarrow \boldsymbol{B}''$ *induces a quasi-isomorphism* $\overline{\mathcal{C}ob}_\bullet(\boldsymbol{B}'') \longrightarrow \overline{\mathcal{C}ob}_\bullet(\boldsymbol{B}')$. *Let* $\boldsymbol{N}'$ *(resp.,* $\boldsymbol{N}''$*) be a DG-module over* $\boldsymbol{B}'$ *(resp.,* $\boldsymbol{B}''$*). Then a quasi-isomorphism* $\boldsymbol{N}' \longrightarrow \boldsymbol{N}''$ *compatible with* f *induces a quasi-isomorphism* $\overline{\mathcal{C}ob}_\bullet(\boldsymbol{B}'',\boldsymbol{N}'') \longrightarrow \overline{\mathcal{C}ob}_\bullet(\boldsymbol{B}',\boldsymbol{N}')$.

The decreasing filtrations induced by G on the algebra $A = H_0\mathcal{C}ob_\bullet(\boldsymbol{B})$ and on the A-module $M = H_0\mathcal{C}ob_\bullet(\boldsymbol{B},\boldsymbol{N})$ are the A_+-adic filtrations $G^lA = A^l_+$ and $G^lM = A^l_+M$. Furthermore, it is easy to see that taking completions of the cobar-complexes leads to the A_+-adic completions on H_0:

$$H_0\overline{\mathcal{C}ob}_\bullet(\boldsymbol{B}) \simeq A\,\hat{}\, = \varprojlim A/A^i_+,$$

$$H_0\overline{\mathcal{C}ob}_\bullet(\boldsymbol{B},\boldsymbol{N}) \simeq M\,\hat{}\, = \varprojlim M/A^i_+M.$$

Remark. We see that the completions $A\,\hat{}$ and $M\,\hat{}$ are determined by the quasi-isomorphism classes of $\boldsymbol{B}$ and $\boldsymbol{N}$ (cf. Remark 2 in section 6). In fact, it is not difficult to prove a more general statement. For any augmented algebra A, the A_+-adic completion $A^\frown$ is determined by the class of the differential coalgebra $\mathcal{B}ar_\bullet(A)$ up to morphisms inducing isomorphisms of H_1 and H_2. For any A-module M, the completion $M^\frown$ is determined by the class of the differential comodule $\mathcal{B}ar_\bullet(A, M)$ up to morphisms inducing isomorphisms of H_0 and H_1.

Now let A be a Koszul quadratic-linear algebra and let $\boldsymbol{B}$ be the dual DG-algebra. Then $\mathcal{C}ob_\bullet(\boldsymbol{B})$ is a free DG-resolution of A (see Proposition 7.2(ii)). Therefore, the homology spaces $H_i\mathcal{C}ob_\bullet(\boldsymbol{B})$ can be viewed as nonabelian derived functors of the A_+-adic completion functor $A \to A^\frown$. Combining Remark 1 of section 6 with Proposition 8.1 we see that these homology spaces are invariants of the algebra A (not depending on a Koszul quadratic-linear structure). The example below shows that these invariants are quite nontrivial. Similarly, the homology $H_*\mathcal{C}ob_\nu(\boldsymbol{N})$ is an invariant of an A-module M admitting a nonhomogeneous Koszul structure (where $\boldsymbol{N}$ is Koszul dual to M).

Example. Let $A = U\mathfrak{g}$ be the enveloping algebra of a semisimple Lie algebra $\mathfrak{g}$. Then the dual DG-algebra is the standard cohomological complex $C^\bullet(\mathfrak{g})$. The inclusion of the subalgebra of invariant forms defines a quasi-isomorphism of $C^\bullet(\mathfrak{g})$ with its cohomology algebra $H^*(\mathfrak{g})$. It follows that the algebra $H_*\overline{\mathcal{C}ob}_\bullet(C^\bullet(\mathfrak{g})) \simeq H_*\overline{\mathcal{C}ob}_\bullet(H^*(\mathfrak{g}))$ is isomorphic to the symmetric algebra with generators in even degrees $\geqslant 2$ (it is isomorphic to the center of $U\mathfrak{g}$). Thus, in this case higher derived functors of the A_+-completion carry interesting information, unlike the usual A_+-completion $H_0\overline{\mathcal{C}ob}_\bullet(C^\bullet(\mathfrak{g})) \simeq U\mathfrak{g}^\frown = \Bbbk$. On the other hand, if M is a nontrivial irreducible $\mathfrak{g}$-module then the dual DG-module $\boldsymbol{N} = C^\bullet(\mathfrak{g}, M^*)$ (see Example 2 in section 5) is acyclic. Thus, in this case $H_*\overline{\mathcal{C}ob}_\bullet(C^\bullet(\mathfrak{g}), C^\bullet(\mathfrak{g}, M^*)) = 0$.

CHAPTER 6

Families of quadratic algebras and Hilbert series

In this chapter we consider families of quadratic algebras. The main result is the Koszul Deformation Principle (Theorem 2.1) stating that in a neighborhood of a Koszul algebra flatness of a deformation in grading components of degree 3 implies its flatness in (any finite number of) higher degrees. We deduce from this the finiteness of the number of Hilbert series of Koszul algebras with a fixed number of generators and give an explicit upper bound on this number (see section 3). Furthermore, in section 7 we prove that there exists a uniform bound on this number over all ground fields. In section 4 we determine which generic quadratic algebras are Koszul and in section 5 consider some examples of possible Koszul Hilbert series (when the number of generators and relations is small).

1. Openness of distributivity

Let W be a vector space over $\Bbbk$ and let $R_1(x), \dots, R_m(x) \subset W$ be algebraic families of subspaces (of fixed dimensions) parametrized by an algebraic variety X over $\Bbbk$. In other words, we have a collection of morphisms from X to the corresponding Grassmannians of subspaces in W.

Proposition 1.1. *Assume that the dimensions of pairwise intersections $R_i(x) \cap R_j(x)$ are locally constant functions on X. Then the set*

$$U = \{x \in X \mid (R_1(x), \dots, R_m(x)) \text{ is distributive}\}$$

is open in X. Moreover, the dimensions of all subspaces obtained from $R_i(x)$ using the operations of sum and intersection are locally constant functions on U.

Proof: First, assume that $m = 3$. Consider the inclusion

$$R_1(x) \cap (R_2(x) + R_3(x)) \supset R_1(x) \cap R_2(x) + R_1(x) \cap R_3(x). \tag{1.1}$$

Assume that it becomes an equality for some $x_0 \in X$. Let $l(x) \geqslant r(x)$ denote the dimensions of the left-hand side and the right-hand side of (1.1), respectively. Then there exists an open neighborhood U_{x_0} of x_0 such that for $x \in U_{x_0}$ one has $l(x) \leqslant l(x_0)$ and $r(x) \geqslant r(x_0)$. Since $l(x_0) = r(x_0)$, we obtain that for $x \in U_{x_0}$ the above inclusion is actually an equality and $l(x) = r(x)$ is constant on U_{x_0}. The statements for other intersections and sums follow easily from this.

Now let us consider the case $m > 3$. Assume that $(R_1(x_0), \dots, R_m(x_0))$ is distributive for some $x_0 \in X$. Let T be a word in the free lattice generated by R_i. We claim that there exists an open neighborhood U_T of x_0 such that the dimension of $T(x)$ is constant on U_T. We can argue by induction in the length of T. If the length of T is at most 2 then this is true by our assumption. Now let $T = T_1 * (T_2 * T_3)$, where $*$ denotes intersection or sum. The induction assumption implies that the dimensions of $T_i(x)$ and $T_i(x) \cap T_j(x)$ are constant on some open neighborhood of x_0. Applying the case $m = 3$ considered before we derive the existence of a neighborhood U_T of x_0 such that the dimension of $T(x)$ is constant on U_T. This finishes the proof of our claim. Recall that by Theorem 6.3 to check distributivity of a finitely generated lattice one has to check only a finite number of distributivity conditions. Hence, there exists an open neighborhood of x_0 on which the collection $(R_1(x), \dots, R_m(x))$ is distributive. Since such a lattice is finite, the dimensions of all elements of this lattice are constant near x_0. □

2. Deformations of Koszul algebras

The following important result is essentially due to V. Drinfeld [**43**]. Recall that a quadratic algebra A is said to be *n-Koszul* if $\mathrm{Ext}_A^{ij}(\Bbbk, \Bbbk) = 0$ for $i < j \leqslant n$, or, equivalently, the relation lattice in $A_1^{\otimes n}$ is distributive (see section 4 of chapter 2). The advantage of the notion of n-Koszulness is that it is given by a finite number of conditions (unlike Koszulness).

THEOREM 2.1. *Let $A(x) = \{V, I(x)\}$ be a family of quadratic algebras with the fixed generating space V and the space of quadratic relations $I(x)$ parametrized by points of an algebraic variety X (so that $x \longrightarrow I(x)$ is a morphism from X to the Grassmannian $\mathbb{G}_r(V^{\otimes 2})$). Assume that $A(x_0)$ is an n-Koszul algebra for some $x_0 \in X$ and that $\dim A_3(x)$ is constant on X. Then there exists an open neighborhood U of x_0 such that for $x \in U$ the algebra $A(x)$ is n-Koszul and $\dim A_i(x) = \dim A_i(x_0)$ for $i \leqslant n$.*

Proof: We have $A_n(x) = V^{\otimes n}/(R_1(x) + \cdots + R_{n-1}(x))$, where $R_i(x) = V^{\otimes i-1} \otimes I(x) \otimes V^{\otimes n-i-1}$. The assumption that $\dim A_3(x)$ is constant on X implies that the dimensions of all pairwise intersections $R_i(x) \cap R_j(x)$ are also constant on X. Hence, we can apply Proposition 1.1. □

Assume for simplicity that the ground field $\Bbbk$ is algebraically closed and uncountable. We will get rid of this assumption in section 7. Let $\mathcal{Q}_{m,s} = \mathbb{G}_{m^2-s}(\Bbbk^{m\otimes 2})$ be the Grassmannian variety of quadratic algebras with $A_1 = \Bbbk^m$ and $\dim A_2 = s$. It has a stratification by the locally closed subvarieties $\mathcal{Q}_{m,s,u}$ consisting of quadratic algebras with $\dim A_3 = u$.

COROLLARY 2.2. *The set of all n-Koszul algebras in $\mathcal{Q}_{m,s,u}$ is open. Therefore, the set of all Koszul algebras in $\mathcal{Q}_{m,s,u}$ is a countable intersection of open subsets. For $i \leqslant n$ the restriction of the function $\dim A_i$ to the set of all n-Koszul algebras in $\mathcal{Q}_{m,s,u}$ is locally constant.*

Proof: Apply Theorem 2.1 to the natural family of quadratic algebras parametrized by $\mathcal{Q}_{m,s,u}$. □

Example. The above corollary implies that if we have a one-parameter family $A(\lambda)$ of quadratic algebras in $\mathcal{Q}_{m,s,u}$ such that all algebras $A(\lambda)$ for $\lambda \neq \lambda_0$ are isomorphic to some fixed algebra A, then Koszulness of $A(\lambda_0)$ implies Koszulness of A and in this case $h_A = h_{A(\lambda_0)}$. This observation is used in [**113**] to prove Koszulness of the Orlik-Solomon algebra of a supersolvable hyperplane arrangement (see [**87**]). This algebra is the quotient of the exterior algebra in generators $x_1, \dots, x_n$ by the relations of the form $x_i x_j - x_i x_k + x_j x_k = 0$ for some set of triples $i < j < k$ (depending on the arrangement). Now one can define the family $A(\lambda)$ by making the change of variables $x_i \to \lambda x_i$ so that the relations in $A(\lambda)$ become

$x_ix_j - \lambda^{k-j}x_ix_k + \lambda^{k-i}x_jx_k = 0$. It turns out that the Hilbert series is constant in this family, so Koszulness of $A(0)$ implies Koszulness of $A(1)$.

COROLLARY 2.3. *There is only a finite number of Hilbert series of Koszul algebras with any fixed number of generators* $m = \dim A_1$.

Proof: All these Hilbert series can be obtained from the algebras corresponding to generic points of irreducible components of $\mathcal{Q}_{m,s,u}$ (more precisely, those of them that are Koszul). Therefore, the number of Hilbert series is bounded above by the total number of irreducible components of the varieties $\mathcal{Q}_{m,s,u}$. □

COROLLARY 2.4. *For every* $m > 0$ *and* $r > 0$ *there is only a finite number of Hilbert series of Koszul modules* M *over Koszul algebras* A*, such that* $\dim A_1 = m$ *and* $\dim M_0 = r$.

Proof: Recall that to a pair (A, M) consisting of a Koszul algebra A and a Koszul (left) module M over it we can associate a Koszul algebra $A_M = A \oplus M(-1)$ (see Corollary 5.5 of chapter 2). Since $h_{A_M}(z) = h_A(z) + zh_M(z)$, our assertion follows from the previous corollary. □

The above corollary has the following nice application.

PROPOSITION 2.5. *Let* A *be a Koszul algebra such that the sequence* $a_n = \dim A_n$ *is bounded. Then this sequence has a periodic tail, i.e., there exists* $N > 0$ *and* $n_0 \geqslant 0$ *such that* $a_{n+N} = a_n$ *for all* $n \geqslant n_0$.

Proof: Consider the sequence of A-modules $A^{[n]} = \bigoplus_{i \geqslant n} A_i$, $n \geqslant 0$. By Proposition 1.1 of chapter 2 each of these modules is Koszul. Since the number of generators of $A^{[n]}$ is equal to a_n, from the above corollary we deduce that there is only a finite number of elements in the set of Hilbert series $h_{A^{[n]}}$. This implies the result. □

Remark 1. As in Theorem 2.1 of chapter 5 the Koszulness condition in Theorem 2.1 can be weakened to the vanishing $\mathrm{Ext}^{3,j}_{A(0)}(\Bbbk, \Bbbk) = 0$ for $j > 3$ (see Remark 1 in section 7 of chapter 5). It follows that one can use this condition in the statements of Corollaries 2.2 and 2.3 instead of Koszulness.

Remark 2. Theorem 2.1 of chapter 5 can be derived from Theorem 2.1 as follows. Let A be the algebra defined by a set of nonhomogeneous quadratic relations $a + \varphi(a) + h(a) = 0$, $a \in I$. Consider the family of quadratic algebras $\widetilde{A}(\lambda)$ with the generator space $\Bbbk t \oplus V$ and the relations

$$(tx = xt,\ x \in V;\ a + \lambda\varphi(a)t + \lambda^2 h(a)t^2,\ a \in I).$$

The analogue of the Jacobi identity implies that this deformation is flat in degree 3. All the algebras $\widetilde{A}(\lambda)$ with $\lambda \neq 0$ are isomorphic to each other, hence they have the same Hilbert series as $\widetilde{A}(0)$ provided that $\widetilde{A}(0)$ is Koszul. Taking the quotient by $t - 1$ we conclude that $\mathrm{gr}^F A$ and $A^{(0)}$ have equal Hilbert series.

Remark 3. It is also natural to consider the characterization of PBW-bases using the third grading component (see sections 2 and 7 of chapter 4) as an analogue of Theorem 2.1, since passing from a filtered object to the associated graded one is somewhat analogous to passing from the generic point of a deformation to the special one. For *commutative* PBW-bases, this analogy can be made more precise: as was explained in Example 1 of section 8 of chapter 4, for any commutative quadratic algebra with a fixed ordered set of generators the corresponding commutative monomial algebra can be obtained as a limit of a family of quadratic algebras isomorphic to the original one (see also [**72**]). The same is true for $\mathbb{Z}$-PBW-bases if we consider deformations in the class of $\mathbb{Z}$-algebras (cf. section 10 of chapter 4).

Furthermore, consider the "variety" $\mathcal{K}_h$ of Koszul algebras with the fixed space of generators $\mathbb{k}^m$ and the Hilbert series $h(z) = 1 + mz + \cdots$. Since the algebraic group $\mathrm{GL}(m)$ is connected, its action preserves irreducible components of $\mathcal{K}_h$. Therefore, it makes sense to consider the component(s) containing a given Koszul algebra A as invariants of the isomorphism class of A.

Now for the usual *noncommutative* PBW-bases, it is natural to ask whether the associated monomial algebra will always belong to the same irreducible component as the original PBW-algebra. The following example shows that the answer is "no". Consider the quadratic algebra A with 2 generators x_1, x_2 and 2 relations $(x_1+x_2)^2 = (x_1-x_2)^2 = 0$. Then x_1, x_2 are PBW-generators of A in the standard order and the corresponding monomial algebra is $x_2^2 = 0$, $x_2x_1 = 0$. It is easy to verify that these algebras belong to different irreducible components (see Example 2 in section 5).

On the other hand, we do not know any examples when $\mathcal{K}_h$ is not *connected.*

Remark 4. A generalization of Corollary 2.3 was recently obtained by Piontkovskii [**94**]. Namely, he proved that for given integers n, a, b, c the set of Hilbert series of graded algebras A with $A_0 = \mathbb{k}$, with $\leqslant n$ generators of degree $\leqslant a$, relations concentrated in degrees $\leqslant b$, and $\mathrm{Tor}_3^A(\mathbb{k},\mathbb{k})$ concentrated in degrees $\leqslant c$, is finite.

3. Upper bound for the number of Koszul Hilbert series

Using the results of section 2 it is not difficult to give an explicit bound for the number of Hilbert series of Koszul algebras with fixed $\dim A_1$.

LEMMA 3.1. *Let X be an algebraic variety, U and W vector spaces, $\mathcal{E} \subset U \otimes \mathcal{O}_X$ a vector subbundle of rank e, and $\rho\colon \mathcal{E} \longrightarrow W \otimes \mathcal{O}_X$ a morphism of vector bundles over X. Then the closed subvariety $X_k = \{x \in X : \mathrm{rk}\,\rho_x < k\}$ can be presented as an intersection of divisors from the linear system $|\det^{-1}\mathcal{E}|$, where $\det\mathcal{E} = \bigwedge^e \mathcal{E}$.*

Proof: Consider the composition φ of the following maps:

$$\bigwedge^e \mathcal{E} \longrightarrow \bigwedge^{e-k} \mathcal{E} \otimes \bigwedge^k \mathcal{E} \longrightarrow \bigwedge^{e-k} U \otimes \bigwedge^k W \otimes \mathcal{O}.$$

It is clear that $\mathrm{rk}\,\rho_x < k \iff \varphi_x = 0$. Hence, X_k is the set of common zeros of the space of sections $\bigwedge^{e-k} U^* \otimes \bigwedge^k W^* \longrightarrow H^0(X, \det^{-1}\mathcal{E})$. $\square$

PROPOSITION 3.2. *The number of Hilbert series of Koszul algebras with* $\dim A_1 = m$ *and* $\dim A_2 = s$, *where* $s \geqslant 1$, *is bounded above by*

$$m^3(2m)^{s(m^2-s)}(s(m^2-s))!\frac{1!\cdots(s-1)!}{(m^2-s)!\cdots(m^2-1)!} \leqslant (2ms)^{s(m^2-s)}.$$

Proof: Clearly, we can assume that $1 < s \leqslant m^2/2$. Let us apply the above lemma to $X = \mathcal{Q}_{m,s} = \mathbb{G}_{m^2-s}((\mathbb{k}^m)^{\otimes 2})$, $W = V^{\otimes 3}$, $U = V^{\otimes 3} \oplus V^{\otimes 3}$ and the subbundle $\mathcal{E} \subset U \otimes \mathcal{O}_X$ with the fiber $I \otimes V \oplus V \otimes I$ over $I \in \mathbb{G}_{m^2-s}((\mathbb{k}^m)^{\otimes 2})$. We obtain that the subvariety $\overline{\mathcal{Q}}_{m,s,u}$ is cut out by a linear subspace in the projective embedding of X defined by the linear system $|\mathcal{O}_X(2m)|$, where $\mathcal{O}_X(1)$ is the ample line bundle corresponding to the Plücker embedding (with the fiber $\det^{-1} I$ over I). By the refined Bezout's theorem (see [58], Thm. 12.3) this implies that the projective degree of $\overline{\mathcal{Q}}_{m,s,u}$ in this embedding is bounded above by

$$\deg_{\mathcal{O}(2m)} X = (2m)^{s(m^2-s)}(s(m^2-s))!\frac{1!\cdots(s-1)!}{(m^2-s)!\cdots(m^2-1)!},$$

where we used the formula for the degree of the Grassmannian in the Plücker embedding (see e.g. [58], Example 14.7.11(iii)). Now we observe that

$$\frac{1!\cdots(s-1)!}{(m^2-s)!\cdots(m^2-1)!} =$$
$$\frac{1}{(m^2-s)!^s\binom{m^2-s+1}{m^2-s}\cdots\binom{m^2-1}{m^2-s}} \leqslant \frac{1}{(m^2-s)!^s(m^2-s+1)^{s-1}}.$$

Therefore, we have

$$\deg_{\mathcal{O}(2m)} X \leqslant \frac{(2m)^{s(m^2-s)}(s(m^2-s))!}{(m^2-s)!^s(m^2-s+1)^{s-1}}.$$

By Corollary 2.2 the number of Hilbert series in question is bounded above by $m^3 \deg_{\mathcal{O}(2m)} X$, so to finish the proof it suffices to check the following inequality:

$$\frac{m^3(s(m^2-s))!}{(m^2-s+1)^{s-1}(m^2-s)!^s} \leqslant s^{s(m^2-s)}, \tag{3.1}$$

where $1 < s \leqslant m^2/2$. Assume first that $s > 2$. Then one can easily check that $m^3 \leqslant (m^2-s+1)^{s-1}$, so (3.1) follows from the well-known inequality

$$\frac{(s(m^2-s))!}{(m^2-s)!^s} \leqslant s^{s(m^2-s)}.$$

In the case $s = 2$ we can rewrite (3.1) in the form

$$\frac{m^2}{m^2-1} \leqslant \frac{2^{2(m^2-2)}(m^2-2)!^2}{(2m^2-4)!m},$$

where $m \geqslant 2$. Note that for $m = 2$ this becomes an equality. It remains to note that the right-hand side grows with m while the left-hand side decreases with m. □

Remark. The number of Hilbert series of PBW-algebras is obviously bounded above by 2^{m^2}, which is substantially less than the bound m^{m^4} for Koszul algebras, though still very large. One can see from the following table (calculated using a computer) that the number of PBW-series does grow very fast:

m	1	2	3	4	5
PBW series	2	7	46	803	50650

4. Generic quadratic algebras

In this section we consider the variety of quadratic algebras with fixed $\dim A_1 = m$ and $\dim A_2 = s$ (that can be identified with a certain Grassmannian). By a generic quadratic algebra (with given m and s) we mean a point of a countable intersection of nonempty Zariski open subsets of this variety. Clearly, a generic quadratic algebra has coefficient-wise minimal Hilbert series. It follows that the open stratum $\mathcal{Q}_{m,s,u} \subset \mathcal{Q}_{m,s}$ corresponds to the minimal value of $u = \dim A_3$. The next result is due to D. J. Anick [8].

PROPOSITION 4.1. *The minimal possible value of* $\dim A_3$ *for quadratic algebras with* $\dim A_1 = m$ *and* $\dim A_2 = s$ *is given by the formula:*

$$u_{min} = \begin{cases} 0, & \text{if } s \leqslant \frac{m^2}{2}, \\ 2ms - m^3, & \text{if } s \geqslant \frac{m^2}{2}. \end{cases}$$

Proof: Because of the duality it is sufficient to consider the first case. We have to show that there exists a quadratic algebra with $s = \lfloor m^2/2 \rfloor$ and $\dim A_3 = 0$. Here are explicit examples (taken from [**8**]):

$$k\{x_1, \dots, x_n, y_1, \dots, y_n\}/(x_i y_j,\, x_i x_j - y_i y_j,\, i, j = 1, \dots, n)$$

for $m = 2n$ and

$$k\{x_1, \dots, x_n, y_0, \dots, y_n\}/(x_i y_{j-1},\, x_i x_j - y_{i-1} y_j,$$
$$x_i y_n - y_i y_0,\, y_0 x_j - y_n y_j,\, y_0^2,\, i, j = 1, \dots, n)$$

for $m = 2n + 1$. □

Remark 1. It is an open problem to find the minimal value of $u = \dim A_3$ for *Koszul* algebras with fixed $m = \dim A_1$ and $s = \dim A_2$. Conjecturally, it coincides with the minimum for PBW-algebras. For $s \leqslant m^2/4$ this is true (with $u = 0$) by Thm. 3.1 of [**40**] stating that $H(z) = 1 + mz + sz^2$ is the Hilbert series of a Koszul algebra iff it is the Hilbert series of a PBW-algebra iff $H(-z)^{-1}$ has positive coefficients iff $s \leqslant m^2/4$ (this also follows from the proof of Proposition 4.2 below). Moreover, in this case there exists a commutative Koszul algebra with $u = 0$.

On the other hand, the *maximal* value of $\dim A_3$ for one-generated algebras is obviously attained on PBW-algebras, since one has $\dim A_3^0 \geqslant \dim A_3$ for the quadratic monomial algebra A^0 associated with A. Thus, to find this maximal value one has to find the maximum of the sum of all entries in M^2 among all $0 - 1$ matrices M of size $m \times m$ and with s entries equal to 1 (cf. section 6 of chapter 4). This maximal value was found independently in the works [**6, 5, 120**]. For example, one has $u \leqslant s^{3/2}$ for $s \geqslant m^2/2$ and $u \leqslant (s^2 + 4s - 1)/4$ (unless $s = 4$ when $u \leqslant 8$) for any m. In addition, the problem of determining the minimal value of the sum of all entries in M^2 for a $0 - 1$ matrix M of a given size and a given number of 1's was solved in [**120**]. Hence, this paper gives the minimal value of $\dim A_3$ for a PBW-algebra with given $\dim A_1 = m$ and $\dim A_2 = s$. In particular, it is shown in [**120**] that as $m \to \infty$ and $s/m^2 \to \sigma \in \mathbb{R}$, the minimal value of u/m^3 (for a PBW-algebra) tends to a piecewise algebraic function of σ with break points at $\sigma = (n \pm 1)/2n$ for $n = 2, 3, \dots$ (corresponding to the symmetric and exterior algebras with n generators).

The main tool in studying generic relations is the Golod–Shafarevich inequality (see Proposition 2.3 of chapter 2). The following well-known result is a typical application of this inequality.

PROPOSITION 4.2. *A generic quadratic algebra with* $\dim A_1 = m$ *and* $\dim A_2 = s$ *is Koszul iff one of the inequalities holds:*

$$s \leqslant \frac{m^2}{4} \quad \text{or} \quad s \geqslant \frac{3m^2}{4}.$$

Proof: It suffices to consider the case $s \leqslant m^2/2$. Then for a generic quadratic algebra we have $h_A(z) = 1 + mz + sz^2$. It is easy to check that the power series $(1 + mz + sz^2)^{-1}$ has nonnegative coefficients iff $s \leqslant m^2/4$. Hence, a generic algebra with $m^2/4 < s \leqslant m^2/2$ is not Koszul. For $s \leqslant m^2/4$ we claim that $h_{A^!}(z) = (1 - mz + sz^2)^{-1}$ for a generic quadratic algebra. Indeed, this value of the Hilbert series is attained at the monomial algebra corresponding to a bichromatic graph $S \subset [1, \lfloor m/2 \rfloor] \times [\lfloor m/2 \rfloor + 1, m]$. It remains to use Proposition 2.3 of chapter 2. □

Remark 2. We have seen that the Hilbert series of a generic quadratic algebra is equal to $1+mz+sz^2$ for $s \leqslant m^2/2$ and to $(1+mz+(m^2-s)z^2)^{-1}$ for $s \geqslant 3m^2/4$. It is a natural *conjecture*[1] that the Golod–Shafarevich estimate is attained for all s, i.e., the Hilbert series of a generic quadratic algebra is equal to $|(1-mz+(m^2-s)z^2)^{-1}|$. This would imply that a generic algebra with $s < 3m^2/4$ is finite-dimensional (more precisely, that $A_n = 0$ iff $s/m^2 \leqslant 1 - \frac{1}{4}\cos^{-2}\frac{\pi}{n+1}$). However, the analogous statement for non-quadratic one-generated graded algebras in not true [**8**].

Remark 3. The simplest way to see the difference between the classes of PBW and Koszul algebras is to consider the analogue of Proposition 4.2 for PBW-algebras. We claim that a generic algebra A with $\dim A_1 = m$ and $\dim A_2 = s$ admits a PBW-basis iff $s \leqslant m - 1$ or $s \geqslant m^2 - m + 1$. One may assume that $s \leqslant m^2/2$, so that $\dim A_3 = 0$. Then the PBW condition for a set of generators $x_1, \ldots, x_m$ means that the corresponding set S does not contain any (i, j) and (j, k) simultaneously. In particular, we must have $(1, 1) \notin S$, that is $x_1^2 \in I$. Clearly, a generic subspace of relations $I \subset V^{\otimes 2}$ intersects the m-dimensional cone $\{x^2 : x \in V\}$ at a nonzero point iff $s = \operatorname{codim} I \leqslant m - 1$. The reader will easily check that for $s \leqslant m - 1$ a generic algebra does admit a set of PBW-generators with the corresponding set $S = \{(1, 2), \ldots, (1, s + 1)\}$.

A similar argument shows that a generic algebra with $2m-2 < s < m^2-2m+2$ has no $\mathbb{Z}$-PBW-basis in any order (see section 10 of chapter 4).

Remark 4. The analogue of Proposition 4.2 for commutative algebras states that a generic commutative quadratic algebra with $\dim A_1 = m$ and $\dim A_2 = s$ is Koszul iff $s \leqslant m^2/4$ or $s \geqslant \binom{m+1}{2} - m$ (see [**57**]).

5. Examples with small $\dim A_1$ and $\dim A_2$

In this section we consider examples of the stratification of the variety of quadratic algebras by the dimension of A_3. By a "component" we mean an "irreducible component". We use the notation $m = \dim A_1$, $s = \dim A_2$, $u = \dim A_3$.

1. $m \geqslant 2$, $s = 1$. The stratification has the form $\mathcal{Q}_{m,1} = \mathcal{Q}_{m,1,0} \sqcup \mathcal{Q}_{m,1,1}$, where the closed subvariety $\mathcal{Q}_{m,1,1}$ consists of quadratic algebras isomorphic to $B^!$, where $B = \Bbbk\{x_1, \ldots, x_m\}/(x_1^2)$. All algebras in $\mathcal{Q}_{m,1}$ are Koszul. If in addition $\Bbbk$ is quadratically closed (contains all square roots) then all of them are PBW-algebras. The possible Hilbert series of algebras in $\mathcal{Q}_{m,1}$ are

$$1 + mz + z^2,$$
$$1 + mz + z^2 + z^3 + z^4 + z^5 + \ldots$$

For $m = 2$, any algebra from $\mathcal{Q}_{2,1,0}$ is isomorphic to $\Bbbk\{x, y\}/(xy - \lambda yx)$ with $\lambda \in \mathbb{P}^1_{\Bbbk}$, or to $\Bbbk\{x, y\}/(xy - yx + x^2)$.

[1]The argument of Zvyagina (Zapiski Nauchn. Sem. LOMI **155**, 1986) concerning this problem is incorrect.

In the dual case $m = 2$, $s = 3$, the Hilbert series are

$$1 + 2z + 3z^2 + 4z^3 + 5z^4 + 6z^5 + \dots$$
$$1 + 2z + 3z^2 + 5z^3 + 8z^4 + 13z^5 + \dots$$

2. $m = 2$, $s = 2$. We have $\mathcal{Q}_{2,2} = \mathcal{Q}_{2,2,0} \sqcup \mathcal{Q}_{2,2,1} \sqcup \mathcal{Q}_{2,2,2}$. The closed subvariety $\mathcal{Q}_{2,2,2}$ consists of three 2-dimensional irreducible components. The algebras corresponding to their generic points are $\Bbbk\{x, y\}/(x^2, y^2)$, $\Bbbk\{x, y\}/(x^2, xy - \lambda yx)$, and $\Bbbk\{x, y\}/(xy, yx)$, and any algebra from $\mathcal{Q}_{2,2,2}$ is isomorphic to one of these. The first and the third components intersect the second one (at $\lambda = -1$ and $\lambda = 1$, respectively), but do not intersect each other. Geometrically, this is a union of three surfaces $\mathbb{P}^1 \times \mathbb{P}^1$, where the lines $\mathbb{P}^1 \times \{\mp 1\}$ in the second component are identified with the diagonals in the first and the third components.

The locally closed subvariety $\mathcal{Q}_{2,2,1}$ is a disjoint union of two 3-dimensional components consisting of algebras isomorphic to $\Bbbk\{x, y\}/(x^2, y(x + y))$ and to $\Bbbk\{x, y\}/(xy, y(x + y))$, respectively. Their closures contain the first two and the last two components of $\mathcal{Q}_{2,2,2}$, respectively. The open subset $\mathcal{Q}_{2,2,0}$ consists of algebras isomorphic to $\Bbbk\{x, y\}/(xy, x^2 + \lambda yx + y^2)$, $\lambda^2 \neq 1$.

A quadratic algebra from $\mathcal{Q}_{2,2}$ is Koszul iff it belongs to $\mathcal{Q}_{2,2,2}$. The Hilbert series of any such algebra is equal to

$$1 + 2z + 2z^2 + 2z^3 + 2z^4 + 2z^5 + \dots$$

3. $m \geqslant 3$, $s = 2$. This case was analyzed in [**15**]. The stratification has the same form $\mathcal{Q}_{m,2} = \mathcal{Q}_{m,2,0} \sqcup \mathcal{Q}_{m,2,1} \sqcup \mathcal{Q}_{m,2,2}$. The closed subvariety $\mathcal{Q}_{m,2,2}$ consists of Koszul algebras of the form $A \sqcap (\Bbbk \oplus V)$, where $A \in \mathcal{Q}_{2,2,2}$ and $\dim V = m - 2$, so it has three $2m - 2$-dimensional components (see the case $m = 2$). The subvariety $\mathcal{Q}_{m,2,1}$ consists of two intersecting components of dimensions $3m - 3$ and $m^2 + m - 3$. A generic algebra on the first component is quadratic dual to $\Bbbk\{x_1, \dots, x_m\}/(x_1x_2, x_2x_3)$. Other algebras from this component form two non-Koszul isomorphism classes: that of the algebra B dual to $\Bbbk\{x_1, \dots, x_m\}/(x_1^2, x_1x_2 + x_3x_1)$ and of the algebra $A \sqcap (\Bbbk \oplus V)$, where $A = \Bbbk\{x, y\}/(x^2, y(x+y))$ (A corresponds to the first component of $\mathcal{Q}_{2,2,1}$). The second component of $\mathcal{Q}_{m,2,1}$ consists of the algebras A such that $A^!$ has a relation $x^2 = 0$. All algebras of this component are Koszul except for those isomorphic to B (constituting the intersection with the first component) and except for algebras arising from the second component of $\mathcal{Q}_{2,2,1}$ (by the direct sum with $\Bbbk \oplus V$). All algebras on $\mathcal{Q}_{m,2,0}$ are Koszul except for those arising from $\mathcal{Q}_{2,2,0}$ and the isomorphism classes of 6 algebras listed in [**15**]. The Hilbert series of Koszul algebras with $\dim A_1 = m$ and $\dim A_2 = 2$ are

$$1 + mz + 2z^2$$
$$1 + mz + 2z^2 + z^3$$
$$1 + mz + 2z^2 + z^3 + z^4 + z^5 + \dots$$
$$1 + mz + 2z^2 + 2z^3 + 2z^4 + 2z^5 + \dots$$

4. $m = 3$, $s = 3$. In this case (and in the next one) we do not know either a description of the stratification or the list of Hilbert series of Koszul algebras. The possible values of u are $u = 0, \dots, 5$. For Koszul algebras one has $u \neq 0$. We can only list the Hilbert series of PBW-algebras (together with a possible set

$S \subset \{1,2,3\}^2$):

$$\begin{aligned}
&1 + 3z + 3z^2 + z^3 && (12, 23, 13)\\
&1 + 3z + 3z^2 + 2z^3 + 2z^4 + 2z^5 + 2z^6 + \dots && (12, 22, 13)\\
&1 + 3z + 3z^2 + 3z^3 + 3z^4 + 3z^5 + 3z^6 + \dots && (11, 22, 33)\\
&1 + 3z + 3z^2 + 4z^3 + 4z^4 + 4z^5 + 4z^6 + \dots && (12, 22, 23)\\
&1 + 3z + 3z^2 + 4z^3 + 5z^4 + 6z^5 + 7z^6 + \dots && (11, 12, 22)\\
&1 + 3z + 3z^2 + 5z^3 + 8z^4 + 13z^5 + 21z^6 + \dots && (11, 12, 21).
\end{aligned}$$

5. $m = 3$, $s = 4$. For quadratic algebras we can have $u = 0, \dots, 8$. There exist Koszul algebras with $u = 4, \dots, 8$. There are no Koszul algebras with $u = 0$, 1, or 2. We do not know whether there exists a Koszul algebra with $\dim A_1 = 3$, $\dim A_2 = 4$ and $\dim A_3 = 3$. Here are the Hilbert series of PBW-algebras:

$$\begin{aligned}
&1 + 3z + 4z^2 + 4z^3 + 4z^4 + 4z^5 + \dots && (12, 13, 23, 32)\\
&1 + 3z + 4z^2 + 5z^3 + 6z^4 + 7z^5 + \dots && (11, 12, 22, 33)\\
&1 + 3z + 4z^2 + 5z^3 + 7z^4 + 9z^5 + 12z^6 + 16z^7 + 21z^8 + 28z^9 + \dots && (12, 23, 32, 31)\\
&1 + 3z + 4z^2 + 6z^3 + 8z^4 + 10z^5 + 12z^6 + 14z^7 + 16z^8 + 18z^9 + \dots && (11, 12, 22, 23)\\
&1 + 3z + 4z^2 + 6z^3 + 8z^4 + 12z^5 + 16z^6 + 24z^7 + 32z^8 + 48z^9 + \dots && (12, 21, 13, 31)\\
&1 + 3z + 4z^2 + 6z^3 + 9z^4 + 13z^5 + 19z^6 + 28z^7 + 41z^8 + 60z^9 + \dots && (11, 12, 23, 31)\\
&1 + 3z + 4z^2 + 6z^3 + 9z^4 + 14z^5 + 22z^6 + 35z^7 + 56z^8 + 90z^9 + \dots && (11, 12, 21, 33)\\
&1 + 3z + 4z^2 + 6z^3 + 10z^4 + 16z^5 + 26z^6 + 42z^7 + 68z^8 + 110z^9 + \dots && (11, 12, 21, 23)\\
&1 + 3z + 4z^2 + 7z^3 + 11z^4 + 18z^5 + 29z^6 + 47z^7 + 76z^8 + 123z^9 + \dots && (11, 12, 21, 13)\\
&1 + 3z + 4z^2 + 8z^3 + 16z^4 + 32z^5 + 64z^6 + 128z^7 + 256z^8 + \dots && (11, 12, 21, 22).
\end{aligned}$$

6. Koszulness is not constructible

The following examples from the paper [**56**] demonstrate that the set of all Koszul algebras with 3 generators and 3 relations is not constructible:

$$\begin{cases} zy = \lambda yz \\ xz = xy \\ zx = yx, \quad \lambda \in \Bbbk^* \end{cases} \qquad \text{or} \qquad \begin{cases} zy = yz + \lambda y^2 \\ xz = xy \\ zx = 0, \quad \lambda \in \Bbbk. \end{cases}$$

For the first algebra one has $xy^{n+1}x = xzy^nx = \lambda^n xy^n zx = \lambda^n xy^{n+1}x$, so the Hilbert series differs from the generic one (equal to $(1+z)(1-2z+z^3)^{-1}$) iff λ is a root of unity. The Hilbert series of the second algebra is equal to $(1-z)^{-3}$ unless $\lambda^{-1} = 1, 2, 3, \dots$. We claim that in both cases algebras corresponding to non-exceptional values of λ are Koszul. To prove this note that these algebras admit a Γ-grading (in the sense of section 7 of chapter 4) with values in the free group with two generators α, β. Namely, set $A_{1,\alpha} = \langle x \rangle$ and $A_{1,\beta} = \langle y, z \rangle$. Since $\dim A_{1,\alpha} = 1$, it is enough to check that the relation lattice in $A_{1,\alpha} \otimes A_{1,\beta}^{\otimes n+1} \otimes A_{1,\alpha}$ is distributive. But in the gradings $\alpha\beta\beta$, $\beta\beta\beta$ and $\beta\beta\alpha$ one has $I \otimes V \cap V \otimes I = 0$, hence Koszulness follows from the calculation of the Hilbert series (cf. Proposition 2.3). Taking $\lambda = -1$ in the second example we obtain a quadratic algebra A over $\mathbb{Z}$ such that $A \otimes_{\mathbb{Z}} \mathbb{Q}$ is Koszul but $A \otimes_{\mathbb{Z}} (\mathbb{Z}/p\mathbb{Z})$ is not Koszul for any prime. Note that the quadratic dual $A^!$ (defined over $\mathbb{Z}$) is a free $\mathbb{Z}$-module. Substituting a new generator u instead of x in the third relation we obtain a family with the generic series $(1-4z+3z^2)^{-1}$.

Let us outline a more general approach to such examples following [**56**]. Suppose we are given a collection of linear automorphisms $M_1,\dots,M_k\colon V\longrightarrow V$ of a vector space V and two subspaces $V_l,\ V_r\subset V$. Consider the quadratic algebra A with the generator space $\langle x_1,\dots,x_k,a_l,a_r\rangle\oplus V$ and the relations $a_lV_l=0$, $x_iv=M_i(v)x_i$ for $v\in V$, and $V_ra_r=0$. We claim that the Hilbert series of A is given by the formula

$$h_A(z)^{-1}=1-mz+rz^2-\sum_{n=0}^{\infty}z^{n+3}\sum_{i_1,\dots,i_n=1}^{k}\dim V_l\cap M_{i_1}\cdots M_{i_n}V_r.$$

This can be easily verified if all these intersections are zero. One can reduce to this case iteratively taking quotients of A by strongly free elements (see the end of section 5 of chapter 2). Moreover, a straightforward computation shows that $\operatorname{gl}\dim A\leqslant 3$ and

$$\dim\operatorname{Ext}_A^{3,3+n}(\Bbbk,\Bbbk)=\sum_{i_1,\dots,i_n=1}^{k}\dim V_l\cap M_{i_1}\cdots M_{i_n}V_r.$$

Therefore, A is Koszul iff $V_l\cap M_{i_1}\cdots M_{i_n}V_r=0$ for any $i_1,\dots,i_n$, $n\geqslant 1$, and $A_3=0$ iff $V_l\cap V_r=0$. There exists also a construction with the same properties for which $A^!$ is commutative [**8**]. Furthermore, it is shown in [**9**] that for any system F of (quasi-)polynomial Diophantine equations in nonnegative integer variables $z_1,\dots,z_k$ with coefficients in $\Bbbk$ there exists a construction of linear operators $M_1,\dots,M_k$ on a vector space $V\supset V_l,\ V_r$ (with $\dim V$ bounded by the "size" of F) such that the right-hand side of the above formula is equal to the number of solutions of F with $z_1+\cdots+z_k=n$.

Thus, the set of Koszul algebras can be non-constructible. However, we do not know whether there exists an irreducible component of some $\mathcal{Q}_{m,s,u}$ that contains a Koszul algebra but does not contain a nonempty Zariski open subset consisting of Koszul algebras.

7. Families of quadratic algebras over schemes

Our goal in this section is to prove that there is a finite number of Hilbert series of Koszul algebras with fixed $\dim A_1$ *over all ground fields uniformly.* [2]

PROPOSITION 7.1. *Let Λ be an arbitrary ring, W a Λ-module, and $R_1,\dots,R_n\subset W$ a collection of submodules generating a distributive lattice Ω of submodules in W. Assume that for any subset $I\subset[1,n]$ the Λ-module $W/\sum_{i\in I}R_i$ is projective. Then for any $S,T\in\Omega$ such that $S\subset T$ the Λ-module T/S is projective. Moreover, there exists a direct sum decomposition $W=\bigoplus_{\alpha\in A}W_\alpha$ over Λ such that each R_i is the sum of a set of submodules W_α.*

Proof: Every submodule $T\in\Omega$ can be presented as the intersection $T_1\cap\dots\cap T_k$ of submodules of the form $\sum_{i\in I}R_i$. Let us prove by induction in k that the quotient

[2]We are grateful to J. Bernstein for posing this question.

module W/T is projective. Indeed, if the modules W/T_1, W/T_2, and $W/(T_1+T_2)$ are projective then it follows from the exact sequence

$$0 \longrightarrow W/(T_1 \cap T_2) \longrightarrow W/T_1 \oplus W/T_2 \longrightarrow W/(T_1+T_2) \longrightarrow 0$$

that the module $W/(T_1 \cap T_2)$ is also projective. It remains to use the exact triple $0 \to T/S \to W/S \to W/T \to 0$ to deduce projectivity of T/S. The second assertion is clear from the proof of Proposition 7.1 of chapter 1. □

PROPOSITION 7.2. *Let Λ be a Noetherian local ring with the residue field $\Bbbk$ and let $R_1, \ldots, R_n \subset W$ be a collection of submodules in a finitely generated Λ-module. Assume that*

(1) *all of the Λ-modules W, W/R_i, and $W/(R_i+R_j)$ are free, where $i,j \in [1,n]$;*
(2) *the collection of subspaces $(R_i \otimes_\Lambda \Bbbk)$, $i \in [1,n]$, in $W \otimes_\Lambda \Bbbk$ is distributive (they are indeed subspaces since the modules W/R_i are free).*

Then the collection of submodules (R_i) generates a distributive lattice and for any submodule $T \subset W$ from this lattice the quotient module W/T is free.

Proof: It suffices to consider the case $n=3$. The case $n>3$ will follow exactly as in the proof of Proposition 1.1. Consider the complex of free Λ-modules $C_2 \longrightarrow C_1 \longrightarrow C_0$ given by

$$W \longrightarrow W/R_1 \oplus W/R_2 \oplus W/R_3 \longrightarrow W/(R_1+R_2) \oplus W/(R_1+R_3) \oplus W/(R_2+R_3).$$

We have $H_0(C_\bullet) = W/(R_1+R_2+R_3)$ and $H_2(C_\bullet) = R_1 \cap R_2 \cap R_3$. Furthermore, the triple (R_1,R_2,R_3) is distributive iff $H_1(C_\bullet)=0$ (see Proposition 7.2 of chapter 1). The analogous statements hold for the complex $C_\bullet \otimes_\Lambda \Bbbk$ and the collection of subspaces $(R_i \otimes_\Lambda \Bbbk)$ since the tensor product commutes with quotients. In particular, we have $H_1(C_\bullet \otimes_\Lambda \Bbbk) = 0$. Now consider the hyperhomology spectral sequence for the functor $\mathrm{Tor}_*^\Lambda(-,\Bbbk)$ and the complex $C_\bullet$:

$$E^2_{p,q} = \mathrm{Tor}_p^\Lambda(H_q(C_\bullet),\Bbbk) \Longrightarrow H_{p+q}(C_\bullet \otimes_\Lambda \Bbbk)$$

with the differentials $d^r\colon E^r_{p,q} \longrightarrow E^r_{p-r,q+r-1}$. This sequence implies that $\mathrm{Tor}_1^\Lambda(H_0(C_\bullet),\Bbbk) \subset H_1(C_\bullet \otimes_\Lambda \Bbbk) = 0$ and therefore the Λ-module $H_0(C_\bullet) = W/(R_1+R_2+R_3)$ is free. It follows that $\mathrm{Tor}_2^\Lambda(H_0(C_\bullet),\Bbbk)=0$ and $H_1(C_\bullet)\otimes_\Lambda \Bbbk \simeq H_1(C_\bullet \otimes_\Lambda \Bbbk) = 0$. Hence, $H_1(C_\bullet)=0$ and the collection (R_i) is distributive. It remains to apply Proposition 7.1. □

THEOREM 7.3. *Let Λ be a Noetherian local ring with the residue field $\Bbbk$ and let*

$$\mathcal{A} = \Lambda \oplus \mathcal{A}_1 \oplus \mathcal{A}_2 \oplus \cdots = \mathbb{T}_\Lambda(\mathcal{A}_1)/(\mathcal{I}), \qquad \mathcal{I} \subset \mathcal{A}_1 \otimes_\Lambda \mathcal{A}_1$$

be a quadratic algebra over Λ with finitely generated Λ-modules $\mathcal{A}_i$. Then the following two conditions are equivalent:

(a) *the Λ-modules $\mathcal{A}_1$, $\mathcal{A}_2$ and $\mathcal{A}_3$ are free and the quadratic algebra $\mathcal{A} \otimes_\Lambda \Bbbk$ over the field $\Bbbk$ is Koszul;*
(b) *for any n the collection of submodules $(\mathcal{A}_1^{\otimes_\Lambda i-1} \otimes_\Lambda \mathcal{I} \otimes_\Lambda \mathcal{A}_1^{\otimes_\Lambda n-i-1})$, $i \in [1,n-1]$, generates a distributive lattice in the Λ-module $\mathcal{A}_1^{\otimes n}$ (n-th tensor power over Λ) and all the quotients of this module by submodules from this lattice are free over Λ.*

If these conditions are satisfied then all the grading components $\mathcal{A}_n$ are free Λ-modules. If in addition Λ is a domain with the field of quotients $\mathbb{K}$ then the algebra $\mathcal{A} \otimes_\Lambda \mathbb{K}$ is Koszul. The same statements are true if we consider these properties in internal degree $\leqslant N$.

Proof: This follows easily from Proposition 7.2. □

For a closed point y of a scheme Y we denote by $\Bbbk(y)$ the corresponding residue field. Also, for a coherent sheaf $\mathcal{F}$ over Y we set $\mathcal{F}(y) = \mathcal{F} \otimes_{\mathcal{O}_Y} \Bbbk(y)$.

LEMMA 7.4. *Let Y be a Noetherian scheme, $\rho\colon \mathcal{E} \longrightarrow \mathcal{F}$ a morphism of locally free sheaves of finite ranks over Y. Let also k be an integer. Assume that for any closed point $y \in Y$ one has $\dim_{\Bbbk(y)} \operatorname{coker}(\rho)(y) \leqslant f - k$, where $f = \operatorname{rk} \mathcal{F}$. Then the sheaf $\operatorname{coker}(\rho)$ is locally free of rank $f-k$ iff the morphism $\wedge^{k+1}\rho : \wedge^{k+1}\mathcal{E} \longrightarrow \wedge^{k+1}\mathcal{F}$ is identically zero.*

Proof: "Only if": If the sheaf $\mathcal{F}/\rho(\mathcal{E})$ is locally free of rank $f-k$ then the map ρ factors through a locally free sheaf $\rho(\mathcal{E})$ of rank k and therefore $\wedge^{k+1}\rho = 0$.

"If": Since $\dim_{\Bbbk(y)}(\mathcal{F}/\rho(\mathcal{E}))(y) \leqslant f-k$, locally in Y one can choose a set of k sections $s_1, \ldots, s_k$ of the sheaf $\rho(\mathcal{E})$ such that their images in $\mathcal{F}(y)$ are linearly independent over $\Bbbk(y)$. Since $\wedge^{k+1}\rho = 0$, it is easy to see that the sections $s_1, \ldots, s_k$ generate $\rho(\mathcal{E})$. The desired statement follows immediately. □

LEMMA 7.5. *Let X be a Noetherian scheme and let $\rho\colon \mathcal{E} \longrightarrow \mathcal{F}$ be a morphism of locally free sheaves of finite rank. Then for any integer k there exists a locally closed subscheme $X_k \subset X$ with the following universal property: a morphism of Noetherian schemes $f\colon Y \longrightarrow X$ factors through X_k iff the pull-back $f^* \operatorname{coker}(\rho)$ is locally free of rank k.*

Proof: The condition $\dim_{\Bbbk(x)} \operatorname{coker}(\rho)(x) \leqslant k$ defines an open subscheme in X. Inside this subscheme the condition $\wedge^{f-k+1}\rho = 0$ defines the closed subscheme X_k. The universal property follows from Lemma 7.4. □

In the following proposition for coherent sheaves $\mathcal{F}$ and $\mathcal{G}$ on a scheme Y we denote their tensor product over $\mathcal{O}_Y$ simply by $\mathcal{F} \otimes \mathcal{G}$.

PROPOSITION 7.6. *For any integers $m, s, u \geqslant 0$ there exists a scheme $\mathcal{Q} = \mathcal{Q}^{\mathbb{Z}}_{m,s,u}$ of finite type over $\operatorname{Spec} \mathbb{Z}$ together with a subsheaf $\mathcal{I}_\mathcal{Q} \subset (\mathcal{O}^m_\mathcal{Q})^{\otimes 2}$ of the free sheaf of rank m^2 such that*
(i) the sheaf $(\mathcal{O}^m_\mathcal{Q})^{\otimes 2}/\mathcal{I}_\mathcal{Q}$ is locally free of rank s,
(ii) the sheaf $(\mathcal{O}^m_\mathcal{Q})^{\otimes 3}/(\mathcal{O}^m_\mathcal{Q} \otimes \mathcal{I}_\mathcal{Q} + \mathcal{I}_\mathcal{Q} \otimes \mathcal{O}^m_\mathcal{Q})$ is locally free of rank u,
and the scheme $\mathcal{Q}^{\mathbb{Z}}_{m,s,u}$ has the following universal property: for any Noetherian scheme Y together with a subsheaf $\mathcal{I}_Y \subset \mathcal{O}_Y^{\otimes 2}$ with the properties similar to (i) and (ii) there exists a unique morphism $f\colon Y \longrightarrow \mathcal{Q}^{\mathbb{Z}}_{m,s,u}$ such that $\mathcal{I}_Y$ coincides with the pull-back of $\mathcal{I}_\mathcal{Q}$ as a subsheaf in $(\mathcal{O}^m_Y)^{\otimes 2} = f^*(\mathcal{O}^m_\mathcal{Q})^{\otimes 2}$.

Proof: Note that the scheme $\mathcal{Q}$ should be empty unless $s \leqslant m^2$. Recall that for any integers $0 \leqslant r \leqslant n$ there exists the *Grassmannian scheme* $\mathbb{G}_{r,n}$ of finite type over $\operatorname{Spec} \mathbb{Z}$ together with a locally free subsheaf $\mathcal{I}_\mathbb{G} \subset \mathcal{O}^n_\mathbb{G}$ of rank r with a locally free quotient sheaf $\mathcal{O}^n_\mathbb{G}/\mathcal{I}_\mathbb{G}$ such that the scheme $\mathbb{G}_{r,n}$ is universal in the category of all schemes Y equipped with subsheaves $\mathcal{I}_Y \subset \mathcal{O}^n_Y$ with the same properties (see [**84**]). It remains to consider the scheme $X = \mathcal{Q}^{\mathbb{Z}}_{m,s} = \mathbb{G}_{m^2-s,m^2}$ and apply Lemma 7.5 to the morphism $\rho\colon \mathcal{O}^m_\mathcal{Q} \otimes \mathcal{I}_\mathcal{Q} \oplus \mathcal{I}_\mathcal{Q} \otimes \mathcal{O}^m_\mathcal{Q} \longrightarrow (\mathcal{O}^m_\mathcal{Q})^{\otimes 3}$ to obtain the locally closed subscheme $\mathcal{Q}^{\mathbb{Z}}_{m,s,u} \subset \mathcal{Q}^{\mathbb{Z}}_{m,s}$. □

COROLLARY 7.7. *For any integers* $m, s, u \geqslant 0$ *and any field* $\Bbbk$ *there exists a natural bijective correspondence between the set of quadratic algebras* A *over* $\Bbbk$ *with the fixed generators space* $A_1 = \Bbbk^m$ *such that* $\dim A_2 = s$, $\dim A_3 = u$ *and the set of all* $\Bbbk$*-points of the scheme* $\mathcal{Q} = \mathcal{Q}^{\mathbb{Z}}_{m,s,u}$. *More precisely, a point* $f\colon \operatorname{Spec}\Bbbk \longrightarrow \mathcal{Q}$ *corresponds to the quadratic algebra* $A = \{\Bbbk^m, I\}$ *with* $I = f^*\mathcal{I}_{\mathcal{Q}} \subset \Bbbk^m = f^*\mathcal{O}^m_{\mathcal{Q}}$. *Furthermore,* n*-Koszul algebras* A *correspond to* $\Bbbk$*-points that belong to an open subscheme* ${}_n\mathcal{K}^{\mathbb{Z}}_{m,s,u} \subset \mathcal{Q}^{\mathbb{Z}}_{m,s,u}$. *This subscheme has a natural decomposition into a disjoint union of open subschemes* ${}_n\mathcal{K}^{\mathbb{Z}}_{m,s,u} = \coprod_{h(z)} {}_n\mathcal{K}^{\mathbb{Z}}_h$ *numbered by power series* $h(z) = 1 + mz + sz^2 + uz^3 + \cdots$ *considered modulo* z^{n+1} *such that* $\Bbbk$*-points in* ${}_n\mathcal{K}^{\mathbb{Z}}_h$ *correspond to* n*-Koszul algebras* A *with* $h_A(z) \bmod z^{n+1} = h(z)$.

Proof: To prove the first statement apply Proposition 7.6 to $Y = \operatorname{Spec}\Bbbk$. The remaining assertions follow easily from Theorem 7.3. □

COROLLARY 7.8. *The total number of Hilbert series of Koszul algebras with fixed* $\dim A_1 = m$ *over all fields* $\Bbbk$ *is finite.*

Proof: The number of Koszul Hilbert series cannot exceed the number of irreducible components of the schemes $\mathcal{Q} = \mathcal{Q}^{\mathbb{Z}}_{m,s,u}$. Indeed, if a point $f\colon \Bbbk \longrightarrow \mathcal{Q}$ corresponding to a Koszul algebra A belongs to an irreducible component with the general point $\eta\colon \mathbb{K} \longrightarrow \mathcal{Q}$ then it follows from Corollary 7.7 that the quadratic algebra B over the field $\mathbb{K}$ corresponding to η is also Koszul and has the same Hilbert series $h_A(z) = h_B(z)$. □

CHAPTER 7

Hilbert series of Koszul algebras and one-dependent processes

In this chapter we describe a relation (first observed in [**100**]) between Hilbert series of Koszul algebras and one-dependent stochastic sequences of 0's and 1's. The connection is based on certain polynomial inequalities satisfied by dimensions of grading components of arbitrary Koszul algebras (see section 2). Our interest in this relation is due to its potential relevance for the conjecture on rationality of Hilbert series of Koszul algebras (see sections 1 and 8). It is also interesting that some of the features of the theory of Koszul algebras have analogues for one-dependent processes. For example, there is a class of one-dependent processes playing the role similar to that of PBW-algebras (see section 5). Also, one can define modules over one-dependent processes (see section 10) and consider operations similar to those studied in chapter 3 (see section 6).

1. Conjectures on Hilbert series of Koszul algebras

Conjecture 1. *The Hilbert series of a Koszul algebra A is a rational function. In particular, the growth of A is either polynomial or exponential.*

This conjecture is true for PBW-algebras and for commutative PBW-algebras (see sections 6 and 8 of chapter 4). Perhaps the most important philosophical reason to believe Conjecture 1 is provided by the finiteness of Hilbert series of Koszul algebras with fixed $\dim A_1$ (see sections 2 and 7 of chapter 6) It is well known that neither this nor Conjecture 1 is true for general quadratic algebras (see [**8, 56**] and section 6 of chapter 6). Among partial confirmations of Conjecture 1 let us mention the result of Piontkovskii [**93**] stating that algebras admitting a Koszul filtration have rational Hilbert series.

Remark 1. It is well known that any power series with integral coefficients defining a meromorphic function on the entire complex plane is rational (see [**32**]). Hence, to prove Conjecture 1 it suffices to show that the Hilbert series of a Koszul algebra admits a meromorphic continuation to $\mathbb{C}$. It is clear that the Hilbert series of a one-generated algebra is holomorphic in the disk $|z| < m^{-1}$, where $m = \dim A_1$. In section 8 we will show that the Hilbert series of a Koszul algebra admits a meromorphic continuaton to the disk $|z| < 2m^{-1}$ (see Corollary 8.3).

Note that Conjecture 1 would also imply that every Koszul module over a Koszul algebra has a rational Hilbert series (by Corollary 5.5 of chapter 2). One can hope that the following stronger version of this assertion is true.

Conjecture 1bis. *Let A be a Koszul algebra. Then there exists a polynomial $q(z)$ depending only on A such that for every Koszul A-module M one has $h_M(z) = p(z)/q(z)$ for some polynomial $p(z)$.*

Finally, let us point out that Conjecture 1 would imply rationality of Hilbert series for a much wider class of graded algebras.

PROPOSITION 1.1. *Assume that Conjecture 1 holds. Then the Hilbert series of any graded algebra of finite rate (see section 3 of chapter 3) is rational.*

Proof: Note that if $\mathrm{rate}(A) \leqslant d$ then the algebra $A^{(d)}$ is Koszul (by Theorem 3.1). Consider the decomposition $A = \bigoplus_{r=0}^{d-1} A^{(d,r)}$, where $A_i^{(d,r)} = A_{r+di}$. It suffices to show that each $A^{(d,r)}$ has rational Hilbert series. For $r = 0$ this follows from the Koszulness of $A^{(d)}$. On the other hand, applying Proposition 3 of chapter 3 to the A-module $A^{[r]} = \bigoplus_{i=0}^{\infty} A_{r+i}$ for $r > 0$ we derive that $A^{(d,r)}$ is a Koszul $A^{(d)}$-module. Hence, its Hilbert series is also rational. □

We have seen that there are examples of Koszul algebras with Hilbert series different from that of any PBW-algebra (see section 8 of chapter 4). Nevertheless, all of the known Hilbert series of Koszul algebras are quotients of polynomials of degree $\leqslant \dim A_1$. The following particular case of this observation may be of independent interest.

Conjecture 2. *Any Koszul algebra A of finite global homological dimension d has $\dim A_1 \geqslant d$. By duality this is equivalent to the following statement: for a Koszul algebra B with $B_{d+1} = 0$ and $B_d \neq 0$ one has $\dim B_1 \geqslant d$.*

This conjecture holds trivially for $d \leqslant 2$. To verify it in the case $d = 3$ we just have to rule out the possibility $\dim A_1 = 2$ which is not difficult since all the Hilbert series of Koszul algebras with two generators are known (see section 5 of chapter 6). It is also easy to check that Conjecture 2 holds for PBW-algebras: the proof reduces to the fact that all vertices in a path of maximal length in an oriented graph are distinct. The second statement of the conjecture is not true for non-Koszul quadratic algebras: for the quadratic algebra

$$B = \Bbbk\{x, y\}/(x^2, y(x+y))$$

one has $h_B(z) = 1 + 2z + 2z^2 + z^3$. Also, the first statement of Conjecture 2 is not true for general one-generated graded algebras: the algebra $A = \{x, y\}/(x^2y, xy^2)$ with two generators and two cubic relations has global dimension 3 and the Hilbert series $(1-z)^{-2}(1-z^2)^{-1}$ [**119**].

Another simple observation can be formulated as follows.

Conjecture 3. *Let A be a Koszul algebra. If both algebras A and $A^!$ have polynomial growth then the Hilbert series of A coincides with that of the tensor product of a symmetric algebra and an exterior algebra:*

$$h_A(z) = \frac{(1+z)^a}{(1-z)^b} \qquad \text{for some } a, b \geqslant 0.$$

For a finite-dimensional algebra A Conjecture 3 follows easily from Conjecture 2. The first nontrivial case to check would be that of the series $h(z) = 1 + 3z + 4z^2 + 3z^3 + z^4 = (1 + z + z^2)(1 + z)^2$. Both conjectures imply that there should be no Koszul algebras with $h(z)$ as the Hilbert series. We do not know whether this is true (see also Example 5 in section 5 of chapter 6).

Here is a proof of Conjecture 3 for PBW-algebras. According to Proposition 6.3 of chapter 4 we should consider a subset $S \subset [1, m]^2$ such that both graphs G_S and $G_{\bar{S}}$ have no intersecting cycles. Assume that G_S has a cycle of length a. Since G_S has no intersecting cycles, the only edges between the corresponding a vertices are the ones constituting the cycle. Since $G_{\overline{S}}$ also has no intersecting cycles, it follows that $a \leqslant 2$. Thus, the only cycles in G_S are loops and 2-cycles (cycles of length 2).

Note that if there are no edges connecting i and j in G_S (we say in this case that $\{i, j\}$ is a disconnected pair of vertices) then i and j form a 2-cycle in $G_{\overline{S}}$. Since 2-cycles in G_S cannot intersect, we derive that for every vertex $i \in [1, m]$ there is at most one other vertex j such that $\{i, j\}$ is a disconnected pair of vertices in G_S. Note also that for every disconnected pair $\{i, j\}$ in G_S we should have loops at both vertices i and j.

Let us call a one- or two-element subset in $[1, m]$ a *quasi-cycle* if it is either a disconnected pair of vertices, or a vertex with a loop that does not belong to any such pair, or a 2-cycle, or a vertex that does not belong to any cycle. It is clear that every vertex belongs to exactly one quasi-cycle and the sets of quasi-cycles for G_S and for $G_{\overline{S}}$ are the same. By the construction there is exactly one edge in G_S between any two vertices from different quasi-cycles. It is not difficult to see that all the edges joining two fixed quasi-cycles start in one of them, and that this relation defines a total order on the set of quasi-cycles.

It remains to observe that our set S coincides with the subset of $[1, m]^2$ associated with the natural PBW-basis in the tensor product of the monomial quadratic algebras with $\leqslant 2$ generators corresponding to the quasi-cycles (see section 4 of chapter 4). Therefore, the Hilbert series of S is equal to the product of those Hilbert series. □

Remark 2. It is also easy to verify Conjecture 3 for commutative and skew-commutative algebras A. Indeed, a quadratic algebra A is skew-commutative (resp., commutative) iff the quadratic dual algebra $A^!$ is the universal enveloping algebra of a quadratic Lie algebra (resp., Lie superalgebra), so this follows from Examples 2 and 3 in section 2 of chapter 2.

Remark 3. Our Conjecture 3 is also reminiscent of the well-known conjecture on Hilbert series of Artin–Schelter *regular* algebras [**12**] (see also [**79**]). Recall that a graded algebra is called *regular*, if it is Gorenstein, of finite homological dimension and has polynomial growth. It is expected that the Hilbert series of a regular algebra A coincides with that of a polynomial algebra with homogeneous generators in appropriate degrees:

$$h_A(z) = \prod_{i=1}^{d} (1 - z^{n_i})^{-1}.$$

If A is a Koszul regular algebra this would follow immediately from either Conjecture 2 or Conjecture 3 (to deduce it from Conjecture 2 observe that roots of $h_{A^!}(-z)$ have absolute value 1 and their sum is equal to $\dim A_1$).

2. Koszul inequalities

In this section we show that the numbers $a_i = \dim A_i$ for a Koszul algebra A satisfy a certain system of polynomial inequalities. In section 4 we will use these inequalities to associate with A a *one-dependent stochastic sequence* of 0's and 1's.

The inequalities that we want to prove generalize inequalities

$$a_{i+j} \leqslant a_i a_j$$

that hold for every one-generated algebra. For every collection of indices $i_1, \dots, i_r \geqslant 1$ (where $r \geqslant 1$) let us consider the polynomial

$$\Phi_{i_1,\dots,i_r}(a_\bullet) = \Phi_{i_1,\dots,i_r}(a_1, a_2, \dots, a_{i_1+\dots+i_r}) :=$$
$$a_{i_1}\dots a_{i_r} - a_{i_1+i_2}a_{i_3}\dots a_{i_r} - a_{i_1}a_{i_2+i_3}a_{i_4}\dots a_{i_r} - \dots - a_{i_1}\dots a_{i_{r-2}}a_{i_{r-1}+i_r}$$
$$+ a_{i_1+i_2+i_3}a_{i_4}\dots a_{i_r} + \dots + a_{i_1}\dots a_{i_{r-3}}a_{i_{r-2}+i_{r-1}+i_r}$$
$$+ a_{i_1+i_2}a_{i_3+i_4}a_{i_5}\dots a_{i_r} + \dots + a_{i_1}\dots a_{i_{r-4}}a_{i_{r-3}+i_{r-2}}a_{i_{r-1}+i_r} - \dots \pm a_{i_1+\dots+i_r}.$$

The summation here is taken over all partitions of the segment $[1, r]$ into nonempty subsegments $[1, t_1]$, $[t_1+1, t_2]$, ..., $[t_{s-1}+1, r]$ and the sign is $(-1)^{r-s}$.

THEOREM 2.1. *If A is an m-Koszul algebra then the numbers $a_i = \dim A_i$ for $i \geqslant 1$ satisfy inequalities*

$$\Phi_{n_1,\dots,n_r}(A) := \Phi_{n_1,\dots,n_r}(a_\bullet) \geqslant 0 \tag{2.1}$$

for every $n_1, \dots, n_r \geqslant 1$ such that $n_1 + \dots + n_r \leqslant m$.

Proof: Consider the complex of Proposition 8.3 of chapter 2. Since it is exact, we get

$$\Phi_{n_1,\dots,n_r}(a_\bullet) = \dim V(n_1, \dots, n_r) \geqslant 0.$$

□

Remark 1. One can get a more direct proof of the above theorem using Backelin's criterion of m-Koszulness (see section 4 of chapter 2) and the following elementary observation. Suppose we are given a finite set S and a collection of subsets $S_1, \dots, S_m \subset S$. For every $J \subset [1, m]$ let b_J be the number of elements in $S_J := \bigcap_{j\in J} S_j$. Then for every $I \subset [1, m]$ one should have

$$\sum_{J\supset I} (-1)^{|J|-|I|} b_J \geqslant 0.$$

Indeed, the set S can be partitioned into a disjoint union of the atomic subsets

$$T_I = (\bigcap_{i\in I} S_i) \cap (\bigcap_{i\in I} S_i^c),$$

where S_i^c denotes the complement to S_i in S, I runs through subsets of $[1, m]$. The subsets S_j are recovered from the partition into T_I's by the formula $S_j = \bigcup_{I:j\in I} T_I$. Using the exclusion-inclusion formula one can easily express cardinalities $f_I := |T_I|$ in terms of (b_J):

$$f_I = \sum_{J\supset I} (-1)^{|J|-|I|} b_J. \tag{2.2}$$

This implies the above inequalities.

Remark 2. Inequalities (2.1) with $r = 1$ are true for any graded algebra, the inequalities with $r = 2$ hold for a one-generated algebra A, and those with $r = 3$ are satisfied for a quadratic algebra. This observation can be generalized further: the inequality $\Phi_{i_1,\dots,i_r}(A) \geqslant 0$ holds provided that $\mathrm{Ext}_A^{ij}(\Bbbk, \Bbbk) = 0$ for all $i \leqslant r-1$ and for $i < j \leqslant i_1 + \dots + i_r$. Moreover, if one has $i_s \geqslant c$ for all s, then it suffices to consider $(i-1)c < j-1$ instead of $i < j$ in the condition above. This follows from the $\mathbb{Z}$-algebra version of Backelin's theorem on the cohomology of Veronese subalgebras (see [**16**] and section 3 of chapter 3).

There is a nice determinantal formula for polynomials $\Phi_{i_1,\dots,i_r}$.

PROPOSITION 2.2. *For every* $i_1, \dots, i_r \geqslant 1$ *one has*

$$\Phi_{i_1,\dots,i_r}(a_\bullet) = \det X(i_1, \dots, i_r),$$

where $X(i_1, \dots, i_r) = (x_{mn})$ *is the* $r \times r$ *matrix with the entries*

$$x_{mn} = \begin{cases} a_{i_{m,n}}, & m \leqslant n, \text{ where } i_{m,n} = \sum_{s=m}^{n} i_s, \\ 1, & m = n+1, \\ 0, & m > n+1. \end{cases}$$

Proof: We use induction in r. If $r = 1$ then the assertion is trivial. To prove the induction step we use the following recursion formula:

$$\Phi_{i_1,\dots,i_r}(a_\bullet) = a_{i_1}\Phi_{i_2,\dots,i_r}(a_\bullet) - \Phi_{i_1+i_2,i_3,\dots,i_r}(a_\bullet). \tag{2.3}$$

The similar formula for $\det X(i_1, \dots, i_r)$ is immediately obtained by expanding the determinant in the first column. □

The determinants appearing in the above formula for $\Phi_{i_1,\dots,i_r}$ are actually minors of the infinite Toeplitz matrix $X = (a_{j-i})_{i,j\geqslant 0}$, where we set $a_i = 0$ for $i < 0$. More precisely, $X(i_1, \dots, i_r)$ corresponds to the $r \times r$ minor of X with rows $(0, i_1, i_1 + i_2, \dots, i_1 + \dots + i_{r-1})$ and columns $(i_1, i_1 + i_2, \dots, i_1 + \dots + i_r)$.

The reason we think this determinantal formula is of interest is because of the relation with total positivity. Recall that a sequence (a_n) is called *totally positive* (or *Polya frequency sequence*) if the matrix X is totally positive, i.e., *all* its minors are nonnegative. It is well known (see [**71**]) that in this case the Hilbert series $h(z) = \sum_{n\geqslant 0} a_n z^n$ meromorphically extends to the entire complex plane (and hence is rational in the case when a_n are integers). Moreover, in this case poles (resp., zeros) of $h(z)$ are positive (resp., negative) real numbers. However, it is not true that for every Koszul algebra the sequence (a_n) is totally positive because the above condition on poles and zeros is not necessarily satisfied (even for PBW-algebras). In [**42**] Davydov proved that total positivity holds for a quadratic algebra associated with an R-matrix satisfying the Hecke condition $(R+1)(R-q) = 0$.

It is not true that every sequence $a_\bullet = (a_n)$ of positive integers satisfying inequalities (2.1) comes from a Koszul algebra. In fact, there are some extra non-homogeneous inequalities that a_n should satisfy. Here is the simplest example.

PROPOSITION 2.3. *Assume that* A *is a one-generated algebra and let* $a_n = \dim A_n$. *Then*

$$a_k \leqslant a_i a_j$$

for $\max(i, j) \leqslant k \leqslant i + j$.

Proof: This is proved using Gröbner basis techniques. Let $x_1, \dots, x_m$ be a basis of A_1. Let $\mathcal{N}$ be the set of *normal monomials* in x_i's. By definition, $\mathcal{N}$ consists of all monomials $x^\alpha \in A$ such that x^α is not a linear combination (in A) of $x^{\alpha'}$ with $\alpha' < \alpha$ in lexicographical order (cf. section 1 of chapter 4). Let $\mathcal{N}_n \subset \mathcal{N}$ be the set of monomials of degree n in $\mathcal{N}$. It is clear that $\mathcal{N}_n$ is a basis of A_n, hence, $|\mathcal{N}_n| = a_n$. Consider the map

$$\mathcal{N}_k \longrightarrow \mathcal{N}_i \times \mathcal{N}_j : x^\alpha \longmapsto ((x^\alpha)_{\leqslant i}, (x^\alpha)_{>k-j}),$$

where $(x^\alpha)_{\leqslant i}$ (resp., $(x^\alpha)_{>k-j}$) denotes the truncation of the monomial x^α to its initial segment of length i (resp., its final segment of length j). It is easy to see that this map is injective for $\max(i, j) \leqslant k \leqslant i + j$. Hence,

$$|\mathcal{N}_k| \leqslant |\mathcal{N}_i| \cdot |\mathcal{N}_j|. \quad \square$$

Example. The series $h(z) = 1 + 9z + z^2 + 2z^3$ cannot be the Hilbert series of a Koszul algebra since the inequality $a_3 \leqslant a_2^2$ does not hold for it. However, we claim that it satisfies all inequalities (2.1). Indeed, this follows from the fact that

$$h(8z) = 1 + 72z + 64z^2 + 1024z^3$$

is the Hilbert series of the monomial algebra corresponding to the oriented graph with 72 vertices

$$\{u_1, \dots, u_{32}, v, w_1, \dots, w_{32}, s_1, \dots, s_7\}$$

and 64 edges

$$\{(u_1, v), \dots, (u_{32}, v), (v, w_1), \dots, (v, w_{32})\}.$$

3. Koszul duality and inequalities

For a sequence of numbers $a_\bullet = (a_0 = 1, a_1, a_2, \dots)$ let us define the dual sequence $a_\bullet^! = (1, a_1^!, a_2^!, \dots)$ by the equality

$$\sum_{n \geqslant 0} a_n^! z^n = (\sum_{n \geqslant 0} a_n(-z)^n)^{-1}.$$

Thus, if $a_\bullet$ consists of dimensions of grading components of a Koszul algebra A then $a_\bullet^!$ corresponds to the Koszul dual algebra $A^!$.

PROPOSITION 3.1. *For every $n \geqslant 1$ one has*

$$a_n^! = \Phi_{1^n}(a_\bullet),$$

where $1^n = (1, \dots, 1)$ (1 repeated n times).

Proof: It suffices to prove the following identity for every $n \geqslant 1$:

$$\sum_{i=0}^{n} (-1)^i a_{n-i} \Phi_{1^i}(a_\bullet) = 0.$$

This identity is easily derived from the recursion formula (2.3) applied to the sequences 1^n, $(2, 1^{n-2})$, $(3, 1^{n-3})$, etc. □

The next proposition is analogous to Proposition 8.5 of chapter 2. Recall that $(i_1, \dots, i_r) \longmapsto J_{i_1, \dots, i_r}$ denotes the correspondence between collections of positive numbers $(i_1, \dots, i_r)$ such that $i_1 + \dots + i_r = n$ and subsets of $[1, n-1]$ considered in Lemma 8.2 of ch. 2.

PROPOSITION 3.2. *For any sequence of numbers $a_\bullet = (a_n)$ one has*

$$\Phi_{i_1, \dots, i_r}(a_\bullet) = \Phi_{j_1, \dots, j_s}(a_\bullet^!),$$

where $a_\bullet^!$ is the dual sequence, $J_{j_1, \dots, j_s}$ is the complement to $J_{i_1, \dots, i_r}$ in $[1, n-1]$.

One can prove this by mimicking the proof of Proposition 8.5 of chapter 2. We will give another proof using one-dependent sequences in section 4.

COROLLARY 3.3. *For any quadratic m-Koszul algebra A and the quadratic dual algebra $A^!$ one has*

$$\Phi_{i_1, \dots, i_r}(A) = \Phi_{j_1, \dots, j_s}(A^!),$$

where $J_{j_1, \dots, j_s} = J^c_{i_1, \dots, i_r}$, provided that $i_1 + \dots + i_r = j_1 + \dots + j_s = n \leqslant m$.

4. One-dependent processes

In this section we show how to associate a one-dependent stationary stochastic sequence of 0's and 1's with a Koszul algebra. Such a sequence is determined by a certain collection of probablities that can be viewed as a function on the free algebra in two variables. We will give a definition first and then explain its origin in probablity theory.

Definition. A *one-dependent process* is a linear functional $\phi : \mathbb{R}\{x_0, x_1\} \to \mathbb{R}$ such that
(i) $\phi(1) = 1$,
(ii) $\phi(m) \geqslant 0$ for every monomial m in x_0 and x_1, and
(iii)

$$\phi(f \cdot (x_0 + x_1) \cdot g) = \phi(f) \cdot \phi(g) \tag{4.1}$$

for all $f, g \in \mathbb{R}\{x_0, x_1\}$.

Let us point out some easy corollaries from conditions (i)–(iii).

PROPOSITION 4.1. *For every one-dependent process one has*
(a) $\phi(f \cdot (x_0 + x_1)^n \cdot g) = \phi(f) \cdot \phi(g)$, *where* $n \geqslant 1$, $f, g \in \mathbb{R}\{x_0, x_1\}$;
(b) $\phi = \phi \circ \iota$, *where* $\iota : \mathbb{R}\{x_0, x_1\} \to \mathbb{R}\{x_0, x_1\}$ *is the anti-involution sending* $x_{i_1} \dots x_{i_r}$ *to* $x_{i_r} \dots x_{i_1}$.

Proof: (a) This follows from (4.1) by induction.
(b) Let us prove by induction that for every monomial m of length n one has $\phi(\iota(m)) = \phi(m)$. We can use the second induction $\deg_{x_0}(m)$. If $m = x_1^n$ then the assertion is clear. Otherwise, we can write $m = m' x_0 x_1^k$. Then $\iota(m) = x_1^k x_0 \iota(m')$, hence

$$\phi(\iota(m)) = \phi(x_1^k(x_1 + x_0)\iota(m')) - \phi(x_1^{k+1}\iota(m')) = \phi(x_1^k)\phi(m') - \phi(m' x_1^{k+1}),$$

where we used the induction assumption for m' and for $m' x_1^{k+1}$. Therefore,

$$\phi(m) = \phi(m'(x_0 + x_1)x_1^k) - \phi(m' x_1^{k+1}) = \phi(x_1^k)\phi(m') - \phi(m' x_1^{k+1}) = \phi(\iota(m)).$$

□

Now let us explain a more standard definition of a one-dependent process in probablity theory. A 0-1-*valued stochastic sequence* is a sequence of random 0-1-valued variables $(\xi_i)_{i \in \mathbb{Z}}$. Such a sequence is determined by the set of probabilities

$$0 \leqslant P\{\xi_m = \varepsilon_m, \xi_{m+1} = \varepsilon_{m+1}, \dots, \xi_n = \varepsilon_n\} \leqslant 1, \tag{4.2}$$

where $m \leqslant n$, $\varepsilon_i \in \{0, 1\}$. These numbers should satisfy the obvious compatibility conditions:

$$P\{\xi_{m+1} = \varepsilon_{m+1}, \dots, \xi_n = \varepsilon_n\} =$$
$$P\{\xi_m = 0, \xi_{m+1} = \varepsilon_{m+1}, \dots, \xi_n = \varepsilon_n\} + P\{\xi_m = 1, \xi_{m+1} = \varepsilon_{m+1}, \dots, \xi_n = \varepsilon_n\},$$

$$P\{\xi_m = \varepsilon_m, \dots, \xi_{n-1} = \varepsilon_{n-1}\} =$$
$$P\{\xi_m = \varepsilon_m, \dots, \xi_{n-1} = \varepsilon_{n-1}, \xi_n = 0\} + P\{\xi_m = \varepsilon_m, \dots, \xi_{n-1} = \varepsilon_{n-1}, \xi_n = 1\}.$$

A stochastic sequence (ξ_i) is called *one-dependent* if the collection of all ξ_i with $i < k$ is *independent* from the collection of all ξ_i with $i > k$ for any $k \in \mathbb{Z}$— "the future is independent of the past provided that nothing is known about the present". Equivalently, probabilities (4.2) should satisfy the following system of identities:

$$\begin{aligned}(4.3)\quad & P\{\xi_m = \varepsilon_m, \dots, \xi_{k-1} = \varepsilon_{k-1}, \xi_k = 0, \xi_{k+1} = \varepsilon_{k+1}, \dots, \xi_n = \varepsilon_n\} + \\ & P\{\xi_m = \varepsilon_m, \dots, \xi_{k-1} = \varepsilon_{k-1}, \xi_k = 1, \xi_{k+1} = \varepsilon_{k+1}, \dots, \xi_n = \varepsilon_n\} = \\ & \qquad P\{\xi_m = \varepsilon_m, \dots, \xi_{k-1} = \varepsilon_{k-1}\} \cdot P\{\xi_{k+1} = \varepsilon_{k+1}, \dots, \xi_n = \varepsilon_n\},\end{aligned}$$

where $m \leqslant k \leqslant n$. Note that the cases $k = m$ and $k = n$ correspond precisely to the compatibility conditions between elementary probabilities (4.2).

A stochastic sequence is called *stationary* if the probabilities are invariant under the shift $\xi_i \longmapsto \xi_{i+1}$. In this case the elementary probablities are invariant under the simultaneous shift of m and n.

We claim that a *one-dependent process* can be viewed as a generating function for the set of probablities associated with a stationary 0-1-valued one-dependent stochastic sequence. Indeed, if we set

$$\phi(x_{\varepsilon_1} \dots x_{\varepsilon_n}) = P\{\xi_1 = \varepsilon_1, \dots, \xi_n = \varepsilon_n\}$$

then we immediately see that the condition of one-dependence (4.3) is equivalent to (4.1). Note that given a one-dependent process $\phi : \mathbb{R}\{x_0, x_1\} \to \mathbb{R}$ one can realize it canonically by a stochastic sequence (ξ_n) on a probabilistic space Ω. Namely, let Ω be the space of all infinite sequences $(\varepsilon_n)_{n\geqslant 1}$ of 0's and 1's and let $\xi_n : \Omega \to \{0, 1\}$ be the projection onto the n-th component. Let $S(\varepsilon_1, \dots, \varepsilon_n) \subset \Omega$ be the subset of all sequences starting with $(\varepsilon_1, \dots, \varepsilon_n)$. Then we set $P(S(\varepsilon_1, \dots, \varepsilon_n)) = \phi(x_{\varepsilon_1} \dots x_{\varepsilon_n})$ and extend it by σ-additivity. Correctness of this is guaranteed by the identity

$$\phi(fx_0) + \phi(fx_1) = \phi(f)$$

that follows from property (i) in the definition of a one-dependent process.

Now the crucial observation is that to an arbitrary sequence of numbers $a_\bullet = (a_n)$ satisfying Koszul inequalities (2.1) (and hence to any Koszul algebra) one can assign a one-dependent process. To explain this connection let us note that a one-dependent process is determined by the sequence of numbers $\alpha_n := \phi(x_1^{n-1})$ (i.e., by the probabilities of several units in a row). Indeed, we can rewrite (4.1) as

$$\phi(m_1 x_0 m_2) = \phi(m_1)\phi(m_2) - \phi(m_1 x_1 m_2) \tag{4.4}$$

for a pair of monomials m_1, m_2. Using this identity we can express $\phi(m)$ for every monomial m in terms of $\phi(m')$ where $\deg_{x_0}(m') < \deg_{x_0}(m)$. Iterating this procedure we can express all numbers $\phi(m)$ as some universal polynomials of α_n. A one-dependent process with prescribed numbers α_n exists iff all these polynomial expressions of α_n's are nonnegative. It is not difficult to check that these expressions coincide with polynomials $\Phi_{i_1,\dots,i_r}(\alpha_1, \alpha_2, \dots, \alpha_n)$ (where $\alpha_1 = 1$), so we have the following result.

THEOREM 4.2. *Assume that $a_1 > 0$. A sequence of numbers $a_\bullet = (a_n)$ satisfies (2.1) iff there exists a one-dependent process $\phi : \mathbb{R}\{x_0, x_1\} \to \mathbb{R}$ such that*

$$\alpha_n = \phi(x_1^{n-1}) = a_n / a_1^n \tag{4.5}$$

for all $n \geqslant 1$. For such a one-dependent process one has

(4.6)

$$\phi(x_{\varepsilon_1} \dots x_{\varepsilon_{n-1}}) = \Phi_{i_1,\dots,i_r}(1, \alpha_2, \dots, \alpha_n) = \frac{\Phi_{i_1,\dots,i_r}(a_1, a_2, \dots, a_n)}{a_1^n},$$

where in the notation of Lemma 8.2 of chapter 2, $J_{i_1,\dots,i_r} = \{j \in [1, n-1] : \varepsilon_j = 1\}$.

Proof: Clearly, it is enough to prove (4.6). Let us denote by $m_{i_1,\dots,i_r}$ the monomial of degree $n-1$ that has x_1 exactly in places corresponding to $J = J_{i_1,\dots,i_r}$. It suffices to show that the quantities $\phi(m_{i_1,\dots,i_r})$ satisfy the recursion similar to (2.3). Note that $J = J_{i_1,\dots,i_r} = [1, i_1 - 1] \cup (i_1 + J')$, where $J' = J_{i_2,\dots,i_r} \subset [1, n - i_1 - 1]$. On the other hand, $J_{i_1+i_2,i_3,\dots,i_r} = J \cup \{i_1\}$. Hence applying (4.4) for the decomposition $m_{i_1,\dots,i_r} = x_1^{i_1-1} x_0 m_{i_2,\dots,i_r}$ we get

$$\phi(m_{i_1,\dots,i_r}) = \alpha_{i_1} \cdot \phi(m_{i_2,\dots,i_r}) - \phi(m_{i_1+i_2,\dots,i_r})$$

as required. □

COROLLARY 4.3. *To every Koszul algebra A one can assign a one-dependent process ϕ_A such that*

$$\phi_A(x_1^{n-1}) = a_n / a_1^n,$$

where $a_n = \dim A_n$.

Remark 1. More explicitly, given a Koszul algebra $A = \{V, R\}$ and an integer n, one can find a realization for the finite part $\xi_1, \dots, \xi_{n-1}$ of the stochastic sequence corresponding to ϕ_A in the following way. Consider a distributing basis $\Omega^{(n)} \subset V^{\otimes n}$ as a finite probabilistic space with the uniform measure $P\{w\} = a_1^{-n}$ and define the random variables ξ_i as follows:

(4.7)

$$\xi_i \colon \Omega^{(n)} \longrightarrow \{0, 1\}, \qquad \xi_i(w) = \begin{cases} 0, & w \in R_i^{(n)} \\ 1, & w \notin R_i^{(n)}, \end{cases} \qquad i = 1, \dots, n-1.$$

Both the compatibility and one-dependence conditions follow from the nature of the subspaces $R_i^{(n)} = V^{\otimes i-1} \otimes R \otimes V^{n-i-1} \subset V^{\otimes n}$.

Remark 2. Equality (4.1) corresponds to the following identity for polynomials $\Phi_{i_1,\dots,i_r}$:

$$\Phi_{i_1,\dots,i_s} \cdot \Phi_{i_{s+1},\dots,i_r} = \Phi_{i_1,\dots,i_r} + \Phi_{i_1,\dots,i_{s-1},i_s+i_{s+1},i_{s+2},\dots,i_r},$$

where $1 \leqslant s < r$. For Koszul algebras the exact sequence of Corollary 8.4 of chapter 2 can be considered as a categorification of this identity.

Remark 3. It was proved in [**59**] that for every one-dependent sequence with $\alpha = P\{\xi_1 = 1\} \geqslant 1/2$ one has

$$P\{\xi_1 = 1, \xi_2 = 1\} \leqslant \alpha^{3/2}.$$

The corresponding inequality $a_3 \leqslant a_2^{3/2}$ holds for an arbitrary one-generated algebra with $a_2 \geqslant a_1^2/2$ (see Remark 1 in section 4 of chapter 6).

Let us denote by $\sigma : \mathbb{R}\{x_0, x_1\} \to \mathbb{R}\{x_0, x_1\}$ the automorphism switching x_0 and x_1. Then for every one-dependent process $\phi : \mathbb{R}\{x_0, x_1\} \to \mathbb{R}$ we define the *dual* one-dependent process

$$\phi^! := \phi \circ \sigma.$$

Note that if (ξ_n) is a one-dependent stochastic sequence corresponding to ϕ then the sequence corresponding to $\phi^!$ is $\xi_n^! := 1 - \xi_n$.

PROPOSITION 4.4. *The construction of Theorem 4.2 is compatible with duality. Namely, if ϕ is the one-dependent process corresponding to a sequence of numbers $a_\bullet$ then the dual process $\phi^!$ corresponds to $a_\bullet^!$.*

Proof: We have for $n \geqslant 2$

$$\phi^!(x_1^{n-1}) = \phi(x_0^{n-1}).$$

Since $J_{1^n} = \emptyset$, using (4.6) we obtain

$$\phi(x_0^{n-1}) = \Phi_{1^n}(1, \alpha_2, \ldots, \alpha_n) = \frac{\Phi_{1^n}(a_1, a_2, \ldots, a_n)}{a_1^n} = \frac{a_n^!}{a_1^n}.$$

□

Now we can give a proof of Proposition 3.2.

Proof of Proposition 3.2. Let us denote $J = J_{i_1,\ldots,i_r} \subset [1, n-1]$ and let m_J be the monomial of degree $n-1$ that has x_1 exactly in places corresponding to J. Applying (4.6) together with Proposition 4.4 we obtain

$$\frac{\Phi_{i_1,\ldots,i_r}(a_\bullet)}{a_1^n} = \phi(m_J) = \phi^!(m_{J^c}) = \frac{\Phi_{j_1,\ldots,j_s}(a_\bullet^!)}{a_1^n},$$

where $J_{j_1,\ldots,j_s} = J^c$. □

Remark 4. One can define a *matrix-valued one-dependent process* to be a linear functional $\phi : \mathbb{R}\{x_0, x_1\} \to \mathrm{Mat}_n(\mathbb{R})$ with values in $n \times n$-matrices over $\mathbb{R}$ with properties similar to (i)-(iii) in the definition of a one-dependent process. We only have to modify (ii) by requiring $\phi(m)$ to have nonnegative entries for every monomial m in x_0 and x_1. Such processes appear naturally when considering Koszul algebras with $A_0 = \Bbbk^n$. We leave for the reader to work out the details.

5. PBW-algebras and two-block-factor processes

In the previous section we showed how to assign a one-dependent process to every Koszul algebra. It turns out that there is a natural class of one-dependent processes containing all sequences associated with PBW-algebras (equivalently, with monomial quadratic algebras). Namely, this is the class of *two-block-factor* processes. In probability theory, a two-block-factor stochastic sequence is a sequence of the form $\xi_n = f(\eta_n, \eta_{n+1})$, where (η_n) is a sequence of independent identically distributed random $\mathbb{R}$-valued variables, f is a 0-1-valued measurable function of two variables. It is clear that every two-block-factor sequence is stationary and one-dependent. Note that without loss of generality one can assume that each η_n is distributed uniformly on the interval $[0,1]$ (since every random $\mathbb{R}$-valued variable η has form $\phi(\eta^u)$, where η^u is uniformly distributed on $[0,1]$ and ϕ is a measurable function). Then f is the characteristic function of a measurable subset $\Delta \subset [0,1]^2$. Thus, we arrive at the following definition.

Definition. Let $\Delta \subset [0,1]^2$ be a measurable subset. The *two-block-factor process* associated with Δ is the one-dependent process ϕ_Δ defined by

$$\phi_\Delta(x_{\varepsilon_1} \ldots x_{\varepsilon_{n-1}}) = \mu_n(\Delta(\varepsilon_1, \ldots, \varepsilon_{n-1})),$$

where

$$\Delta(\varepsilon_1, \ldots, \varepsilon_{n-1}) =$$
$$\{(x_1, \ldots, x_n) \in [0,1]^n \mid (x_i, x_{i+1}) \in \Delta \text{ for } \varepsilon_i = 1 \text{ and } (x_i, x_{i+1}) \notin \Delta \text{ for } \varepsilon_i = 0\},$$

where μ_n is the standard measure on $\mathbb{R}^n$. In particular,

$$\phi_\Delta(x_1^{n-1}) = \mu_n(\Delta^{(n)}),$$

where

$$\begin{aligned}\Delta^{(n)} &= \Delta(1^{n-1}) \\ &= \{(x_1,\ldots,x_n)\in[0,1]^n \mid (x_1,x_2)\in\Delta,(x_2,x_3)\in\Delta,\ldots,(x_{n-1},x_n)\in\Delta\}.\end{aligned}$$

Now assume that we have a subset $S\subset\{1,\ldots,m\}^2$. We can view S as the set of edges of an oriented graph G_S with vertices $\{1,\ldots,m\}$. Let us associate with S the subset

$$\Delta_S=\bigcup_{(i,j)\in S}[\frac{i-1}{m},\frac{i}{m}]\times[\frac{j-1}{m},\frac{j}{m}]\subset[0,1]^2.$$

Let A^S be the quadratic monomial algebra associated with S (such that $x_ix_j=0$ in A^S for $(i,j)\notin S$).

PROPOSITION 5.1. *The one-dependent process ϕ_{A^S} associated with A^S coincides with the two-block-factor process ϕ_{Δ_S} associated with Δ_S.*

Proof: We have

$$\Delta_S^{(n)}=\bigcup_{(i_1,\ldots,i_n)\in S^{(n)}}[\frac{i_1-1}{m},\frac{i_1}{m}]\times\cdots\times[\frac{i_n-1}{m},\frac{i_n}{m}]\subset[0,1]^n,$$

where $S^{(n)}$ was defined in section 1 of chapter 4. It follows that

$$\mu_n(\Delta_S^{(n)})=\frac{|S^{(n)}|}{m^n}=\frac{a_n}{a_1^n},$$

where $a_n=\dim A_n^S$. □

Let us equip the space of linear functionals on $\mathbb{R}\{x_0,x_1\}$ with the topology of pointwise convergence, i.e., we say that $\phi_n\to\phi$ if for every $f\in\mathbb{R}\{x_0,x_1\}$ one has $\phi_n(f)\to\phi(f)$. The set of one-dependent processes is closed with respect to this topology. Now we will show that the set of processes associated with PBW-algebras is dense in the set of all two-block-factors.

PROPOSITION 5.2. *Let ϕ be a two-block-factor process. Then there exists a sequence of quadratic monomial algebras $(A^{S_i})_{i\geqslant1}$ such that $\phi_{A^{S_i}}\longrightarrow\phi$ as $i\to\infty$.*

Proof: By definition we have $\phi=\phi_\Delta$ for some measurable subset $\Delta\subset[0,1]^2$. We can find a sequence of subsets of the form Δ_{S_i}, where $S_i\subset\{1,\ldots,m_i\}^2$, such that Δ_{S_i} tends to Δ with respect to the measure. It is clear that this implies that $\phi_{\Delta_{S_i}}\longrightarrow\phi$. It remains to apply the previous proposition. □

Recall that there exist Koszul algebras with Hilbert series different from those of all PBW-algebras (see section 8 of chapter 4). The first examples of non-two-block-factor one-dependent sequences were found in [**2**]. Two-block-factor and non-two-block-factor sequences are also studied in the monograph [**122**].

6. Operations on one-dependent processes

Recall that the set of Koszul and PBW-algebras is closed under a large set of natural operations (see chapter 3). In this section we are going to consider similar operations for one-dependent processes.

First, let us describe a family of unary operations on one-dependent processes generalizing Veronese powers. Namely, assume we are given a triple of elements

$f_0, f_1, g \in \mathbb{R}\{x_0, x_1\}$, all of them positive linear combinations of monomials, such that

$$f_0 + f_1 = \sum_{i=1}^{d} a_i(x_0 + x_1)^i$$

for some numbers $a_i \geqslant 0$ such that $\sum_i a_i = 1$. Then we can introduce an operation $\phi \longmapsto \phi^g_{f_0,f_1}$ on one-dependent processes by setting

$$\phi^g_{f_0,f_1}(f) := \frac{\phi(\kappa(f))}{\phi(g)^{\deg(f)+1}},$$

where $\deg(f)$ is the total degree of f and $\kappa : \mathbb{R}\{x_0, x_1\} \to \mathbb{R}\{x_0, x_1\}$ is the following linear map:

$$\kappa(x_{\varepsilon_1} x_{\varepsilon_2} \ldots x_{\varepsilon_n}) = g f_{\varepsilon_1} g f_{\varepsilon_2} g \ldots g f_{\varepsilon_n} g. \tag{6.1}$$

Note that this operation is defined on ϕ only if $\phi(g) > 0$. The fact that ϕ_{f_1} is again a one-dependent process follows from the identity

$$\kappa(f \cdot (x_0 + x_1) \cdot f') = \kappa(f)(f_0 + f_1)\kappa(f')$$

and from Proposition 4.1(a). For $f_i = x_{1-i}$ and $g = 1$ we recover the duality: $\phi^! = \phi^1_{x_1,x_0}$. Setting $\phi^{(N)} = \phi^{x_1^{N-1}}_{x_0,x_1}$, where $N > 0$ we get analogues of Veronese powers. One has

$$\phi^{(N)}(x_1^{n-1}) = \frac{\phi(x_1^{Nn-1})}{\phi(x_1^{N-1})^n}.$$

Hence, for a Koszul algebra A one has

$$\phi_{A^{(N)}} = \phi_A^{(N)}.$$

We can generalize the above construction to define a huge family of n-ary operations on one-dependent processes. For simplicity we will consider only binary operations. First, for a pair of one-dependent processes $\phi_1 : \mathbb{R}\{x_0, x_1\} \to \mathbb{R}$ and $\phi_2 : \mathbb{R}\{y_0, y_1\} \to \mathbb{R}$ one can consider the tensor product of the corresponding functionals:

$$\phi_1 \otimes \phi_2 : \mathbb{R}\{x_0, x_1\} \otimes_{\mathbb{R}} \mathbb{R}\{y_0, y_1\} \to \mathbb{R}.$$

Note that $(\phi_1 \otimes \phi_2)(1) = 1$ and in addition we have the following identity:

$$(\phi_1 \otimes \phi_2)(f \cdot (x_0 + x_1)^m (y_0 + y_1)^n g) = (\phi_1 \otimes \phi_2)(f) \cdot (\phi_1 \otimes \phi_2)(g),$$

where $m \geqslant 1$, $n \geqslant 1$. Next, let us fix a triple of positive linear combinations of monomials $f_0, f_1, g \in \mathbb{R}\{x_0, x_1\} \otimes_{\mathbb{R}} \mathbb{R}\{y_0, y_1\}$ such that

$$f_0 + f_1 = (x_0 + x_1)(y_0 + y_1)F(x_0 + x_1, y_0 + y_1)$$

for some polynomial F such that $F(1, 1) = 1$. Then we can define an operation

$$\phi_1 \otimes^g_{f_0,f_1} \phi_2(f) := \frac{\phi_1 \otimes \phi_2(\kappa(f))}{\phi_1 \otimes \phi_2(g)^{\deg(f)+1}},$$

where $\kappa : \mathbb{R}\{x_0, x_1\} \to \mathbb{R}\{x_0, x_1\} \otimes_{\mathbb{R}} \mathbb{R}\{y_0, y_1\}$ is still given by formula (6.1). This operation is well defined provided $(\phi_1 \otimes \phi_2)(g) > 0$. For $g = 1$, $f_1 = x_1 y_1$ and $f_0 = x_0 y_0 + x_1 y_0 + x_0 y_1$ we get an analogue of the Segre product $\phi_1 \circ \phi_2$ that is characterized by the equality

$$(\phi_1 \circ \phi_2)(x_1^n) = \phi_1(x_1^n) \cdot \phi_2(x_1^n).$$

This construction is compatible with the Segre product of algebras: if A and B are Koszul algebras then

$$\phi_{A\circ B} = \phi_A \circ \phi_B.$$

If (ξ_n) and (ξ'_n) are one-dependent stochastic sequences independent from each other then the stochastic sequence realizing their Segre product is simply $\min(\xi_n, \xi'_n)$. On the other hand, if we take $g = 1$, $f_1 = x_1y_0 + x_0y_1$ and $f_0 = x_0y_0 + x_1y_1$, we get an operation realized by the addition of one-dependent sequences (independent from each other).

Another family of operations is motivated by the direct sum of Koszul algebras. We are going to check that if we have two sequences of numbers $a_\bullet = (a_n)_{n\geqslant 1}$ and $b_\bullet = (b_n)_{n\geqslant 1}$ satisfying inequalities (2.1) then the sequence $a_\bullet + b_\bullet = (a_n + b_n)$ also satisfies them. This will lead to a family of operations on one-dependent processes.

LEMMA 6.1. *Let $a^0_\bullet = (a^0_n)_{n\geqslant 1}$ and $a^1_\bullet = (a^1_n)_{n\geqslant 1}$ be two sets of variables. For every $(n_1, \dots, n_r)$ one has*

$$\Phi_{n_1,\dots,n_r}(a^0_\bullet + a^1_\bullet) =$$

$$\sum_{\varepsilon=0,1}\ \sum_{1\leqslant t_1<t_2<\cdots<t_s=r} \Phi_{n_1,\dots,n_{t_1}}(a^\varepsilon_\bullet)\Phi_{n_{t_1+1},\dots,n_{t_2}}(a^{\varepsilon+1}_\bullet)\dots\Phi_{n_{t_{s-1}+1},\dots,n_r}(a^{\varepsilon+s-1}_\bullet)$$

where we set $a^m_\bullet = a^{m \bmod 2}_\bullet$.

Proof: In the case when $a^\varepsilon_n = \dim A^\varepsilon_n$ for some Koszul algebras A^0 and A^1 the assertion follows immediately from the decomposition of

$$V_{A^0\sqcap A^1}(n_1, \dots, n_r) \subset W = (A^0_{n_1} \oplus A^1_{n_1}) \otimes \cdots \otimes (A^0_{n_r} \oplus A^1_{n_1})$$

into a direct sum of subspaces, induced by the natural decomposition of W. It remains to observe that collections of numbers (a_n) arising from Koszul algebras are algebraically independent. Indeed, it is enough to consider direct sums of algebras of polynomials. □

PROPOSITION 6.2. *For every number λ such that $0 < \lambda < 1$ and every pair of one-dependent processes ϕ and ϕ' there exists a one-dependent sequence $\phi \sqcap_\lambda \phi'$ such that*

$$(\phi \sqcup_\lambda \phi')(x_1^{n-1}) = \lambda^n\phi(x_1^{n-1}) + (1-\lambda)^n\phi'(x_1^{n-1}).$$

Proof: Let $\alpha_n = \phi(x_1^{n-1})$, $\beta_n = \phi'(x_1^{n-1})$. Applying Lemma 6.1 to sequences $a_n = \lambda^n\alpha_n$ and $b_n = (1-\lambda)^n\beta_n$ (where $a_1 = \lambda$, $b_1 = 1-\lambda$) we get $\Phi_{n_1,\dots,n_r}(a_\bullet + b_\bullet) \geqslant 0$. It remains to apply Theorem 4.2. □

Finally, let us consider analogues of the tensor product. Given a pair of sequences of numbers $a_\bullet = (a_n)_{n\geqslant 1}$ and $b_\bullet = (b_n)_{n\geqslant 1}$ let us define a sequence $a_\bullet \otimes b_\bullet = (c_n)_{n\geqslant 1}$ by setting

$$c_n = a_n + a_{n-1}b_1 + \cdots + a_1b_{n-1} + b_n.$$

LEMMA 6.3. *In the above situation $\Phi_{n_1,\dots,n_r}(c)$ can be written as a universal polynomial of $(\Phi_{i_1,\dots,i_s}(a))$ and $(\Phi_{j_1,\dots,j_r}(b))$ with nonnegative integer coefficients.*

Proof: The proof is similar to Lemma 6.1: first, one reduces to the case when a and b come from Koszul algebras A and B, and then uses a direct sum decomposition of $V_{A\otimes^0 B}(n_1, \dots, n_r)$. Note that it is more convenient to use the one-sided product $A \otimes^0 B$ rather than the usual tensor product $A \otimes B$. □

As before this leads to the following family of operations on one-dependent processes.

PROPOSITION 6.4. *For every number λ such that $0<\lambda<1$ and every pair of one-dependent processes ϕ and ϕ' there exists a one-dependent process $\phi\otimes_\lambda\phi'$ such that*

$$(\phi\otimes_\lambda\phi')(x_1^{n-1})=\sum_{i=0}^n\lambda^i(1-\lambda)^{n-i}\alpha_i\alpha'_{n-i},$$

where $\alpha_i=\phi(x_1^{i-1})$, $\alpha'_i=\phi'(x_1^{i-1})$ for $i\geqslant 1$ and $\alpha_0=\alpha'_0=1$. □

7. Hilbert space representations of one-dependent processes

In this section we discuss representations of one-dependent sequences in terms of operators in Hilbert spaces considered by de Valk in [**121**].

Let A be a bounded operator on a Hilbert space H and let $x,y\in H$ be a pair of vectors such that $(x,y)=1$. With these data we can associate a pair of operators A_0 and A_1 as follows. Let P_y^x be the operator given by

$$P_y^x(v)=(v,x)y.$$

Then we set $A_1=A$, $A_0=P_y^x-A$. Assume that for every sequence $\varepsilon_1,\dots,\varepsilon_n$ of 0's and 1's one has

$$(A_{\varepsilon_1}\dots A_{\varepsilon_n}y,x)\geqslant 0.$$

Then there exists a one-dependent process ϕ such that

$$\phi(x_{\varepsilon_1}\dots x_{\varepsilon_n})=(A_{\varepsilon_1}\dots A_{\varepsilon_n}y,x).$$

Indeed, this follows from the identity

$$(BP_y^xCy,x)=(By,x)\cdot(Cy,x)$$

that holds for an arbitrary pair of operators B,C. In this situation we say that (H,A,x,y) is a *Hilbert space representation* of a one-dependent process ϕ. The following result is due to de Valk [**121**] (we give a slightly different proof below).

PROPOSITION 7.1. *For every one-dependent process ϕ there exists a Hilbert space representation (H,A,x,x) of Θ.*

Proof: Consider a Hilbert space H with an orthonormal basis $(e_n)_{n\geqslant 1}$. Set

$$x=\frac{1}{2}e_1+\lambda\cdot\sum_{n\geqslant 2}\phi(x_1^{n-1})e_n/n\in H,$$

where λ is a positive constant (the sum converges since $\phi(x_1^{n-1})\leqslant 1$). Let us choose λ such that $(x,x)=1$. Now let A be the unique operator on H such that $A(x)=\frac{2}{\lambda}e_2$ and $A(e_n)=\frac{n+1}{n}e_{n+1}$ for $n\geqslant 2$. It is easy to see that A is bounded. Also, for $n\geqslant 1$ one has

$$(A^nx,x)=\frac{n+1}{\lambda}(e_{n+1},x)=\phi(x_1^n).$$

This implies that (H,A,x,x) is a representation of ϕ. □

Some one-dependent processes admit finite-dimensional representations. In particular, this is true if there exists a $0-1$ sequence $\varepsilon_1,\dots,\varepsilon_N$ with $N\geqslant 1$ such that $P\{\xi_1=\varepsilon_1,\dots,\xi_N=\varepsilon_N\}=0$. Such a sequence is called a *zero-cylinder of length N*. De Valk proves in [**121**] that a one-dependent process with a zero-cylinder of length N admits a representation in the N-dimensional space. It is not known how to characterize one-dependent processes with a finite-dimensional representation. The only result in this direction is Theorem 4.3 of [**121**] asserting that if a one-dependent process admits a two-dimensional representation then it is a two-block-factor process. The structure of minimal zero-cylinders for one-dependent processes is also not known. The conjecture of de Valk states that the only possible minimal zero-cylinders are [101], [010], $[1^N]$ and $[0^N]$.

It is very probable that there exist two-block-factor processes that do not admit a finite-dimensional representation. On the other hand, we have the following result.

PROPOSITION 7.2. *Every two-block-factor process ϕ admits a Hilbert space representation (H, A, x, x), where A is a Hilbert-Schmidt operator.*

Proof: Let $\phi = \phi_\Delta$ for some measurable subset $\Delta \subset [0,1]^2$. Consider $H = L^2([0,1], dx)$ and let $A : H \longrightarrow H$ be the Hilbert-Schmidt operator with the kernel 1_Δ (characteristic function of Δ). Take x to be the constant function 1 on $[0,1]$. It is easy to see that (H, A, x, x) is a representation of ϕ. □

8. Hilbert series of one-dependent processes

With every one-dependent process ϕ we can associate its *Hilbert series*

$$h_\phi(z) := 1 + \sum_{n=1}^{\infty} \phi(x_1^{n-1}) z^n.$$

Clearly, this series converges for $|z| < 1$.

Conjecture. *The series h_ϕ has a meromorphic continuation to the entire complex plane.*

Note that if ϕ_A is a process associated with a Koszul algebra A then

$$h_{\phi_A}(z) = h_A(z/m),$$

where $m = \dim A_1$. Thus, this conjecture would imply Conjecture 1 in section 1 (see Remark 1 of section 1). The above conjecture is true for two-block-factor processes (see Proposition 8.4 below). For general one-dependent processes we can prove only the following weaker version (along with some additional information).

THEOREM 8.1. *The series $h_\phi(z)$ has a meromorphic continuation to the disk $|z| < 2$. Furthermore, $1 + h_\phi(z)$ has no zeros in this disk.*

The proof is based on the following result that can be of independent interest. Let $M \subset \mathbb{R}\{x_0, x_1\}$ be the set of all noncommutative monomials in x_0 and x_1 (including 1).

PROPOSITION 8.2. *Consider the noncommutative generating series*

$$F(x_0, x_1) = \sum_{m \in M} \phi(m) m \in \mathbb{R}\{x_0, x_1\}.$$

Then one has

$$F(x_0, x_1) = [f(x_1 - x_0)^{-1} - x_0]^{-1},$$

where $f(z) = F(0, z) = (h_\phi(z) - 1)/z$.

Proof: Equation (4.1) leads to the following identity in $\mathbb{R}\{x, y\}$:

$$\sum_{m_1, m_2 \in M} \phi(m_1(x_0 + x_1)m_2) m_1 m_2 = F(x_0, x_1)^2. \tag{8.1}$$

Consider the derivations ∂_0 and ∂_1 of $\mathbb{R}\{x_0, x_1\}$ defined by $\partial_i(x_j) = \delta_{ij}$ (and satisfying the Leibnitz identity $\partial_i(ab) = \partial_i(a)b + a\partial_i(b)$). Set $D = \partial_0 + \partial_1$. Then (8.1) can be written as

$$D(F) = F^2.$$

This is equivalent to

$$D(F^{-1}) = -F^{-1}D(F)F^{-1} = -1,$$

or to $D(F^{-1} + x_0) = 0$. It is clear that for any $\widetilde{f}(z) \in 1 + z\mathbb{R}[[z]]$ one has

$$D(\widetilde{f}(x_1 - x_0)) = 0.$$

Therefore, for any such $\widetilde{f}(z)$ the series $[\widetilde{f}(x_1 - x_0) - x_0]^{-1}$ is a unique solution of (8.1) with the initial condition $F(0, z) = \widetilde{f}(z)^{-1}$. It remains to take $\widetilde{f} = f^{-1}$. □

Proof of Theorem 8.1: From the formula of Proposition 8.2 we get

$$F(-t, t) = [f(2t)^{-1} + t]^{-1}. \tag{8.2}$$

Note that $F(-t, t) = \sum_{n \geqslant 0} a_n t^n$, where

$$|a_n| \leqslant \phi((x_0 + x_1)^n) = 1.$$

Hence, $F(-t, t)$ converges for $|t| < 1$. Now (8.2) defines a meromorphic continuation for $f(z)$ in $|z| < 2$. Moreover, it shows that

$$(f(z)^{-1} + z/2)^{-1} = \frac{2(h_\phi(z) - 1)}{z(1 + h_\phi(z))}$$

is holomorphic in this disk. □

COROLLARY 8.3. *If A is a Koszul algebra with* $\dim A_1 = m$ *then the Hilbert series $h_A(z)$ admits a meromorphic continuation to the disk $|z| < 2/m$. Moreover, $1 + h_A(z)$ has no zeros in this disk.*

Note that the result that $1 + h_A(z)$ has no zeros in $|z| < 2/m$ is sharp: for $h_A(z) = 1 + mz$ a zero appears on the boundary of this disk.

PROPOSITION 8.4. *If ϕ is a two-block-factor process then h_ϕ extends meromorphically to the entire complex plane.*

Proof: According to Proposition 7.2 there exists a Hilbert space representation (H, A, x, x) of Θ with A Hilbert-Schmidt. Therefore,

$$h_\phi(z) = 1 + \sum_{n=1}^{\infty} (A^{n-1}x, x)z^n = 1 + z((1 - zA)^{-1}x, x).$$

Since A is compact, the operator-valued function $(1 - zA)^{-1}$ extends meromorphically to the entire complex plane. Therefore, the same is true for $h_\phi(z)$. □

The above proof indicates that one possible approach to the conjecture on meromorphic continuation of h_ϕ is to try to prove the existence of a Hilbert space representation (H, A, x, y) of ϕ such that the operator A is compact.

9. Hermitian construction of one-dependent processes

In this section we describe the construction due to B. Tsirelson (unpublished) of a one-dependent process associated with a quadratic algebra and a Hermitian form on the space of generators.

Let V be a *Hermitian vector space*, i.e., a finite-dimensional complex vector space equipped with a positive-definite Hermitian form $(\cdot,\cdot)$. Recall that for a pair of subspaces $K, L \subset V$ one can define the *angle* between K and L by the formula

$$\langle K, L\rangle = \langle K, L\rangle_V = \mathrm{tr}(P_K P_L),$$

where P_K and P_L are orthogonal projections onto K and L respectively. The angle $\langle K, L\rangle$ is always a nonnegative real number and it is equal to zero iff K and L are orthogonal. In fact, there is another formula for $\langle K, L\rangle$ that shows this. To state it we need the following elementary result.

LEMMA 9.1. *For every pair of subspaces $K, L \subset V$ there exist orthonormal bases $(k_i)_{1\leqslant i\leqslant \dim K}$, $(l_j)_{1\leqslant j\leqslant \dim L}$ of K and L such that $(k_i, l_j) = 0$ for $i \neq j$.*

Proof: If $K \cap L^\perp \neq 0$ then the statement easily reduces to the subspaces $K/K\cap L^\perp$ and L in $V/K \cap L^\perp$. Therefore, we can assume that $K \cap L^\perp = L \cap K^\perp = 0$. In this case the restriction of the projection P_L to K gives an isomorphism $\rho = P_L|_K : K \longrightarrow L$. Let (k_i) be the orthonormal basis of K that diagonalizes the pull-back of the Hermitian form on L by ρ. Setting $l_i = \rho(k_i)/||\rho(k_i)||$ we get the required bases (k_i) and (l_i) in K and L. □

Let us choose bases (k_i) and (l_j) of K and L as in the above lemma. Then one has

$$\langle K, L\rangle = \mathrm{tr}(P_K P_L) = \sum_{j=1}^{\dim L} (l_j, P_K l_j) = \sum_{j=1}^{\min(\dim L, \dim L)} (l_j, (l_j, k_j)k_j)$$
$$= \sum_{j=1}^{\min(\dim K, \dim L)} |(l_j, k_j)|^2.$$

This shows that $\langle K, L\rangle$ is a nonnegative real number. Note that if $K \cap L \neq 0$ then the above bases (k_i) and (l_j) can be chosen in such a way that they both contain the same basis of $K \cap L$. This implies that $\langle K, L\rangle \geqslant \dim K \cap L$. The equality holds iff there exists an orthonormal basis of V containing bases for K and L.

We can equip tensor powers of V with positive-definite Hermitian forms in the standard way. This allows us to consider angles between subspaces in $V^{\otimes n}$ for all $n \geqslant 1$.

Definition. A *Hermitian pair* is a pair (V, R) consisting of a Hermitian vector space V and a subspace $R \subset V^{\otimes 2}$.

Let us define the Hilbert series $h_{V,R}(z) = 1 + \sum_{n\geqslant 1} a_n(V, R)z^n$ associated with a Hermitian pair (V, R) by the formulas

$$a_{2k+1}(V, R) = \langle R^{\otimes k} \otimes V, V \otimes R^{\otimes k}\rangle_{V^{\otimes(2k+1)}},$$

$$a_{2k}(V, R) = \langle R^{\otimes k}, V \otimes R^{\otimes(k-1)} \otimes V\rangle_{V^{\otimes(2k)}}.$$

Note that $a_1(V, R) = \langle V, V\rangle = \dim V$ and $a_2(V, R) = \langle R, V^{\otimes 2}\rangle = \dim R$.

PROPOSITION 9.2. *For every Hermitian pair (V, R) there exists a one-dependent process $\phi_{(V,R)}$ such that*

$$\phi_{V,R}(x_1^{n-1}) = \frac{a_n(V,R)}{(\dim V)^n} \tag{9.1}$$

for all $n \geqslant 1$, i.e.,

$$h_{\phi_{V,R}}(z) = h_{V,R}(z/\dim V).$$

Proof: Let us set

(9.2)
$$\phi(x_{\varepsilon_1} \dots x_{\varepsilon_{n-1}}) = (\dim V)^{-n} \cdot \langle R^{\varepsilon_1} \otimes R^{\varepsilon_2} \otimes \dots, V \otimes R^{\varepsilon_2} \otimes R^{\varepsilon_4} \otimes \dots \rangle_{V^{\otimes n}},$$

where $R^1 = R$ and $R^0 = R^\perp$ (if n is odd then the last factor of the first tensor product is V; if n is even then the last factor of the second tensor product is V). Clearly, this formula is compatible with (9.1), so we just have to check identity (4.3). Let us denote by P_i^1 and P_i^0 the orthogonal projections to $V^{\otimes(i-1)} \otimes R \otimes V^{n-i-1}$ and $V^{\otimes(i-1)} \otimes R^\perp \otimes V^{n-i-1}$, respectively. Note that for $|j-i| \geqslant 2$ the operators P_i^ε and $P_j^{\varepsilon'}$ commute, so we can rewrite (9.2) as

$$\phi(x_{\varepsilon_1} \dots x_{\varepsilon_{n-1}}) = (\dim V)^{-n} \cdot \operatorname{tr}(\prod_{1 \leqslant 2i+1 \leqslant n-1} P_{2i+1}^{\varepsilon_{2i+1}} \cdot \prod_{1 \leqslant 2j \leqslant n-1} P_{2j}^{\varepsilon_{2j}}).$$

Now using the fact that $P_i^0 + P_i^1 = 1$ we obtain

$$\phi(x_{\varepsilon_1} \dots x_{\varepsilon_{k-1}}(x_0 + x_1)x_{\varepsilon_{k+1}} \dots x_{\varepsilon_n}) = (\dim V)^{-n-1} \times$$
$$\operatorname{tr}(\prod_{1 \leqslant 2i+1 \leqslant n-1, 2i+1 \neq k} P_{2i+1}^{\varepsilon_{2i+1}} \cdot \prod_{1 \leqslant 2j \leqslant n-1, 2j \neq k} P_{2j}^{\varepsilon_{2j}}).$$

It remains to observe that the right-hand side is equal to

$$(\dim V)^{-n-1} \times$$
$$\operatorname{tr}(\prod_{1 \leqslant 2i+1 \leqslant k-1} P_{2i+1}^{\varepsilon_{2i+1}} \cdot \prod_{1 \leqslant 2j \leqslant k-1} P_{2j}^{\varepsilon_{2j}}) \cdot \operatorname{tr}(\prod_{k+1 \leqslant 2i+1 \leqslant n-1} P_{2i+1}^{\varepsilon_{2i+1}} \cdot \prod_{k+1 \leqslant 2j \leqslant n-1} P_{2j}^{\varepsilon_{2j}})$$

which leads to the required identity. □

Note that the above construction is compatible with duality: formula (9.2) immediately implies that

$$\phi_{V,R^\perp} = \phi_{V,R}^!.$$

It would be interesting to characterize one-dependent processes associated with Hermitian pairs. It is easy to see that all processes associated with PBW-algebras are in this set. Namely, if $S \subset [1,m]^2$ is a subset then the process associated with the quadratic monomial algebra A^S coincides with $\phi_{\mathbb{C}^n, R_S}$, where $\mathbb{C}^n$ is equipped with the standard Hermitian structure and $R_S = \bigoplus_{(i,j)\in S} \mathbb{C} \cdot (e_i \otimes e_j)$. Indeed, this follows from the fact that $\langle K, L \rangle = \dim K \cap L$ for a pair of subspaces that can be distributed by an orthonormal basis.

Another interesting problem is to check whether for an arbitrary Hermitian pair the Hilbert series $h_{V,R}$ extends meromorphically to the entire complex plane. It seems that this case should be more accessible than the similar conjecture for general one-dependent processes.

10. Modules over one-dependent processes

A *(left) module* over a one-dependent process ϕ is a linear map $\psi : \mathbb{R}\{x_0, x_1\} \longrightarrow \mathbb{R}$ taking nonnegative values on all monomials, such that

$$\psi(f \cdot (x_0 + x_1) \cdot g) = \phi(f) \cdot \psi(g) \tag{10.1}$$

for all $f, g \in \mathbb{R}\{x_0, x_1\}$.

Example. For every monomial m in x_0 and x_1 we have a module ψ_m over ϕ defined by $\psi_m(f) = \phi(f \cdot m)$. Also, if ψ_1 and ψ_2 are modules over ϕ then for any constants $c_1 \geqslant 0$ and $c_2 \geqslant 0$ the map $c_1\psi_1 + c_2\psi_2$ is also a module over ϕ.

The above definition is motivated by the following construction: given a Koszul module M over a Koszul algebra A we can associate with M a module over the one-dependent process ϕ_A using the infinitesimal Hopf module $V_{A,M}$ (see the end of section 8 of chapter 2). Namely, we set

$$\psi_M(m_{J_{n_1,\dots,n_r}}) = \frac{\dim V_{A,M}(n_1, \dots, n_{r-1}, n_r - 1)}{(\dim A_1)^{n_1+\dots+n_r-1}},$$

where $n_1, \dots, n_r \geqslant 1$, the correspondence $(n_1 \dots, n_r) \longmapsto J_{n_1,\dots,n_r} \subset [1, n-1]$ was defined in Lemma 8.2 of chapter 2 (where $n = n_1 + \dots + n_r$), and m_J denotes the monomial of degree $n - 1$ in x_0 and x_1, having x_1 on places corresponding to J.

Let us define the Hilbert series of a module ψ by

$$h_\psi^{mod}(z) = \sum_{n \geqslant 0} \psi(x_1^n) z^n.$$

For example, $h_\phi^{mod}(z) = (h_\phi(z) - 1)/z$. Note that for the module ψ_M associated with a Koszul module M we get the usual Hilbert series of M up to rescaling:

$$h_{\psi_M}^{mod}(z) = h_M(z/m),$$

where $m = \dim A_1$. The next result shows that ψ is completely determined by its Hilbert series. We use the notation from Proposition 8.2.

PROPOSITION 10.1. *Let ψ be a left module over a one-dependent process ϕ. Consider the noncommutative generating function*

$$G(x_0, x_1) = \sum_{m \in M} \psi(m) m \in \mathbb{R}\{x_0, x_1\}.$$

Then one has

$$G(x_0, x_1) = F(x_0, x_1) \cdot g(x_1 - x_0),$$

where $g(z) = G(0, z)/F(0, z) = h_\psi^{mod}(z) z/(h_\phi(z) - 1)$.

Proof: Equation (10.1) leads to the equality

$$D(G) = FG.$$

Since $D(F) = F^2$ this implies that $D(F^{-1}G) = 0$. Hence, $F^{-1}G$ is of the form $g(x_1 - x_0)$. □

Note that the fact that $G(x_0, x_1)$ has nonnegative coefficients is equivalent to certain polynomial inequalities on coefficients of h_ψ^{mod} and h_ϕ that are module analogues of Koszul inequalities considered in section 2.

As in section 8 we can derive the following.

COROLLARY 10.2. *The series $h_{\psi}^{mod}(z)$ has a meromorphic continuation to the disk $|z| < 2$. Moreover, the ratio*

$$\frac{h_{\psi}^{mod}(z)}{1+h_{\phi}(z)}$$

is holomorphic in this disk.

Proof: Since $\psi((x_0+x_1)^n) = \psi(1)$ for $n \geqslant 1$, we obtain that the series $G(-t,t)$ has bounded coefficients. Hence, it converges for $|t| < 1$ (as does $F(-t,t)$). It remains to use the identity $G(-t,t) = F(-t,t)g(2t)$ together with (8.2). □

COROLLARY 10.3. *For a Koszul module M over a Koszul algebra A the Hilbert series $h_M(z)$ admits a meromorphic continuation to the disk $|z| < 2/m$, where $m = \dim A_1$. Moreover, the ratio*

$$\frac{h_M(z)}{1+h_A(z)}$$

is holomorphic in this disk.

APPENDIX A

DG-algebras and Massey products

Definition. A *DG-algebra* is a graded algebra $A = \bigoplus_{n\in\mathbb{Z}} A_n$ equipped with a differential $d_A : A \to A$ such that $d_A(A_n) \subset A_{n+1}$, $d_A^2 = 0$, and the Leibnitz identity is satisfied:

$$d_A(x \cdot y) = d_A(x) \cdot y + (-1)^{\tilde{x}} x \cdot d_A(y),$$

where $x \in A_{\tilde{x}}$.

A *DG-module* M over a DG-algebra A is a graded A-module $M = \bigoplus_{n\in\mathbb{Z}} M_n$ equipped with a differential $d_M : M \to M$ such that $d_M(M_n) \subset M_{n+1}$, $d_M^2 = 0$, and the Leibnitz identity is satisfied:

$$d_M(a \cdot m) = d_A(a) \cdot m + (-1)^{\tilde{a}} a \cdot d_M(m),$$

where $a \in A_{\tilde{a}}$, $m \in M$.

Observe that in section 4 of chapter 5 we consider only nonnegatively graded DG-algebras.

The cohomology $H^*(A) = H^*_{d_A}(A)$ of a DG-algebra A has a natural structure of a graded algebra, and the cohomology $H^*_{d_M}(M)$ of a DG-module M over A has a natural structure of a graded $H^*(A)$-module.

A *morphism of DG-algebras* $f : A \to B$ is a homomorphism of graded algebras such that $fd_A = d_B f$. Such a morphism induces a homomorphism $H^*(f) : H^*(A) \to H^*(B)$. If $H^*(f)$ is an isomorphism then we say that f is a *quasi-isomorphism*.

Let M (resp., N) be a DG-module over a DG-algebra A (resp., B). Assume we are given a morphism of a DG-algebras $f : A \to B$. Then a *morphism of DG-modules* $g : M \to N$ compatible with f is a homomorphism of graded A-modules such that $gd_M = d_N g$. It induces a homomorphism of $H^*(A)$-modules $H^*(g) : H^*(M) \to H^*(N)$. We say that g is a *quasi-isomorphism* if $H^*(g)$ is an isomorphism.

We leave for the reader to give formally dual definitions of a DG-coalgebra and a DG-comodule (and morphisms between them).

Let A be a DG-algebra. Massey products are certain natural multivalued partially defined operations on $H^*(A)$ preserved under quasi-isomorphisms. The simplest example is a triple Massey product $m_3(x_1, x_2, x_3)$, where $x_1 \in H^i(A)$, $x_2 \in H^j(A)$, $x_3 \in H^k(A)$ are cohomology classes satisfying

$$x_1 x_2 = x_2 x_3 = 0.$$

Let us choose cycles $x_1' \in A_i$, $x_2' \in A_j$ and $x_3' \in A_k$ representing x_1, x_2 and x_3, respectively. Since $x_1' x_2'$ is a coboundary, there exists $x_{12} \in A_{i+j-1}$ such that

$$d_A(x_{12}) = x_1' x_2'.$$

Similarly, we can choose $x_{23} \in A_{j+k-1}$ such that

$$d_A(x_{23}) = x_2' x_3'.$$

Now we set

$$\langle x_1, x_2, x_3 \rangle = x_{12} x_3' - (-1)^i x_1' x_{23} \mod \operatorname{im}(d_A).$$

The obtained element in $H^{i+j+k-1}(A)$ depends on the choices made. However, its coset with respect to the subspace $x_1 \cdot H^{j+k-1}(A) + H^{i+j-1} \cdot x_3$ is well defined.

One can also define more general Massey products by replacing the decomposable tensor $x_1 \otimes x_2 \otimes x_3$ with more general tensors. The definition of n-ary Massey products for $n > 3$ is similar but is more involved. They appear as differentials in the spectral sequence associated with the natural filtration on the bar-complex of A (we considered the dual spectral sequence in section 7 of chapter 5; see also [**82**]; Stasheff [**114**] defined similar notions in a more general context of A_∞-algebras). In the case when A is the cobar-complex of a graded algebra the above spectral sequence shows that an algebra is Koszul iff all the higher Massey operations on its cohomology are trivial (see [**104**]).

Bibliography

[1] J. Aaronson, D. Gilat, M. Keane. On the structure of 1 dependent Markov chains. *J. Theoret. Probab.* **5**, #3, 545–561, 1992.

[2] J. Aaronson, D. Gilat, M. Keane, V. de Valk. An algebraic construction of a class of one-dependent processes. *Ann. Probab.* **17**, #1, 128–143, 1989.

[3] M. Aguiar. Infinitesimal Hopf algebras. *New trends in Hopf algebra theory (La Falda, 1999)*, 1–29. Contemp. Math. 267, AMS, Providence, RI, 2000.

[4] M. Aguiar. Infinitesimal bialgebras, pre-Lie and dendriform algebras. *Hopf Algebras*, 1–33. Lect. Notes in Pure and Appl. Math. 237, 2004.

[5] R. Aharoni. A problem in rearrangements of (0, 1) matrices. *Discrete Math.* **30**, #3, 191–201, 1980.

[6] R. Ahlswede, G. O. H. Katona. Graphs with maximal number of adjacent pairs of edges. *Acta Math. Acad. Sci. Hungar.* **32**, #1-2, 97–120, 1978.

[7] D. J. Anick. Noncommutative graded algebras and their Hilbert series. *J. Algebra* **78**, #1, 120–140, 1982.

[8] D. J. Anick. Generic algebras and CW-complexes. *Algebra, topology and algebraic K-theory (Princeton, NJ, 1983)*, 247–321, Ann. of Math. Stud. 113, Princeton Univ. Press, 1987.

[9] D. J. Anick. Diophantine equations, Hilbert series, and undecidable spaces. *Annals of Math.* **122**, #1, 87–112, 1985.

[10] E. Arbarello, M. Cornalba, P. Griffiths, J. Harris. Geometry of algebraic curves I, Springer-Verlag, New York/Berlin, 1985.

[11] S. Arkhipov. Koszul duality for quadratic algebras over a complex. *Funct. Anal. Appl.* **28**, #3, 202–204, 1994.

[12] M. Artin, W. Schelter. Graded algebras of global dimension 3. *Advances in Math.* **6**, #2, 171–216, 1987.

[13] M. Artin, J. Tate, M. Van den Bergh. Some algebras associated to automorphisms of elliptic curves. *The Grothendieck Festschrift*, vol. 1, 33–85, 1990.

[14] L. Avramov, D. Eisenbud. Regularity of modules over a Koszul algebra. *J. Algebra* **153**, #1, 85–90, 1992.

[15] J. Backelin. A distributiveness property of augmented algebras and some related homological results. Ph. D. Thesis, Stockholm, 1981. Available at http://www.matematik.su.se/~joeb/avh/

[16] J. Backelin. On the rates of growth of the homologies of Veronese subrings. *Algebra, topology and their interactions (Stockholm, 1983)*, 79–100, Lecture Notes in Math. 1183, Springer, Berlin, 1986.

[17] J. Backelin. Some homological properties of "high" Veronese subrings. *J. Algebra* **146**, #1, 1–17, 1992.

[18] J. Backelin. Private communication, 1991.

[19] J. Backelin. Koszul duality for parabolic and singular category $\mathcal{O}$. *Represent. Theory* **3**, 139–152, 1999.

[20] J. Backelin, R. Fröberg. Koszul algebras, Veronese subrings and rings with linear resolutions. *Revue Roumaine Math. Pures Appl.* **30**, #2, 85–97, 1985.

[21] D. Bayer, M. Stillman. A criterion for detecting m-regularity. *Inventiones Math.* **87**, #1, 1–11, 1987.

[22] D. Bayer, M. Stillman. On the complexity of computing syzygies. *J. Symbolic Comput.* **6**, #2-3, 135–147, 1988.

[23] A. Beilinson, V. Ginzburg, V. Schechtman. Koszul duality. *J. Geom. Phys.* **5**, #3, 317–350, 1988.

[24] A. Beilinson, V. Ginzburg, W. Soergel. Koszul duality patterns in representation theory. *J. Amer. Math. Soc.* **9**, #2, 473–527, 1996.

[25] A. Beilinson, R. MacPherson, V. Schechtman. Notes on motivic cohomology. *Duke Math. J.* **54**, #2, 679–710, 1987.

[26] G. Bergman. The diamond lemma for ring theory. *Advances in Math.* **29**, #2, 178–218, 1978.

[27] I. Bernstein, I. Gelfand, S. Gelfand. Algebraic vector bundles on P^n and problems of linear algebra. (Russian) Funktsional. Anal. i Prilozhen. **12**, # 3, 66–67, 1978.

[28] R. Bezrukavnikov. Koszul DG-algebras arising from configuration spaces. *Geom. and Funct. Anal.* **4**, #2, 119–135, 1994.

[29] R. Bezrukavnikov. Koszul property and Frobenius splitting of Schubert varieties. Preprint alg-geom/9502021.

[30] A. I. Bondal. Helices, representations of quivers and Koszul algebras. *Helices and vector bundles*, 75–95, London Math. Soc. Lecture Note Ser., 148, Cambridge Univ. Press, Cambridge, 1990.

[31] A. I. Bondal, A. E. Polishchuk. Homological properties of associative algebras: the method of helices. *Russian Acad. Sci. Izvestiya Math.* **42**, #2, 219–260, 1994.

[32] E. Borel. Sur une application d'un théorème de M. Hadamard. *Bull. Sc. math.*, 2e série, t.18, 22–25, 1894.

[33] A. Braverman, D. Gaitsgory. The Poincaré–Birkhoff–Witt theorem for quadratic algebras of Koszul type. *J. Algebra* **181**, #2, 315–328, 1996.

[34] W. Bruns, J. Gubeladze, Ngo Viet Trung. Normal polytopes, triangulations, and Koszul algebras. *J. Reine Angew. Math.* **485**, 123–160, 1997.

[35] B. Buchberger. Gröbner bases: An algorithmic method in polynomial ideal theory. *Multidimensional Systems Theory* (N. K. Bose ed.), 184–232, Reidel, Dordrecht, 1985.

[36] B. Buchberger, F. Winkler, eds., Gröbner bases and applications, Cambridge Univ. Press, 1998.

[37] D. C. Butler. Normal generation of vector bundles over a curve. *J. Diff. Geom.* **39**, #1, 1–34, 1994.

[38] A. Conca. Gröbner bases for spaces of quadrics of low codimension. *Adv. in Appl. Math.* **24**, #2, 111–124, 2000.

[39] A. Conca, M. E. Rossi, G. Valla. Gröbner flags and Gorenstein algebras. *Compositio Math.* **129**, #1, 95–121, 2001.

[40] A. Conca, N. V. Trung, G. Valla. Koszul property for points in projective spaces. *Math. Scand.* **89**, #2, 201–216, 2001.

[41] A. Connes. Noncommutative differential geometry. *Publ. Math. IHES* **62**, 257–360, 1985.

[42] A. A. Davydov. Totally positive sequences and R-matric quadratic algebras. *Journal of Math. Sciences* **100**, #1, 1871–1876, 2000.

[43] V. G. Drinfeld. On quadratic quasi-commutational relations in quasi-classical limit. *Mat. Fizika, Funkc. Analiz*, 25–34, "Naukova Dumka", Kiev, 1986. English translation in: *Selecta Math. Sovietica* **11**, #4, 317–326, 1992.

[44] L. Ein, R. Lazarsfeld. Syzygies and Koszul cohomology of smooth projective varieties of arbitrary dimension. *Inventiones Math.* **111**, #1, 51–67, 1993.

[45] D. Eisenbud, S. Goto. Linear free resolutions and minimal multiplicity. *J. Algebra* **88**, #1, 89–133, 1984.

[46] D. Eisenbud, A. Reeves, B. Totaro. Initial ideals, Veronese subrings, and rates of algebras. *Advances in Math.* **109**, #2, 168–187, 1994.

[47] P. Etingof, V. Ostrik. Module categories over representations of $SL_q(2)$ and graphs. Preprint math.QA/0302130.

[48] B. Feigin, A. Odesskii. Sklyanin's elliptic algebras. *Funct. Anal. Appl.* **23**, #3, 207–214, 1990.

[49] B. Feigin, B. Tsygan. Cyclic homology of algebras with quadratic relations, universal enveloping algebras, and group algebras. *K-theory, arithmetic and geometry (Moscow, 1984–1986)*, 210–239, Lecture Notes in Math. 1289, Springer, Berlin, 1987.

[50] M. Finkelberg, A. Vishik. The coordinate ring of general curve of genus $\geqslant 5$ is Koszul. *J. Algebra* **162**, 535–539, 1993.

[51] G. Floystad. Koszul duality and equivalences of categories, preprint math.RA/0012264.

[52] R. Fröberg. Determination of a class of Poincaré series. *Math. Scand.* **37**, #1, 29–39, 1975.

[53] R. Fröberg. A study of graded extremal rings and of monomial rings. *Math. Scand.* **51**, #1, 22–34, 1982.

[54] R. Fröberg. Rings with monomial relations having linear resolutions. *J. Pure Appl. Algebra* **38**, #2–3, 235–241, 1985.

[55] R. Fröberg. Koszul algebras. *Advances in commutative ring theory (Fez, 1997)*, 337–350. Dekker, New York, 1999.

[56] R. Fröberg, T. Gulliksen, C. Löfwall. Flat one-parameter family of Artinian algebras with infinite number of Poincaré series. *Algebra, topology and their interactions (Stockholm, 1983)*, 170–191, Lecture Notes in Math. 1183, Springer, Berlin, 1986.

[57] R. Fröberg, C. Löfwall. Koszul homology and Lie algebras with appplication to generic forms and points. *Homology Homotopy Appl.* **4**, no. 2, 227–258, 2002.

[58] W. Fulton, *Intersection theory*, Springer, 1998.

[59] A. Gandolfi, M. Keane, V. de Valk. Extremal two-correlations of two-valued stationary one-dependent processes. *Probab. Theory Related Fields* **80**, #3, 475–480, 1989.

[60] I. M. Gelfand, V. A. Ponomarev. Model algebras and representations of graphs. *Funct. Anal. Appl.* **13**, 157–166, 1979.

[61] V. Ginzburg, M. Kapranov. Koszul duality for operads. *Duke Math. J.* **76**, #1, 203–272, 1994. Erratum in *Duke Math. J.* **80**, #1, 293, 1995.

[62] E. S. Golod. Standard bases and homology. *Algebra—some current trends (Varna, 1986)*, 88–95, Lecture Notes in Math. 1352, Springer, Berlin, 1988.

[63] E. S. Golod, I. R. Shafarevich. On the tower of class fields. *Izvestiya Akad. Nauk. SSSR, Ser. Mat.* **28**, #2, 261–272, 1964 (in Russian).

[64] G. Grätzer. General lattice theory. Birkhäuser, Basel, 1978.

[65] M. Green, R. Lazarsfeld. A simple proof of Petri's theorem on canonical curves, *Geometry Today (Rome, 1984)*, 129–142, Progress in Math. 60, Birkhäuser, Boston, 1985.

[66] R. Hartshorne. Algebraic geometry. Springer, New York, 1977.

[67] S. P. Inamdar, V. B. Mehta. Frobenius splitting of Schubert varieties and linear syzygies. *Amer. Journ. of Math.* **116**, #6, 1569–1586, 1994.

[68] S. A. Joni, G. C. Rota. Coalgebras and bialgebras in combinatorics. *Stud. Appl. Math.* **61**, #2, 93–139, 1979.

[69] B. Jónsson. Distributive sublattices of a modular lattice. *Proc. Amer. Math. Soc.* **6**, #5, 682–688, 1955.

[70] P. Jorgensen, Linear free resolutions over non-commutative algebras. *Compositio Math.* **140**, #4, 1054–1058, 2004.

[71] S. Karlin. Total positivity. Stanford Univ. Press, 1968.

[72] G. Kempf. Some wonderful rings in algebraic geometry. *J. Algebra* **134**, #1, 222–224, 1990.

[73] G. Kempf. Syzygies for points in projective space. *J. Algebra* **145**, #1, 219–223, 1992.

[74] G. Kempf. Projective coordinate rings of abelian varieties. *Algebra analysis, geometry, and number theory (Baltimore, MD, 1988)*, 225–235, Johns Hopkins Univ. Press, Baltimore, MD, 1989.

[75] C. Löfwall. On the subagebra generated by the one-dimensional elements in the Yoneda Ext-algebra. *Algebra, algebraic topology and their interactions (Stockholm, 1983)*, Lecture Notes in Math. 1183, 291–338, Springer, Berlin, 1986.

[76] A. Malkin, V. Ostrik, M. Vybornov, *Quiver varieties and Lusztig's algebra*, preprint math.RT/0403222.

[77] Yu. I. Manin. Some remarks on Koszul algebras and quantum groups. *Ann. Inst. Fourier* **37**, #4, 191–205, 1987.

[78] Yu. I. Manin. Quantum groups and non-commutative geometry. CRM, Université de Montréal, 1988.

[79] Yu. I. Manin. Topics in noncommutative geometry. Princeton Univ. Press, Princeton, 1991.

[80] Yu. I. Manin. Notes on quantum groups and quantum de Rham complexes. *Theoret. and Math. Physics* **92**, #3, 997–1019, 1992.

[81] R. Martínez-Villa. Applications of Koszul algebras: the preprojective algebra. *Representation Theory of Algebras (Cocoyoc, 1994)*, 487–504, CMS Conf. Proc., vol. 18, AMS, Providence, RI, 1996.

[82] J. P. May. The cohomology of augmented algebras and generalized Massey products for DGA-algebras. *Trans. Amer. Math. Soc.* **122**, 334–340, 1966.

[83] T. Mora. An introduction to commutative and noncommutative Gröbner bases. *Theoret. Comput. Science* **134**, #1, 131–173, 1994.

[84] D. Mumford. Lectures on curves on an algebraic surface. Princeton Univ. Press, Princeton, 1966.

[85] D. Mumford. Varieties defined by quadratic relations, *Questions on Algebraic Varieties (C.I.M.E., III Ciclo, Varenna, 1969)*, 29–100, Edizioni Cremonese, Rome, 1970.

[86] R. Musti, E. Buttafuoco. Sui subreticoli distributivi dei reticoli modulari. *Boll. Unione Mat. Ital.* **(3) 11**, #4, 584–587, 1956.

[87] P. Orlik, H. Terao. Arrangements of hyperplanes. Springer-Verlag, Berlin, 1992.

[88] S. Papadima, S. Yuzvinsky. On rational $K[\pi, 1]$ spaces and Koszul algebras. *J. Pure Appl. Algebra* **144**, #2, 157–167, 1999.

[89] G. Pareschi. Koszul algebras associated to adjunction bundles. *J. Algebra* **157**, #1, 161–169, 1993.

[90] G. Pareschi, B. Purnaprajna. Canonical ring of a curve is Koszul: A simple proof. *Illinois J. of Math.* **41**, #2, 266-271, 1997.

[91] I. Peeva, V. Reiner, B. Sturmfels. How to shell a monoid. *Math. Ann.* **310**, #2, 379–393, 1998.

[92] D. Piontkovskii. On the Hilbert series of Koszul algebras. *Funct. Anal. Appl.* **35**, #2, 133–137, 2001.

[93] D. Piontkovskii. Noncommutative Koszul filtrations. Preprint math.RA/0301233.

[94] D. Piontkovskii. Sets of Hilbert series and their applications. Preprint math.RA/0502149. Fundam. Prikl. Mat. 10, 143-156, 2004.

[95] D. Piontkovskii. Algebras associated to pseudo-roots of noncommutative polynomials are Koszul. Preprint math.RA/0405375.

[96] A. Polishchuk. On the Koszul property of the homogeneous coordinate ring of a curve. *J. Algebra* **178**, #1, 122–135, 1995.

[97] A. Polishchuk. Perverse sheaves on a triangulated space. *Math. Res. Letters* **4**, #2-3, 191–199, 1997.

[98] A. Polishchuk. Koszul configurations of points in projective spaces. Preprint math.AG/ 0412441.

[99] L. Positselski. Nonhomogeneous quadratic duality and curvature. *Funct. Anal. Appl.* **27**, #3, 197–204, 1993.

[100] L. Positselski. Koszul inequalities and stochastic sequences. M. A. Thesis, Moscow State University, 1993.

[101] L. Positselski. The correspondence between the Hilbert series of dual quadratic algebras does not imply their having the Koszul property. *Funct. Anal. Appl.* **29**, #3, 1995.

[102] L. Positselski. Koszul property and Bogomolov conjecture. Harvard Ph. D. Thesis, 1998, available at http://www.math.uiuc.edu/K-theory/0296.

[103] L. Positselski, A. Vishik. Koszul duality and Galois cohomology. *Math. Research Letters* **2**, #6, 771–781, 1995.

[104] S. Priddy. Koszul resolutions. *Trans. Amer. Math. Soc.* **152**, 39–60, 1970.

[105] J.-E. Roos. Homology of free loop spaces, cyclic homology, and non-rational Poncaré–Betti series in commutative algebra. *Algebra—some current trends (Varna, 1986)*, 173 189, Lecture Notes in Math. 1352, Springer, Berlin, 1988.

[106] J.-E. Roos. On the characterization of Koszul algebras. Four counter-examples. *Comptes Rendus Acad. Sci. Paris*, ser. I, **321**, #1, 15–20, 1995.

[107] J.-E. Roos. A description of the homological behavior of families of quadratic forms in four variables. *Syzygies and Geometry (Boston, 1995)*, 86–95, Northeastern Univ., Boston, MA, 1995.

[108] M. Rosso. An analogue of P.B.W. theorem and the universal R-matrix for $U_h sl(N+1)$. *Commun. Math. Phys.* **124**, 307–318, 1989.

[109] P. Schenzel. Über die freien Auflösungen extremaler Cohen-Macauley-Ringe. *J. Algebra* **64**, 93–101, 1980.

[110] A. Schwarz. Noncommutative supergeometry and duality. *Lett. Math. Phys.* **50**, #4, 309–321, 1999.

[111] S. Serconek, R. Wilson. The quadratic algebras associated with pseudo-roots of noncommutative polynomials are Koszul algebras. *J. Algebra* **278**, 473–493, 2004.

[112] B. Shelton, C. Tingey. On Koszul algebras and a new construction of Artin-Schelter regular algebras. *J. Algebra* **241**, 789–798, 2001.

[113] B. Shelton, S. Yuzvinsky. Koszul algebras from graphs and hyperplane arrangements. *J. London Math. Soc. (2)* **56**, #3, 477–490, 1997.

[114] J. Stasheff. Homotopy associativity of H-spaces, II. *Trans. Amer. Math. Soc.* **108**, 293–312, 1963.

[115] B. Sturmfels. Gröbner bases and convex polytopes. University Lecture Series, 8. AMS, Providence, RI, 1996.

[116] B. Sturmfels. Four counterexamples in combinatorial algebraic geometry. *J. Algebra* **230**, #1, 282–294, 2000.

[117] J. Tate, M. van den Bergh. Homological properties of Sklyanin algebras. *Inventiones Math.* **124**, #1–3, 619–647, 1996.

[118] B. Tsirelson. A new framework for old Bell inequalities. *Helv. Phys. Acta* **66**, #7-8, 858–874, 1993.

[119] V. A. Ufnarovski. Combinatorial and asymptotic methods in algebra. *Algebra, VI*, 1–196. Encyclopaedia Math. Sci. 57, Springer, Berlin, 1995.

[120] V. de Valk. A problem on 0–1 matrices. *Compositio Math.* **71**, #2, 139–179, 1989.

[121] V. de Valk. Hilbert space representations of m-dependent processes. *Ann. Probab.* **21**, #3, 1550–1570, 1993.

[122] V. de Valk. One-dependent processes: two-block factors and non-two-block factors. *Stichting Mathematisch Centrum, Centrum voor Wiskunde en Informatica, Amsterdam*, 1994.

[123] M. Van den Bergh. Noncommutative homology of some three-dimensional quantum spaces. *K-theory*, **8**, 213–230, 1994.

[124] M. Van den Bergh. A relation between Hochschild homology and cohomology for Gorenstein rings. *Proc. AMS*, **126**. #4, 1345–1348, 1998.

[125] A. M. Vershik. Algebras with quadratic relations. *Spektr. teoriya operatorov i beskonechnomern. analiz*, 32–57, Akad. Nauk Ukrain. SSR, Inst. Mat., Kiev, 1984. English translation in: *Selecta Math. Sovietica* **11**, #4, 293–316, 1992.

[126] M. Vybornov. Mixed algebras and quivers related to cell complexes and Koszul duality, *Math. Res. Letters* **5**, 675–683, 1998.

[127] M. Vybornov. Sheaves on triangulated spaces and Koszul duality. Preprint math.AT/9910150.

Titles in This Series

37 **Alexander Polishchuk and Leonid Positselski,** Quadratic algebras, 2005
36 **Matilde Marcolli,** Arithmetic noncommutative geometry, 2005
35 **Luca Capogna, Carlos E. Kenig, and Loredana Lanzani,** Harmonic measure: Geometric and analytic points of view, 2005
34 **E. B. Dynkin,** Superdiffusions and positive solutions of nonlinear partial differential equations, 2004
33 **Kristian Seip,** Interpolation and sampling in spaces of analytic functions, 2004
32 **Paul B. Larson,** The stationary tower: Notes on a course by W. Hugh Woodin, 2004
31 **John Roe,** Lectures on coarse geometry, 2003
30 **Anatole Katok,** Combinatorial constructions in ergodic theory and dynamics, 2003
29 **Thomas H. Wolff (Izabella Łaba and Carol Shubin, editors),** Lectures on harmonic analysis, 2003
28 **Skip Garibaldi, Alexander Merkurjev, and Jean-Pierre Serre,** Cohomological invariants in Galois cohomology, 2003
27 **Sun-Yung A. Chang, Paul C. Yang, Karsten Grove, and Jon G. Wolfson,** Conformal, Riemannian and Lagrangian geometry, The 2000 Barrett Lectures, 2002
26 **Susumu Ariki,** Representations of quantum algebras and combinatorics of Young tableaux, 2002
25 **William T. Ross and Harold S. Shapiro,** Generalized analytic continuation, 2002
24 **Victor M. Buchstaber and Taras E. Panov,** Torus actions and their applications in topology and combinatorics, 2002
23 **Luis Barreira and Yakov B. Pesin,** Lyapunov exponents and smooth ergodic theory, 2002
22 **Yves Meyer,** Oscillating patterns in image processing and nonlinear evolution equations, 2001
21 **Bojko Bakalov and Alexander Kirillov, Jr.,** Lectures on tensor categories and modular functors, 2001
20 **Alison M. Etheridge,** An introduction to superprocesses, 2000
19 **R. A. Minlos,** Introduction to mathematical statistical physics, 2000
18 **Hiraku Nakajima,** Lectures on Hilbert schemes of points on surfaces, 1999
17 **Marcel Berger,** Riemannian geometry during the second half of the twentieth century, 2000
16 **Harish-Chandra,** Admissible invariant distributions on reductive p-adic groups (with notes by Stephen DeBacker and Paul J. Sally, Jr.), 1999
15 **Andrew Mathas,** Iwahori-Hecke algebras and Schur algebras of the symmetric group, 1999
14 **Lars Kadison,** New examples of Frobenius extensions, 1999
13 **Yakov M. Eliashberg and William P. Thurston,** Confoliations, 1998
12 **I. G. Macdonald,** Symmetric functions and orthogonal polynomials, 1998
11 **Lars Gårding,** Some points of analysis and their history, 1997
10 **Victor Kac,** Vertex algebras for beginners, Second Edition, 1998
9 **Stephen Gelbart,** Lectures on the Arthur-Selberg trace formula, 1996
8 **Bernd Sturmfels,** Gröbner bases and convex polytopes, 1996
7 **Andy R. Magid,** Lectures on differential Galois theory, 1994
6 **Dusa McDuff and Dietmar Salamon,** J-holomorphic curves and quantum cohomology, 1994
5 **V. I. Arnold,** Topological invariants of plane curves and caustics, 1994
4 **David M. Goldschmidt,** Group characters, symmetric functions, and the Hecke algebra, 1993

TITLES IN THIS SERIES

3 **A. N. Varchenko and P. I. Etingof,** Why the boundary of a round drop becomes a curve of order four, 1992

2 **Fritz John,** Nonlinear wave equations, formation of singularities, 1990

1 **Michael H. Freedman and Feng Luo,** Selected applications of geometry to low-dimensional topology, 1989